WILDLIFE ON THE PLANET FURAHA

A speculative biology guide to alien life forms

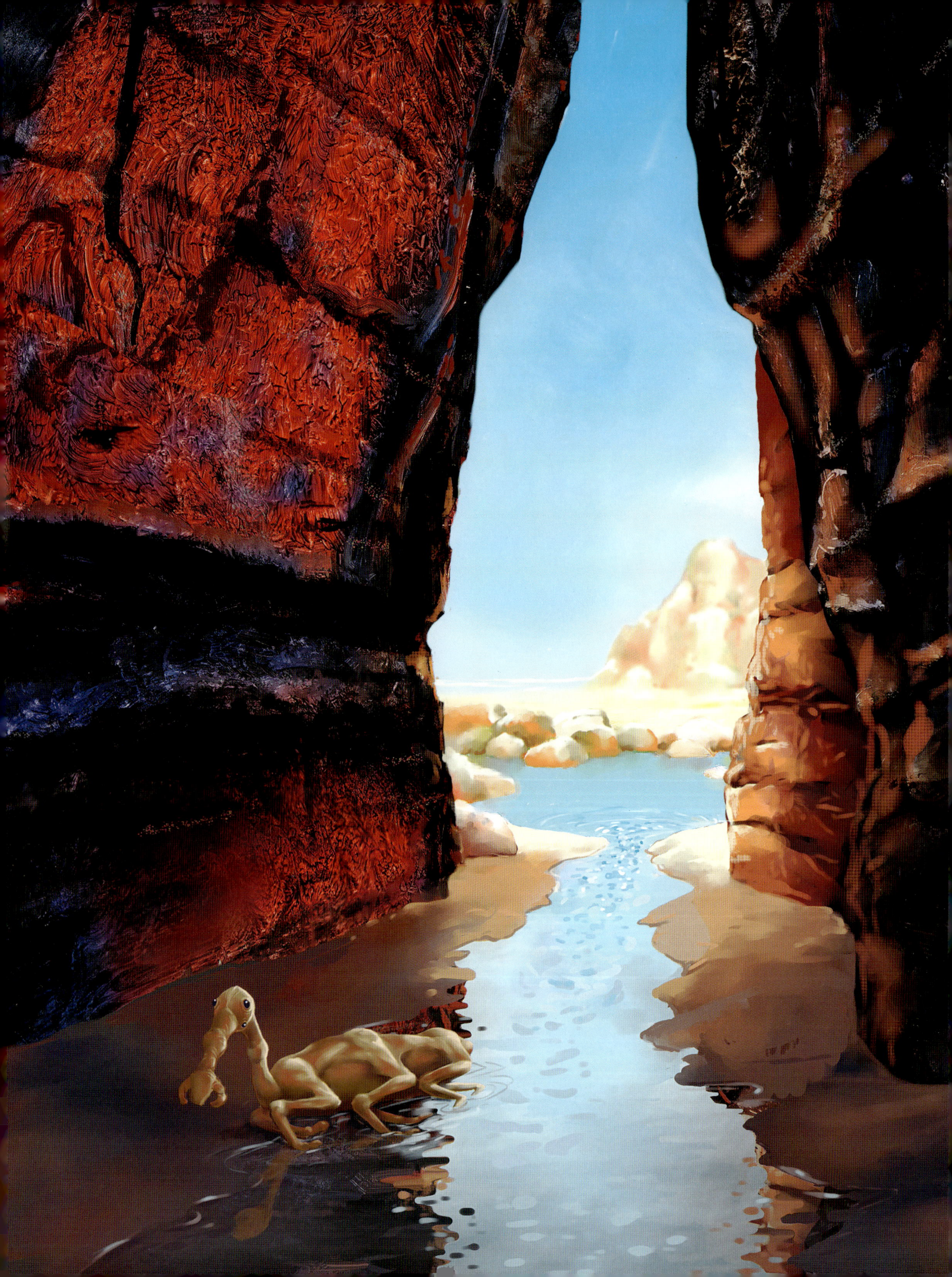

WILDLIFE ON THE PLANET FURAHA

A speculative biology guide to alien life forms

Gert van Dijk

THE CROWOOD PRESS

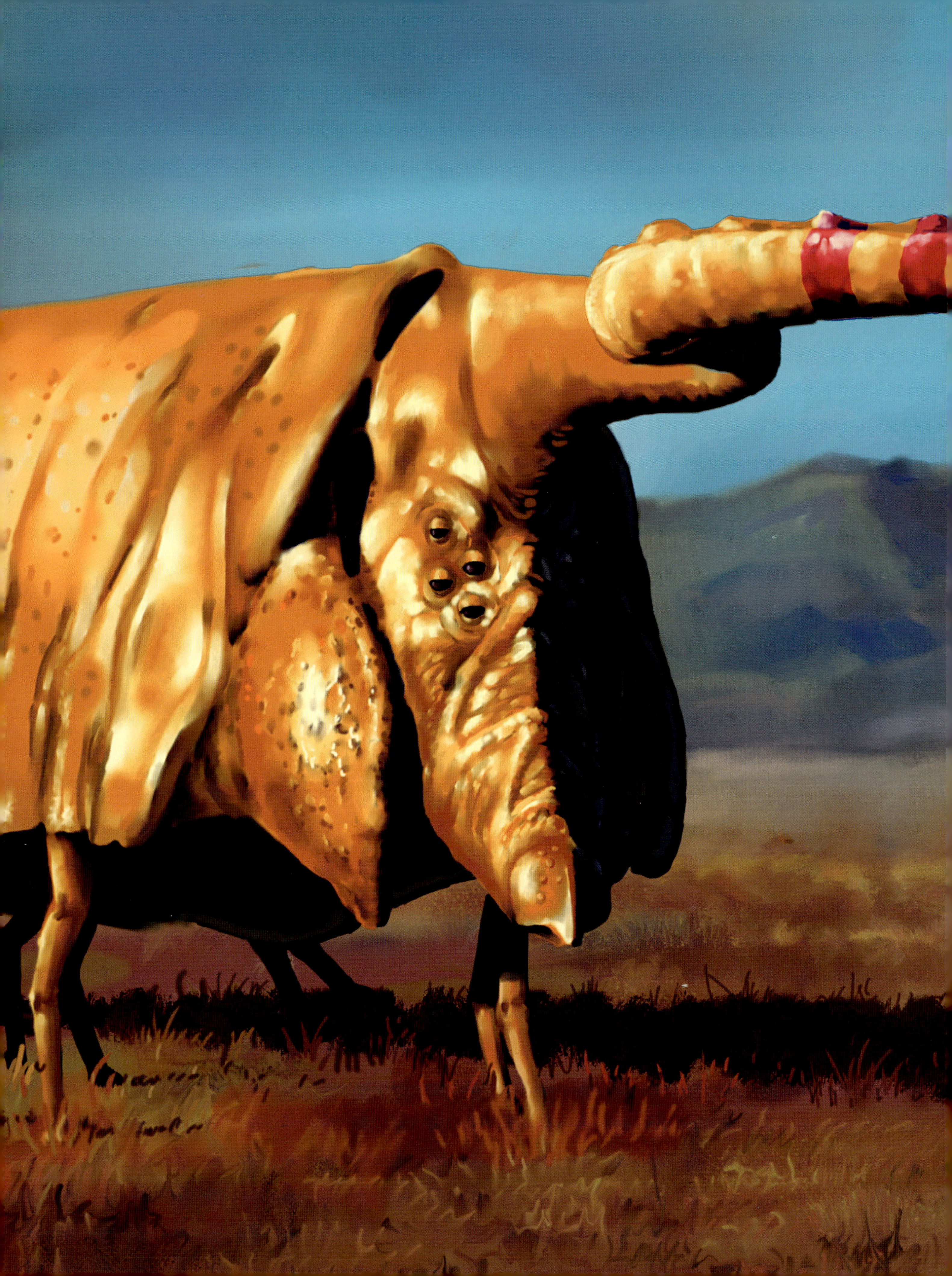

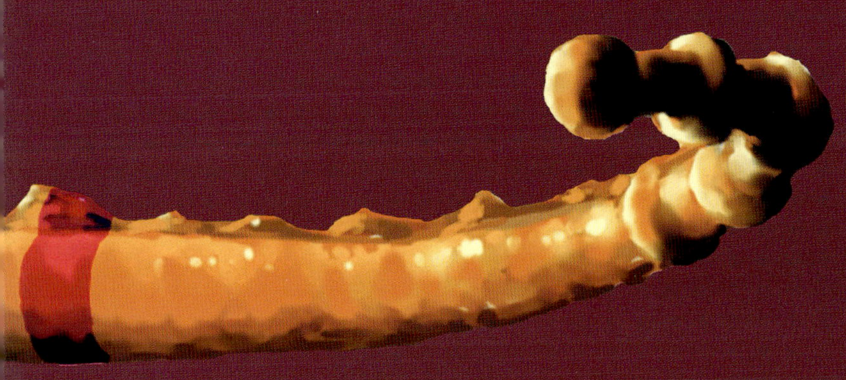

Contents

1 Furaha and its Solar System 7
2 Humanity on Furaha 17
3 Plants 27
4 Mixotrophs 35
5 Various Clades 45
Wadudu 46
Kwals 52
Wardens 57
Cloakfish 65
Polypods 72
Spidrids 77
Tetrapters 84
6 Scalates and Hexapods 91
'Fishes' 93
Hexapods 104
7 Rusps 139
Megarusps 140
Other Rusps 146

Appendix: the Author's 'Prehistory' of the Furaha Project 152
Glossary of Furahan Biological Terms 154
Index 158
Acknowledgements 160

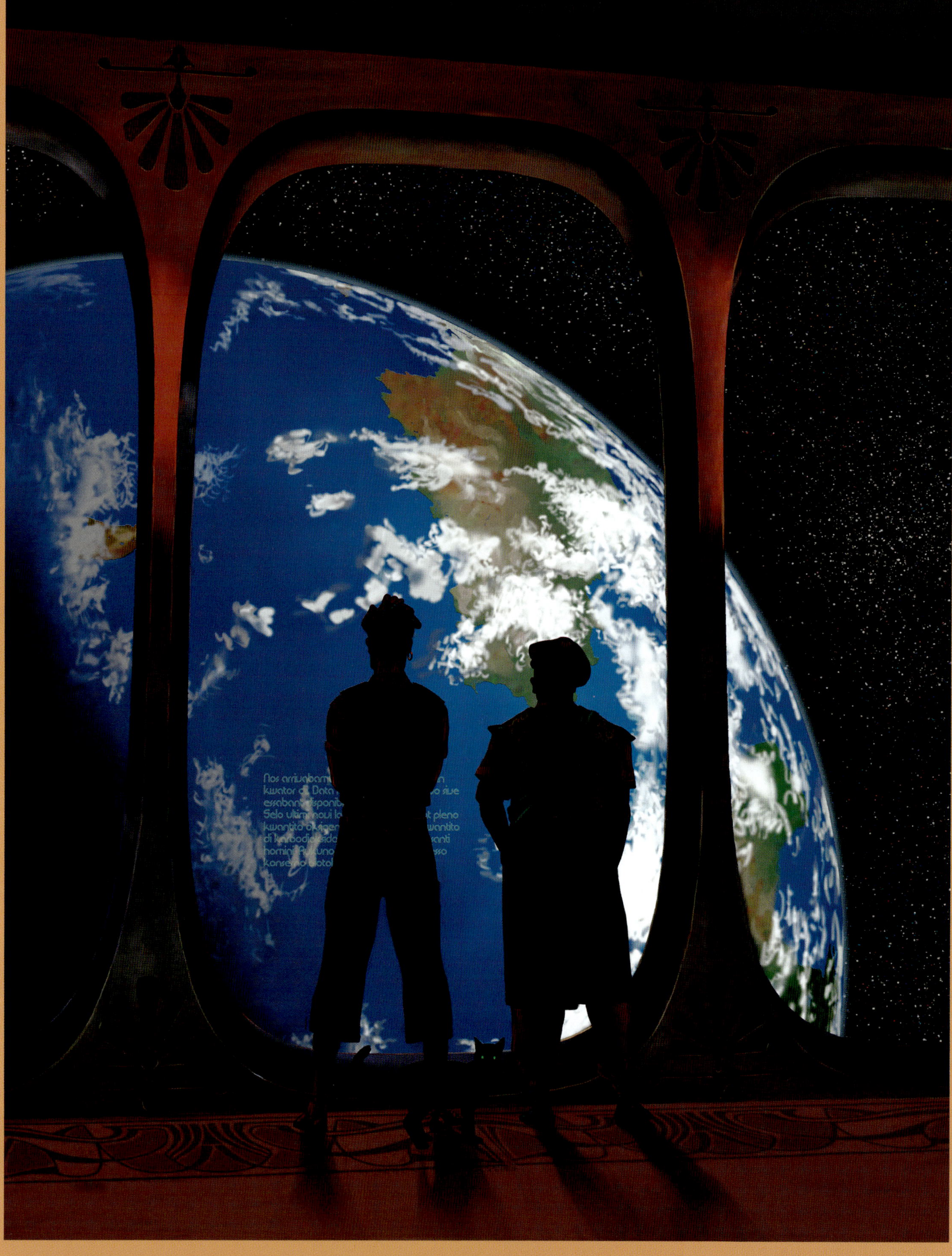
Nos
kwator Data sive
Selo pleno
kwantito
di

Chapter 1

Furaha and Its Solar System

'This planet was utterly unlike the dead cinders we had encountered so far. This was something else, something unique, something vibrant...

This was a living planet.

Bruyningh and I stood together, silently looking at the planet for a long time. Long enough to see it rotate and the clouds move. I think it was there and then that we, without realising it, stopped speaking about 'the planet'. From then on, it was 'the world'.'

From *Worlds Apart: Natural Histories of Furaha and Earth* by Souren Nyoroge

◀ This scene shows Fuscus Filius Bruyningh, on the left, and Souren Nyoroge, on the right, in high orbit around Furaha. At the time, the two men already knew another well but were not yet close friends. Bruyningh had been the force behind the expedition to Furaha, but he was undecided about what to do afterwards. Nyoroge, at heart not an adventurer, had never been enthusiastic about coming and only developed a passion for Furaha much later. Their friendship evolved well after both men came into their own, when Nyoroge had matured into Furaha's first great naturalist and Bruyningh had set up the Institute.

Jua and The Outer Solar System

Five Gas Giants, Many Tiny Moons

The Jua system is a very typical solar system for a G-class star, as far as the number of planets and their composition is concerned. However, the presence of a planet with Earth-like life, a Gaean, is extremely rare.

Jua

The star Nu Phoenicis received its common name Jua, a Swahili word, from the crew of the space vessel Ngonjera, as that crew largely came from East Africa. The Horizonists who later founded the settlement spoke various European languages and might easily have decided to change all earlier Swahili names. They did not, however, and so Nu Phoenicis retained the name Jua, which in Swahili means nothing other than 'star'. In that respect it resembles Earth's sun whose formal name *Sol* is simply the Latin word for 'sun'.

Outer Planets

The outer planets are all gas giants with small rocky cores at their centre, hidden underneath layers and layers of gases compressed to fluids at unimaginable depths.

The innermost of these cold giants, Jitu, is often the closest to Furaha, depending on where it is in its orbit. Jitu is Swahili for

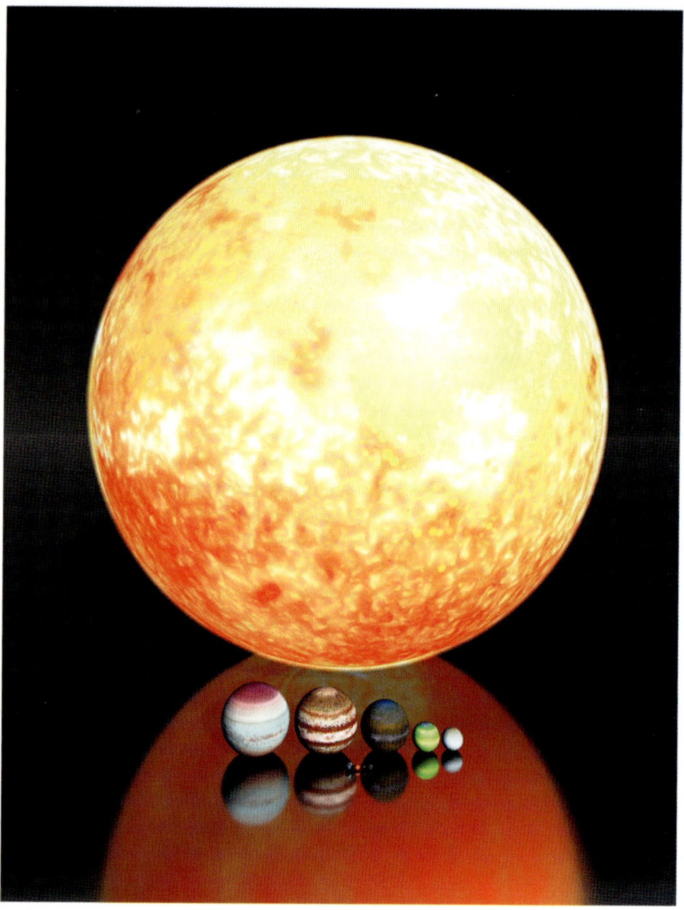

▲ The star Jua and its nine planets are shown at the same scale as far as their sizes are concerned. The planets of the outer and inner systems are shown in two rows before the star Jua.

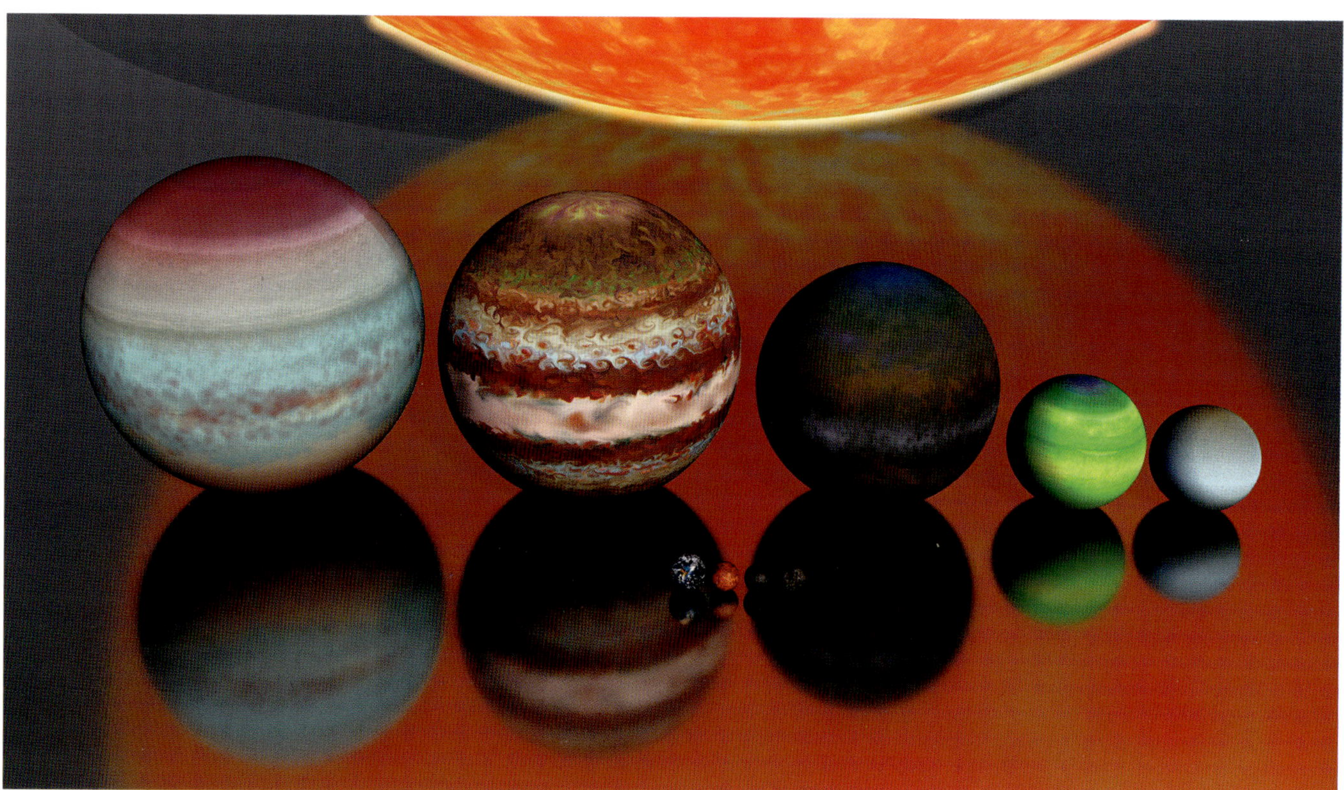

▲ The planets of the outer system are much larger than those of the inner system. The planets are shown from the innermost planets on the far left to the outermost on the far right. From left to right their names are Jitu, Ubaridi, Sititiko, Makamasi and Peke.

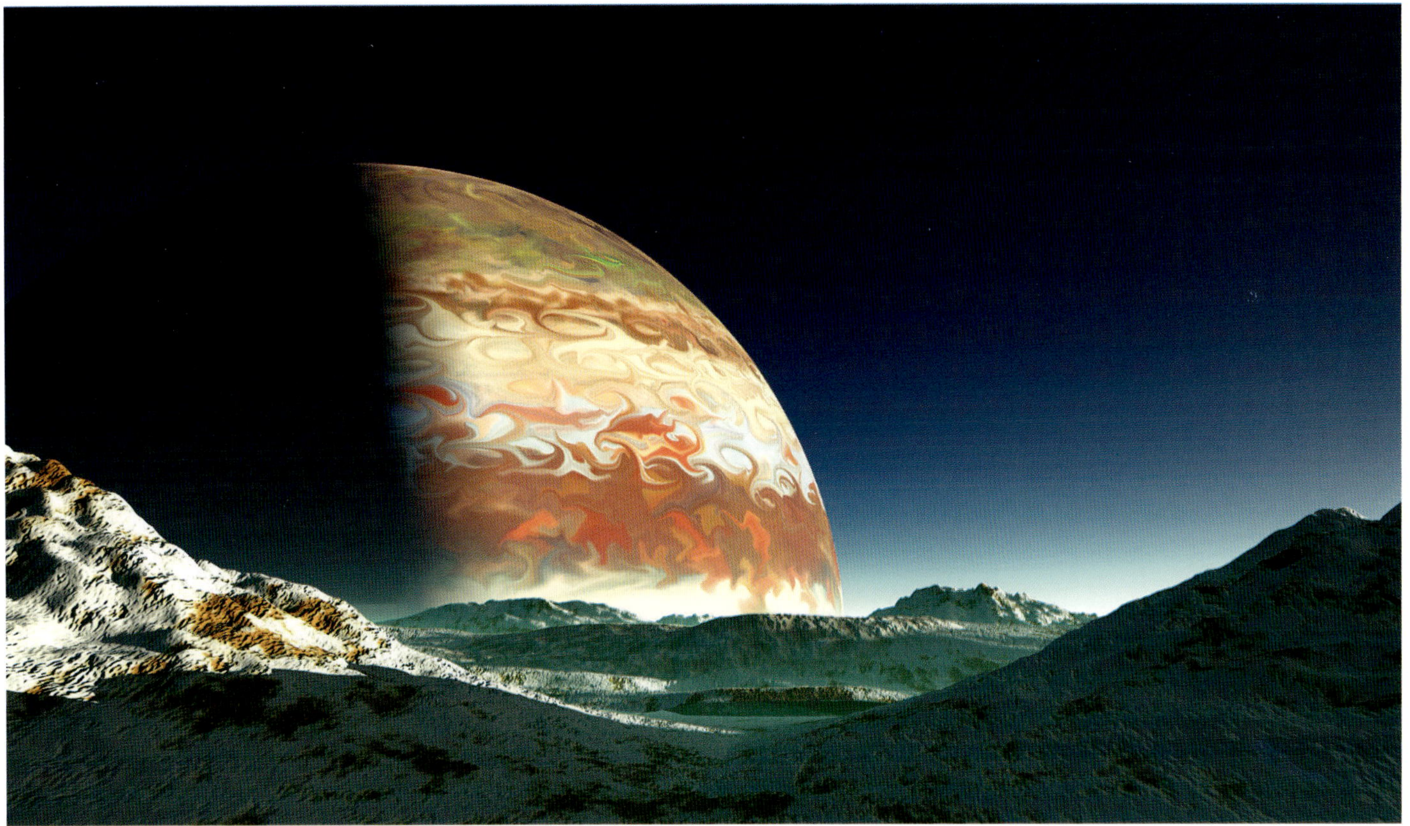

▲ This is an image from one of the dozen or so moons and moonlets that orbit Ubaridi ('coldness'). Ubaridi is the second largest planet in the Jua system, but is much more photogenic than Jitu, the largest. Ubaridi has bands with swirling colours, much like Jupiter in the Sol system. The particular moon this image was taken from, Mchanga, is a cinder, meaning it is mostly a ball of dead rock. Mchanga has just enough of an atmosphere to prevent the sky being all black.

'giant', which is fitting for the largest planet of the system. Jitu circles Jua at a distance of about seven astronomical units (an astronomical unit is defined as the distance between Earth and Sol). Jitu has no distinguishing colours or features, not even from up close.

Next follows Ubaridi (meaning 'coldness'), famous for the swirling colourful bands in its atmosphere.

Sititiko is the next planet out. Like Jitu, its lack of features ensures it does not attract much attention. Still, it is home to many ballooning life forms, known as ballonts, but these ballonts suffer from the common lack of interest of life in gas giants: all ballonts look alike.

Makamasi has an odd greenish tint, and there is no doubt that this hue is what earned its name, as the crew of the Ngonjera dubbed it 'nasal mucus'. As said, the Horizonists who later settled on Furaha should perhaps have paid more attention and renamed this planet, but they did not.

Peke is small as gas giants go, cruising at a distance of 75 to 90 astronomical units away from Jua. Its name means 'isolation' – the Ngonjera crew was certainly right with that name!

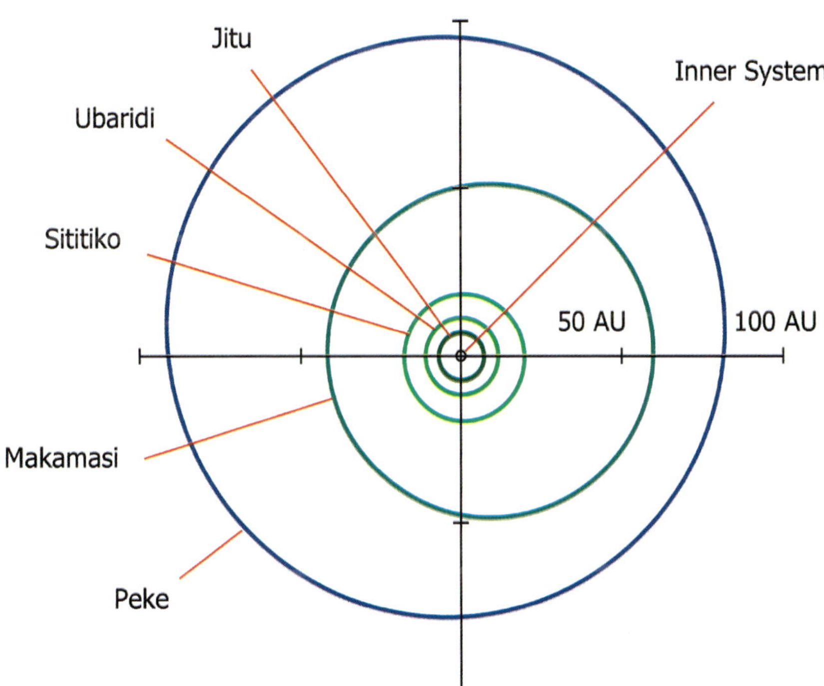

▲ The elliptical paths of the outer planets span a vast distance, with Peke cruising at a distance to Jua equal to 90 times the distance between Earth and its sun. The inner system is also shown, spanning only about 1.5 AU.

The Inner Solar System

Three Cinders, One Cinderella

The whole inner solar system spans less than two astronomical units, meaning it is much smaller than the outer system. Just compare its scale to that of the outer system described previously – the entirety of the inner system is just a tiny circle there. This shows how large a typical solar system is, and yet the Jua solar system is not even a large one.

Furaha is further away from Jua than Earth is from *Sol*, but Jua is a bit hotter than *Sol*. Furaha is, like Earth, comfortably situated in the middle of the 'habitable zone'. This is also known as the 'Goldilocks' zone, where everything is right to bear life: neither too hot nor too cold, but just right.

Discovery of The Jua System

The following quotation was taken from Capitaen Han Lebrouillard's book *Going Out*, which tells the story of what it is like to explore space in a spacer's jargon.

> When you've been in a safe system for some time you get the urge to explore a fresh system again. It's no easy decision: far sensing may show you there are planets in a hab, but that's about it. They're all probably just cinders or cubes. But you don't really know and, in the end, curiosity gets you. So, you say your goodbyes and you drift far enough out of the well to release space-time mooring and fire your Feng. Then you wait for the Boing. It's only when the Boing says you've moored again, and only then, that you find out if the wait was worth it.

As told by an old hand, planet scouring sounds easy, but it is not. Even old spacers welcome the ship's traditional 'Boing', announcing that the moorings have successfully rehooked space-time mooring. Hearing the Boing is a big relief, because no-one knows beforehand how long a trip will take and you do not, and cannot, find out until the wait is over. Greater distances take longer, but there is a large error margin in a trip's duration, both in external and in internal ship time. The 'spacer's nightmare' concerns the tiny chance that a trip may last years or even decades, instead of a few months. After six or nine months on a ship, the uncertainty affects everyone, even born spacers. The spacer phrase 'waiting for the Boing' sums up the slow dread of space travel.

The crew of the vessel Ngonjera had chosen the Jua system without much prior consideration. 'Boing!' said the ship on entering local space near the star Nu Phoenicis and the crew ran to get a look at the first data the ship was gathering.

The Planets

The first data packets to percolate back showed a bunch of joves, or large Jupiter-like planets. Because of their large gravity wells, joves are usually the first planets to be found. The Jua system proved to be one of those rare systems in which the joves were all in the outer system; no hot Jupiters close to Jua wrecked the inner system here.

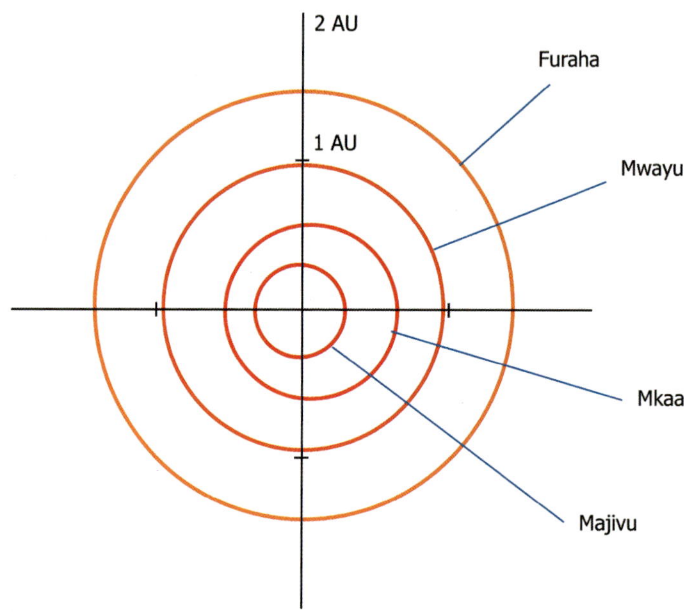

▲ The inner planets are much closer together than those of the outer system. Furaha orbits Jua farther away from Jua than the distance between Earth and its sun. The light from Jua takes almost 12 minutes to reach Furaha.

The first terrestrial planet they found was Majivu, closest to Jua. Then, in quick succession, the ship found two more inner planets: Mkaa and Mwayu. All three were just 'cinders': dead, dry and boring.

Majivu

This planet circles Jua at about one third of an astronomical unit, so its surface is hot enough to melt lead, not that there is any. All traces of an atmosphere blew off aeons ago. The planet's rotation around its axis is becoming tidally locked to its orbit, so in a few million years Majivu will always present the same face to Jua. But for now, it rotates. Its surface roasts one week, only to freeze the next. With its overall grey exterior, it is no wonder that the Ngonjera's crew called it Majivu, meaning 'ashes'.

Mkaa

The next planet out proved to be a brown cinder. Mkaa is too hot wherever it is not too cold. As the universe is full of such places,

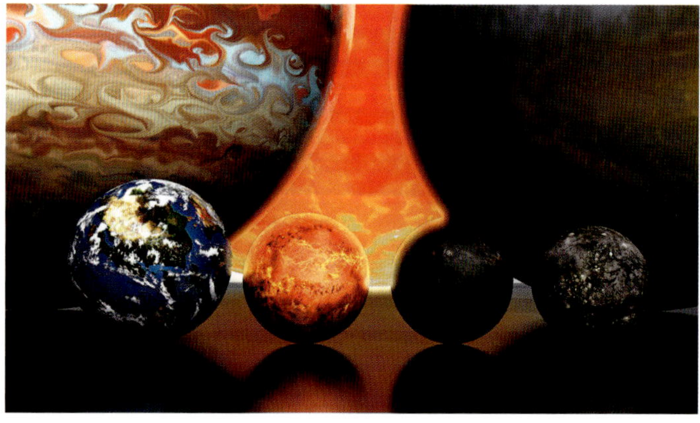

▲ The four inner planets are here shown to scale. The one on the right is Majivu, a planet so close to Jua that its surface is hot enough to melt lead. The next one to the left, Mkaa, may be less hot than Majivu but is just as dead. Mwayu is a proper planet, with a thin atmosphere and a rather Mars-like appearance. The left-most planet is Furaha, a planet with abundant water in fluid form.

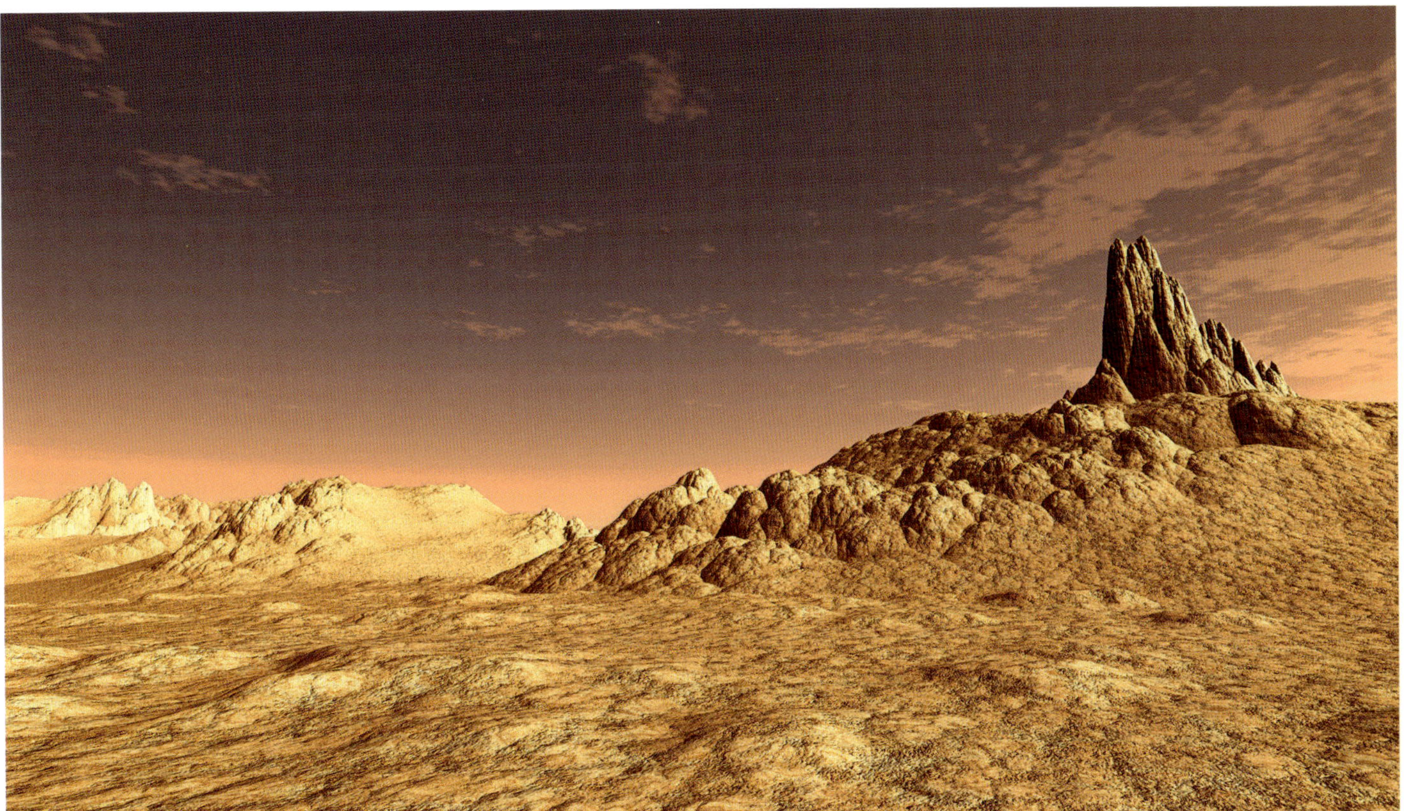

▲ This image shows the surface of Mwayu, one of the three cinders encountered by the vessel Ngonjera. It is somewhat attractive, but you would not want to live there. Mwayu might possibly be terraformed, and the Fogg Corporation expressed an interest to do just that for a while, but so far, their plans have not come to fruition.

the crew spent little time studying this planet, after naming it Mkaa, meaning 'charcoal'.

Mwayu

This planet is the last of the inner cinders. The phrase 'outer cinders', by the way, usually refers to the various lifeless dry moons circling gas giants. The Ngonjera's crew was perhaps a little harsh in dismissing this planet out of hand, as exemplified by the name Mwayu, meaning 'yawn'. After all, it has an atmosphere, albeit a thin and unbreathable one. It also has interesting plains, deserts, stones and sand. Lots of sand. So much sand, in fact, that for once a descriptive term such as 'desert planet' is correct.

Furaha

Only then did the ship find the biggest terrestrial planet, the one farthest out from Jua. Sensors showed nitrogen, as well as free oxygen, so the crew immediately knew they had a 'cinderella' on their hands. Cinderella is the name spacers give to a potential Gaean, as at that stage you do not know whether you are dealing with a maid or a princess.

Still, it was by far the most exciting find of the entire system.

The Ngonjera's crew quickly decided to float towards the cinderella's well. Not long afterwards, incoming spectra told tales of carbon dioxide, with absorption lines hinting at an active biochemistry, so hopes ran higher and higher.

Only when the optical telescopes made out white clouds, shiny seas and bright continents did the crew realise that they had found the spacer's dream. Their cinderella was a true Gaean: a planet with a proper biosphere, in which life is more than a microscopic contaminant.

One Famous Space Farer's Nightmare

Space travel is not for everyone. Some people take it very badly, in fact. Souren Nyoroge, the unwilling traveller who became the first and best cataloguer of Furahan life, found space travel extremely depressing, as is evident from the following excerpt of his famous book *Worlds Apart: Natural Histories of Furaha and Earth*.

> There is no way to grasp the nothingness of space. It cannot be done. The size of Earth can just be imagined, but holding its image next to that of a Jovian forces you to realise that your entire world, a tiny universe unto itself, is small. Indeed, it is as nothing. It IS nothing. In turn, that stupendous Jovian is nothing compared to the magnificence of the sun. And we know that our bright sun is nothing but a speck next to a red giant.
>
> You ask me to describe the size of nothingness between stars, to catch it in words? I cannot, and neither can you. Waiting for time and space to come together again made me feel I was nothing. The emptiness first impressed me, but gradually it wore me down and then depressed me. I questioned whether we should be out there, but what does *there* even mean if there is nothing?
>
> To those who have been there and know space, I admit without hesitation that I was afraid. So very afraid. To those who mock me I say, go and look into the nothingness yourself, and only then return and speak to me about fear. Who measures up to nothingness? It is all hubris...

Closing In on Furaha

Orbit, Moons, Tides

A Year on Furaha

Furaha is further from Jua than Earth is from Sol. It is 1.44 times further, to be precise. Furaha also takes longer to complete one orbit of Jua, so one Furaha year lasts 1.60 Earth years. For Earth travellers, it may be confusing that a 30-year-old Furahan citizen would be 48 years old on Earth. During the long Furahan year, the planet rotates 551.1 times around its axis, so there are 551 days in a Furahan year. Luckily for humans on Furaha, a Furaha day lasts only slightly longer than an Earth one, so human biological clocks adapt easily to the new rhythm.

Seasons

While Furaha's orbit is of course elliptical, as it is for all planets, its ellipse is so close to a circle as to make little difference. This also holds for Earth, so for both planets the seasons do not depend on a changing distance from the sun, as they do on many other worlds. Furahan and Terran seasons instead depend on the degree of tilt of the planet's axis. The planet's axis points towards or away from the sun depending on where the planet is in its orbit and that is what causes seasons.

The polar zones are smaller on Furaha than they are on Earth, which is one reason Furaha's ice caps are not very impressive. Another reason is that there are no big land masses on or near a pole, as is the case for Antarctica and Greenland on Earth. When present, such masses help form a stable ice core. Still, floating sea ice forms a modest ice cap each winter on either polar ocean on Furaha. The sea ice disappears nearly completely every summer.

Earth and Furaha as Siblings

Furaha is the larger of the two planets: Earth's radius is 6378km and Furaha's radius is 6941km, or 1.09 times larger. Some therefore call Furaha Earth's big brother, but size is not everything. Furaha formed about one gigayear after Earth did, so it may with equal reason be called Terra's younger sister. Its mass is 1.36 times larger than that of Earth, resulting in a surface gravity of 1.15 times the one Terrans enjoy at home. That is not a big difference, although some say that people on Furaha are more likely to develop back problems because of the higher gravity.

Furaha's larger radius means that its surface area is 1.29 times bigger than Earth's. While the proportions of land to sea are roughly similar, in Terra's case most of the land is found in the form of a few very large continents along with lots of small islands. On Furaha, there are several intermediately-sized land masses. It is probably for that reason that no-one on Furaha ever bothered to decide whether a specific mass should be called a small continent or a big island, or even to count the number of continents.

Moons: 'Candle' and 'Spark'

Furaha has two moons, named by the crew of the Ngonjera. The names suggest that the crew members were not impressed. The larger one is named Mishunaa (meaning 'candle'). It takes

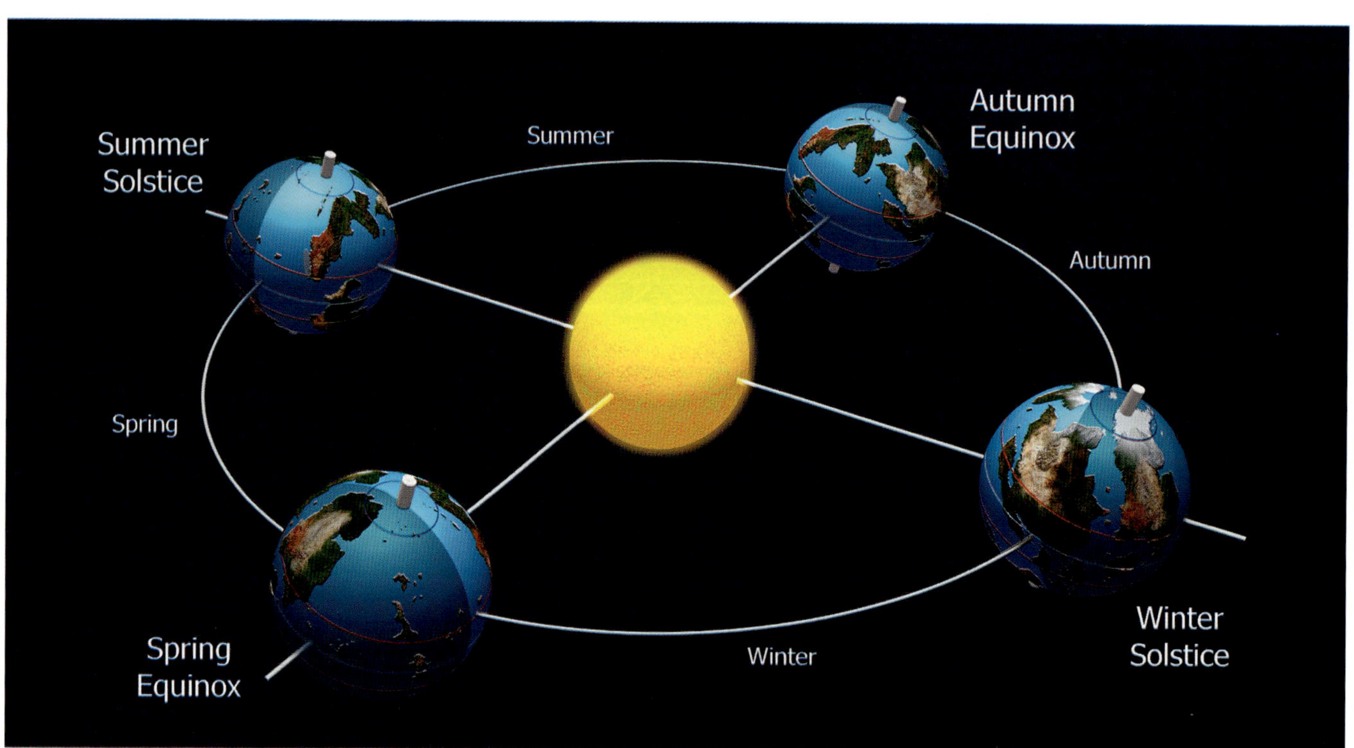

▲ The half of the planet lit by Jua, its day side, is shown with lighter colours then the dark night half. The axis of Furaha is tilted 18.3 degrees compared to the plane in which it circles Jua, less than Earth's 23.5 degrees. One result of this is that seasons tend to be less pronounced than on Earth. Another result is that the part on the planet where the sun can be directly overhead, meaning the tropics, is narrower. The tropics' borders are marked by red lines.

The polar zones, shown as blue circles, are also relatively small. At the summer solstice (top left) the North pole is maximally tilted towards Jua, so the North polar zone stays lit all day, while it is dark in the South polar zone all day. Half a year later, at the winter solstice, the opposite occurs. At the halfway points, day and night last exactly the same time: these are the spring and autumn equinoxes.

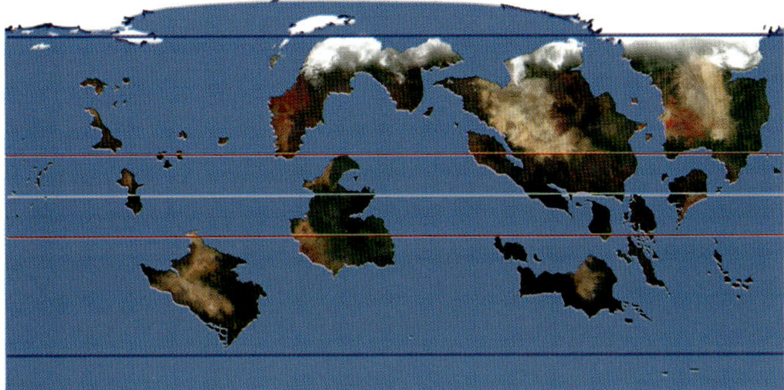

▲ Ice-covered areas in the North show that it is winter in the Northern hemisphere. Sea ice links up with ice extending from the continent *Imparia Septentrionalis.*

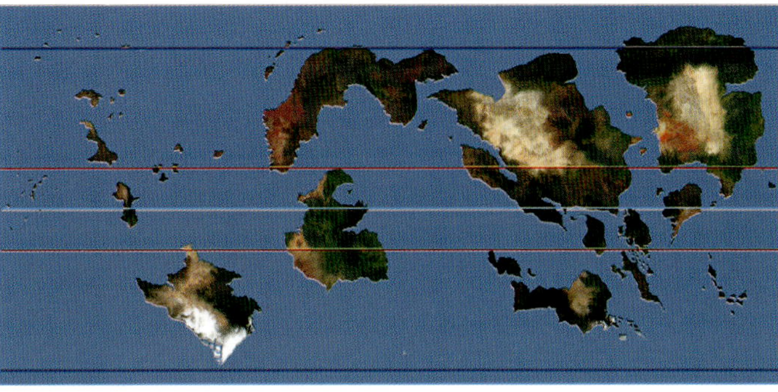

▲ Winter in the Southern hemisphere may show temporary Southern polar ice and always shows ice and snow over *Meralgia.*

11.9 Earth days for Mishunaa to circle Furaha. Its diameter is just over half that of the Earth's moon, Luna, and is closer to Furaha than Luna is to Earth. Seen from Furaha, its angular diameter is nearly 0.4 degrees, which is not much smaller than the 0.5 degrees Luna spans as seen from Earth. The disc of a full Mishunaa moon is still only about 60 per cent the area of a full moon as seen from Earth, so nights tend to be darker on Furaha than they are on Earth.

Poor little Cheche, the other moon, circles Furaha in 22.7 Earth days. It is 7 per cent further away from Furaha than Luna is from Earth, so its angular diameter is only 0.1 degree. The disk of a full Cheche moon is only 4 per cent the area of a full Earth moon, so it does not cast much light during Furahan nights. No wonder the Ngonjera crew called it 'spark'!

Tides

Tides are simply masses of sea water moving around the surface of a planet because of the gravitational attraction of a sun or moon. The tidal force of a planet or sun increases more quickly the nearer it is to the planet with seas. It also increases with its mass, but less strongly so. This is why a small nearby moon can have a stronger effect on tides than the immensely larger, but much more distant, sun.

On Furaha, as on Earth, both the sun and Mishunaa cause tides. As on Earth, Furaha's rotation around its own axis ensures a basic rhythm of two tides per Furahan day. To this rhythm are added the 'monthly' effects of the moons. Note that a 'month' for Mishunaa lasts just less than twelve days. As Mishunaa is smaller than Luna, its tidal effects are smaller than the effects Earth's moon has on Earth's tides. Cheche is so small that it contributes next to nothing to Furaha's tides.

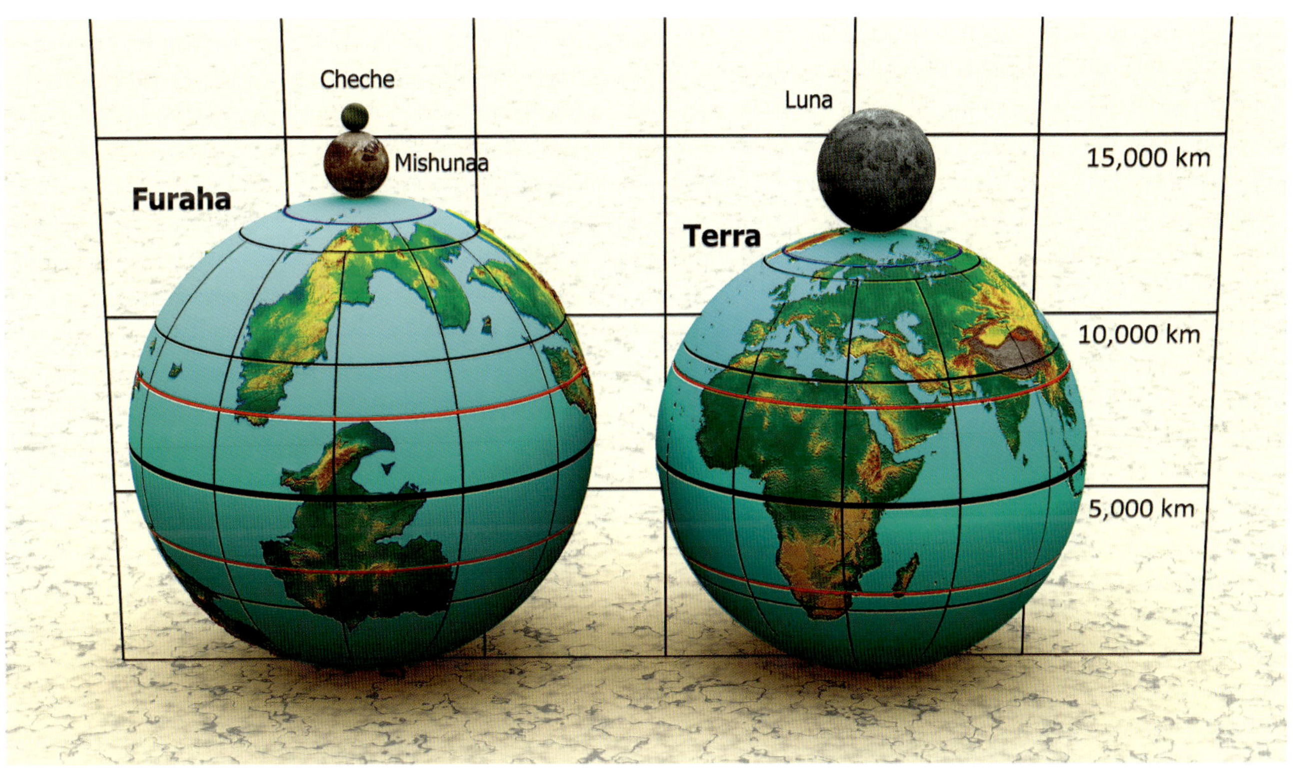

▲ These models show Furaha and postdiluvian Earth along with their moons. The size difference is well visible, as are the different sizes of the tropical and polar zones, indicated with red and blue parallels. To facilitate the size comparison, moons are shown perched on top of their parent planet. Earth's moon Luna is much larger than either of Furaha's moons, Mishunaa ('candle') and Cheche ('spark').

LIFE ADRIFT

COLLIDE, DIVIDE, REPEAT

Universal Traits of Life

We still do not know how life starts on any planet. The discovery of prebiotic bubbling soups on numerous planets suggests that conditions conducive to life are not rare, so apparently life will emerge, given a lucky combination of factors over aeons of time.

We do know that life on Furaha followed the usual major steps of evolution. The first step is that life most readily evolves in a fluid. Furahan life took a long time to leave the oceans and adapt to dry land. The second principle holds that energy for life is either acquired from a non-living source, such as sunlight, or from existing life. The third principle holds that life starts with humble single corpuscles, such as bacteria, before complex multicellular life appears. The final principle is evolution: life forms with better genetic traits are likely to leave more offspring than their less endowed fellows.

Continental Drift and Climate

Furaha has a variety of continents that, as on Earth, drift on the underlying magma. Over many millions of years, parts of continents may split off, following their own course over the surface of the planet. At other times these fragments, or indeed entire continents, collide to form new continents.

This slow dance profoundly influences evolution. The position of the continents influences the world's climate, which in turn strongly influences evolution. But it is not just the place of a continent, near a pole or straddling the equator, that influences climate. Continents can also block or divert oceanic currents, directing cold or warm water to parts of the planet that would otherwise not receive these streams. Mountain ranges can also have strong effects, causing clouds to produce rain on the mountains, leaving no water for rainless deserts beyond.

Collide and Mix

A second major mechanism by which continental drift influences evolution is that each continent functions as a raft carrying its own contingent of species. Some species are not strongly bound to their raft. Coastal fish may survive in open oceans, so they can trek between continents. Floating seeds and flying animals can also jump ship, provided there is land to be found within their range of endurance. But terrestrial species are locked to their raft and will evolve on their continent in isolation. When continents collide, species will come into contact with one another; many things can then happen. A small herbivore may find that its now much bigger world allows it to grow bigger. A top predator may find the new competition too powerful, so it either dies out or evolves quickly, perhaps to survive in a specialist niche. Entire groups may be wiped out following such a clash of species, leaving the fittest to continue.

Break and Separate

If all continents were to collide every ten million years or so, the species they carry would remain recognisable variations on a common theme. This is not the case for continents that remain isolated longer. The longer the isolation lasts, the larger the divergences can be. Australia and New Zealand are good examples of relatively long isolation. Australia broke away from Gondwana about 100 million years ago, after which its marsupial mammal cargo was left to its own devices, missing the placental mammal revolution occurring in the rest of the world. Most other continents on Earth connected with their neighbours fairly recently, except for Antarctica. But Antarctica is too cold for a complex terrestrial ecology, so Earth cannot tell us what a truly long isolation may do.

On Furaha, two land masses remained isolated for very long periods, long enough to result in unique shapes. Meralgia was isolated for 160 million years and the Palaeogaeas have been isolated for an even longer, unknown period.

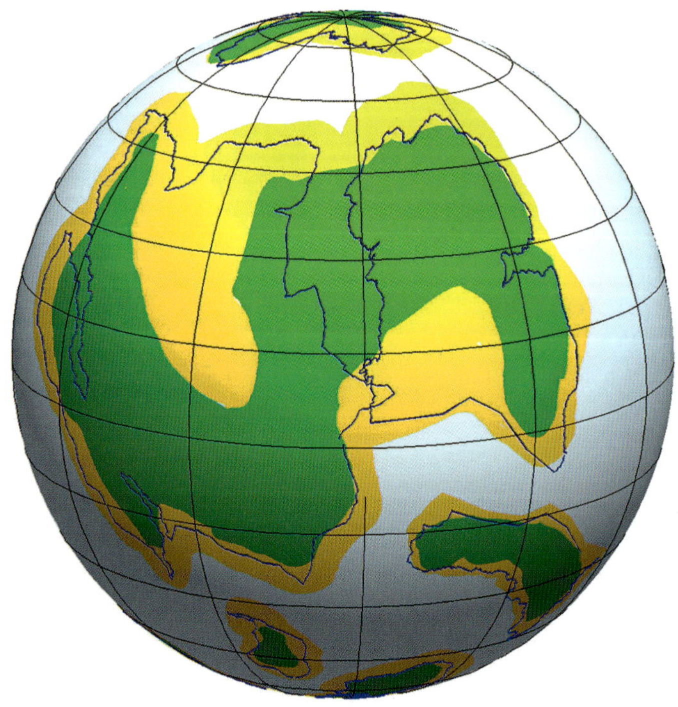

◀ A Reconstruction of Furaha, 200 million years ago
Yellow areas show continental masses and blue lines show present-day coastlines. The two linked landmasses represent the *Imparia* continent. This landmass was, as its name suggests, one gigantic continent rather than two separate ones. A large portion of the continent was covered in shallow seas, because the climate was remarkably warm all over the globe. There is no evidence of glaciation even on rocks that were close to the North or South pole. Areas that were probably dry land are indicated with a green colour.

▶ Present day

This world map, using an Eckert IV projection, shows Furaha at present. Ochre areas indicate continental masses and blue lines demarcate present coastlines. *See* the 'Geography' pages for details on names.

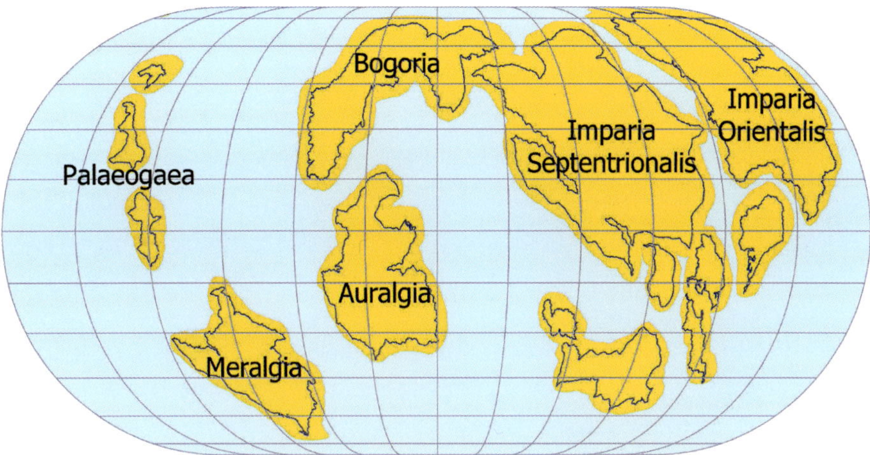

▶ 50 million years ago

The two large *Imparia* continents are closer to one another than in the present but are not fused completely: the breaking of *Imparia* took a very long time. The area named *Arzakia* is drifting north. It will become the southern part of *Imparia Orientalis* and the collision will form a large mountain range. Also, note that *Meralgia* and *Auralgia* are situated more to the east than at present but are still separate. Most of the *Archipelago* forms one land mass in this age.

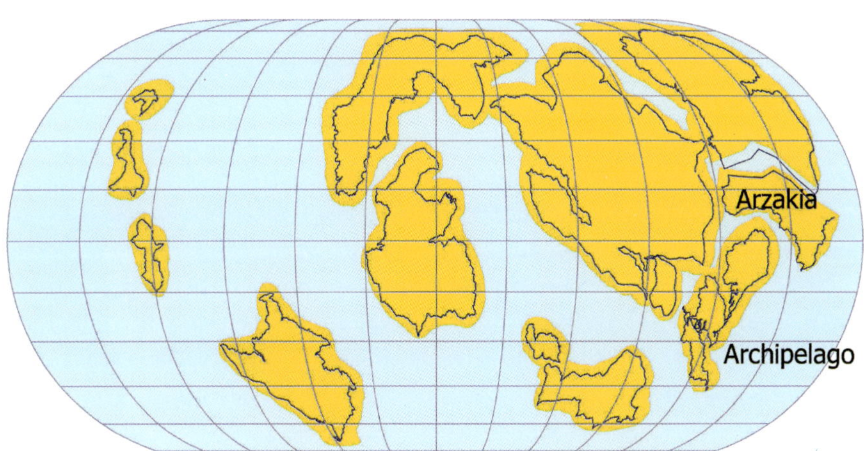

▶ 100 million years ago

The formation of *Bogoria* happened about 55 million years ago. Its two constituent plates are here seen as separate masses: *Bogia* and *Oria*. *Oria* contacted *Imparia* intermittently, depending on the sea level, so exchanges of species happened repeatedly in the last 100 million years. *Auralgia* probably saw some exchanges with *Oria* in this era too, but *Bogia* was isolated until its collision with *Oria*. Unfortunately, very little is known about the palaeontology of Furaha and so we can only guess at *Bogia's* lost species.

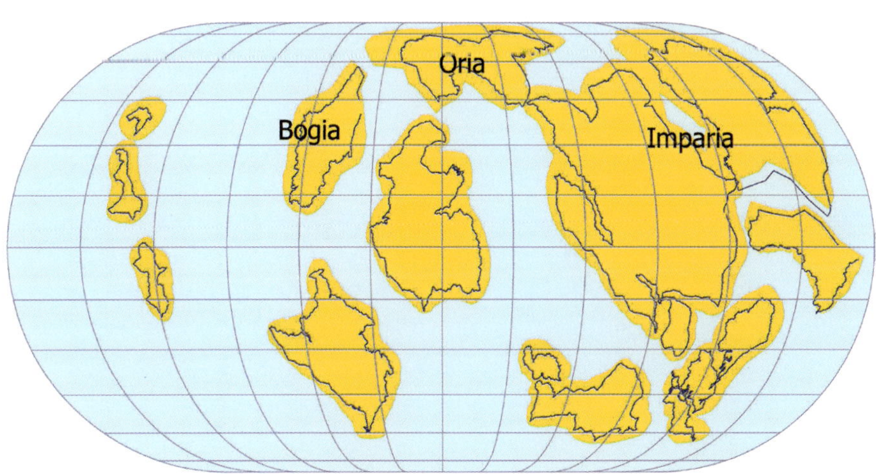

▶ 200 million years ago

Oria straddles the North pole at this time. Likewise, the *Archipelago* sits on the South pole, so both poles were covered by land. *Auralgia* and *Meralgia* were fused this long ago, forming the larger continent of *Panalgia*, that broke up some 160 million years ago. For *Meralgia* this was the last time it was part of a larger ecosystem. The *Palaeogaea's* were isolated all this time and it is not known when their populations last were in contact with those of other land masses. The data on continental drift are quite sketchy for 200 million years ago and no reliable data are available for earlier periods.

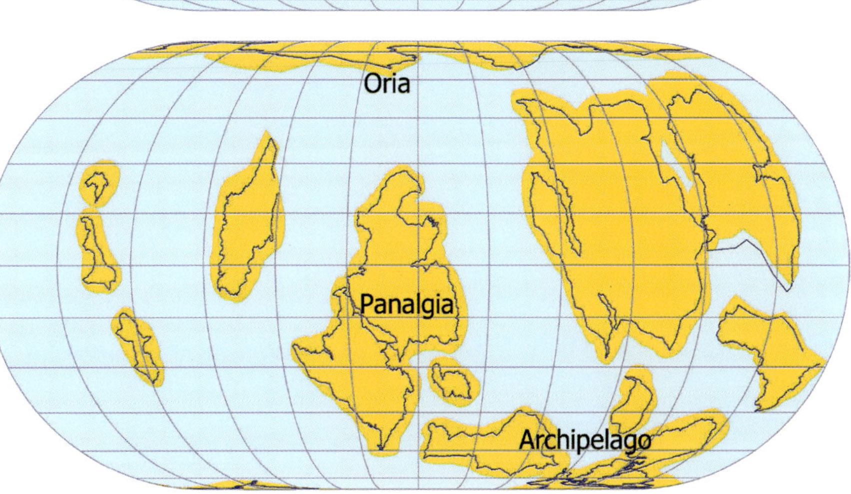

Grover Uytterwaerde

Aetatis suae 38 A.F.

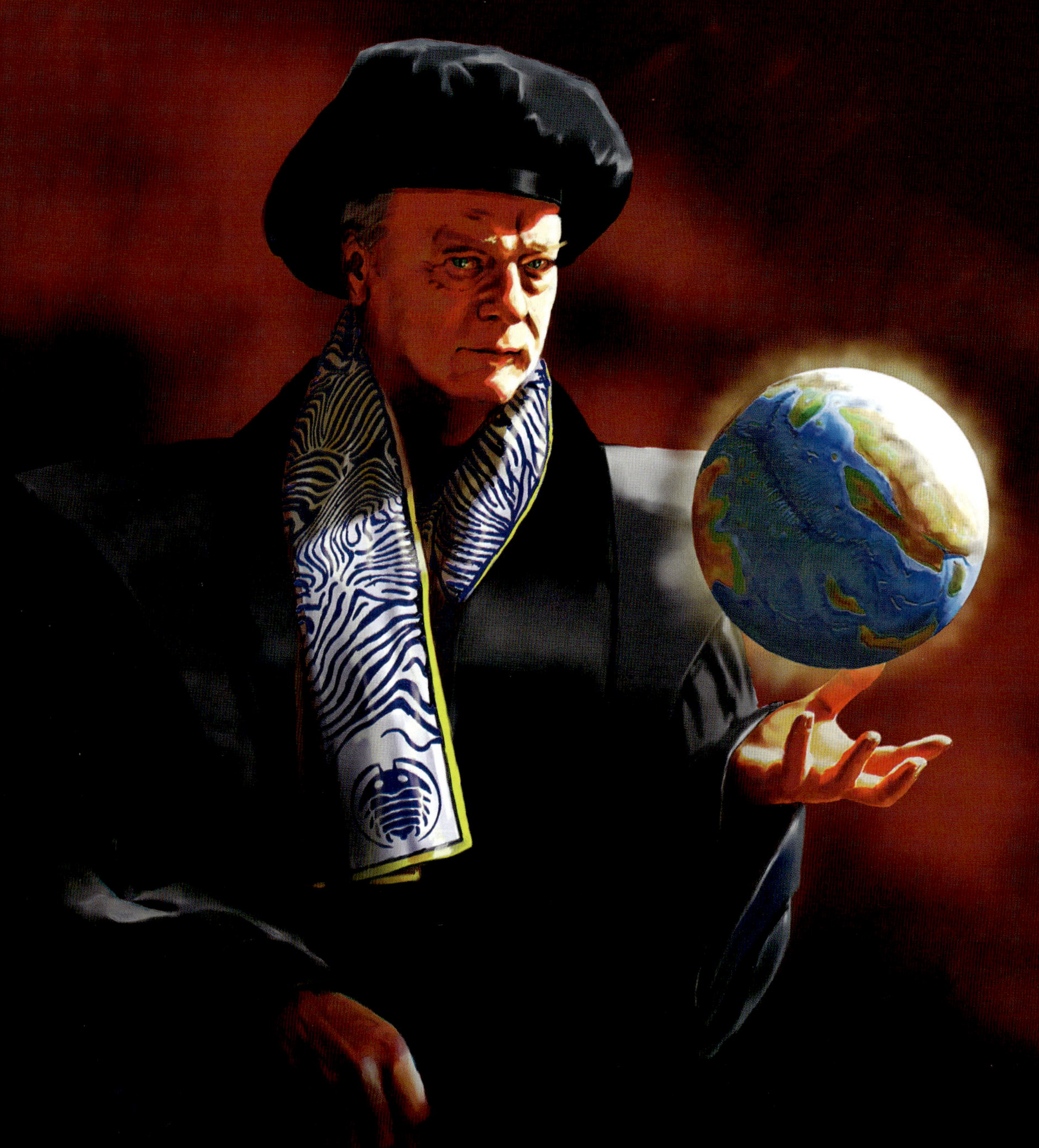

Chapter 2

Humanity on Furaha

'Travelling on Furaha reminded me of the lectures on Terran biotopes I attended as a student. But those virtual trips had not prepared me well for the physicality of the Furahan plains. I had thought that telepresence looked as real as anyone could ever need.

How little had I understood the subject of my studies! I had to stand out there, on the Furahan plains, before I started to realise my mistake. I had to see the horizon shimmer, to feel the heat of Jua on my skin, to hear the calls of the animals, and to *smell* a world. I smelled wet dust after the first drops of rain, I smelled the perfume of bobberpalms on a summer evening, and I smelled the tang of rusps. But before I could register all that, I had to be separated from mere theory to realise, finally, that truly understanding life requires living it.'

From *Worlds Apart: Natural Histories of Furaha and Earth* by Souren Nyoroge

◄ Portraits such as this one were commissioned to commemorate dignitaries in the highest echelons of the Institute and its University. Commonly done in styles of old Earth, the portraits often provide stylised views of the individuals in question.

This portrait shows Profissimus Grover Uytterwaerde, the third *Praeses Rector* of the University of Furaha and the first biographer of Nyoroge. The text states, in classical Latin rather than the common Neolatin, that his age was 38 years in Furahan years (his Earth age would be 61 years). He holds a floating glow-globe, as an obvious symbol of power, and wears the robes and cap denoting *Profissimus* grade and a shawl indicating *Rector* function. The shawl is arranged to display his cartouche, betraying an obvious pride. For about 100 years people embellished clothing and objects with animal and plant cartouches to indicate their clan.

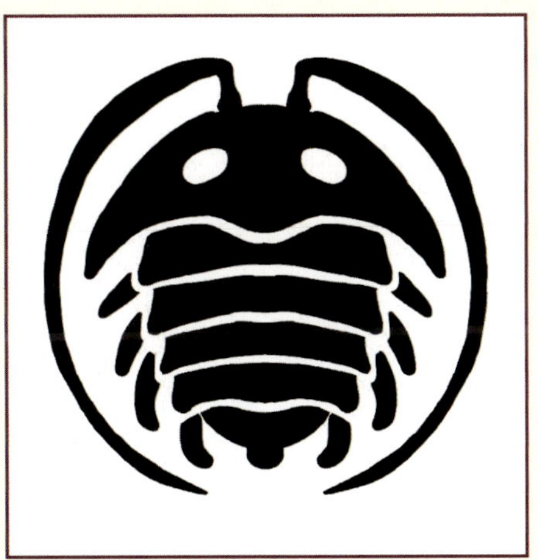

▲ Uytterwaerde's cartouche shows a stylised trilobite, meaning the wearer belongs to the Zoology Faculty. For contemporary citizen-scientists, the image would have conveyed faculty, grade and clan of the depicted person at a glance.

Arriving at Furaha

A Living Planet

Discovery

Once the Ngonjera's crew had identified the fourth planet of the Jua system as the most promising one, the captaincy decided to take a closer look. And so the crew made their ship slipdrift towards that planet, until they reached a high orbit. They shut down the slipdrive and secured the ship as any decent crew would. Only then did they take a good look, first with their instruments and finally with their own eyes. The crew looked with wonder at the planet below: it was glorious.

Similarities and Differences

The planet was a bit larger than Earth, with a higher surface gravity and a denser atmosphere. It was clearly geologically active, with rifts, fractures, volcanoes, subduction zones and all those other things that result from slowly drifting continental plates.

Well over half the planet's surface proved to be covered by water. There was sea ice on one pole. The local day was a bit shorter than Earth's, but there were more days in its year. The tilt of the planet's axis was less that Earth's. It had two moons, one of which was very small and orbited the planet at a considerable distance, while the other one was larger, but not nearly as large as Earth's oversized moon. And so it went: small differences in one direction were offset by small differences in another.

Names and Emotions

The crew enthusiastically started naming major features on the planet. This is how such things always go: there are supposed to be procedures, but people do not bother. The crew of the Ngonjera did not care much for strict procedures anyway. Someone thought of home while looking at this alien planet, and before the matter could be given any serious thought, a continent was named Bogoria, after an East African lake. Apparently, the ship's doctor was invited to come up with some fancy Latin names. This may explain why some continents have odd names, such as *Auralgia* (meaning 'ear pain') or *Meralgia* (meaning 'leg pain'). The crew either did not realise what the names the doctor had chosen meant, or they went along with her jokes.

Rumours flew, and people argued, talked and partied. Before long, there was speculation that humans might live on the surface. But what about the local life forms? For all they knew, the planet could be another Finkelstein's Folly, covered with frothy goo from pole to pole. Of course, there was the tiny chance of finding a rare world with complex animal life, such as Snaiad.

A Living Planet

At long last some robots went down. They sent back breathtaking vistas of white clouds in wide blue skies; the crew saw waves, rivers, deltas, lakes, swamps, mountains and forests. For a while they even dreamed – or feared – that there might be intelligence down there, something never yet encountered. As the robots went through their studies, no dangers emerged that technology could not handle. Then they followed the robots down themselves.

They looked, laughed, wept and gasped, and after half a year of wonder they decided to get back to the oikos and trade their data packets showing life on Furaha, which was the Swahili name they had given their Gaean, meaning 'happiness'.

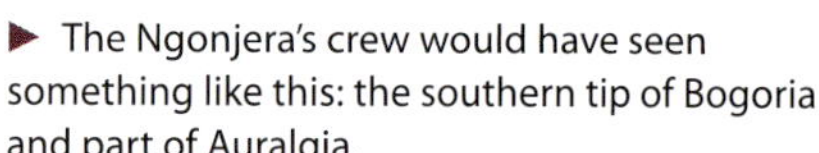

▶ The Ngonjera's crew would have seen something like this: the southern tip of Bogoria and part of Auralgia.

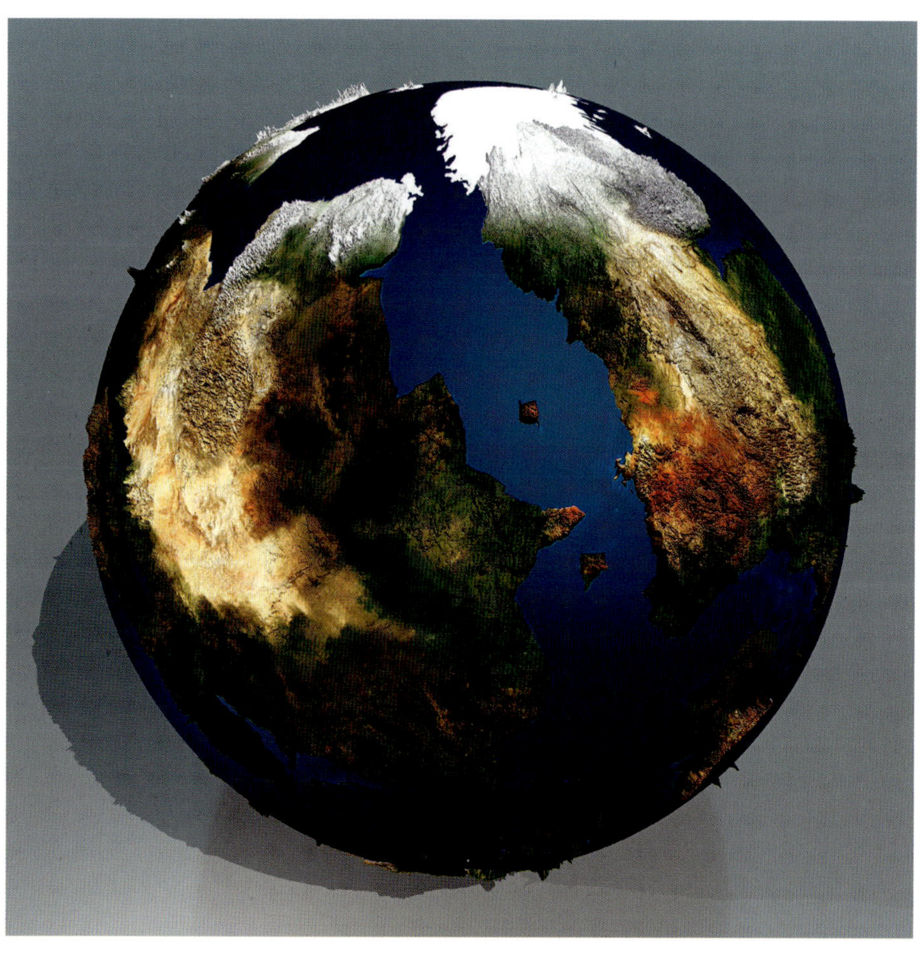

◀ A model of Furaha with accurate colours but exaggerated altitude.

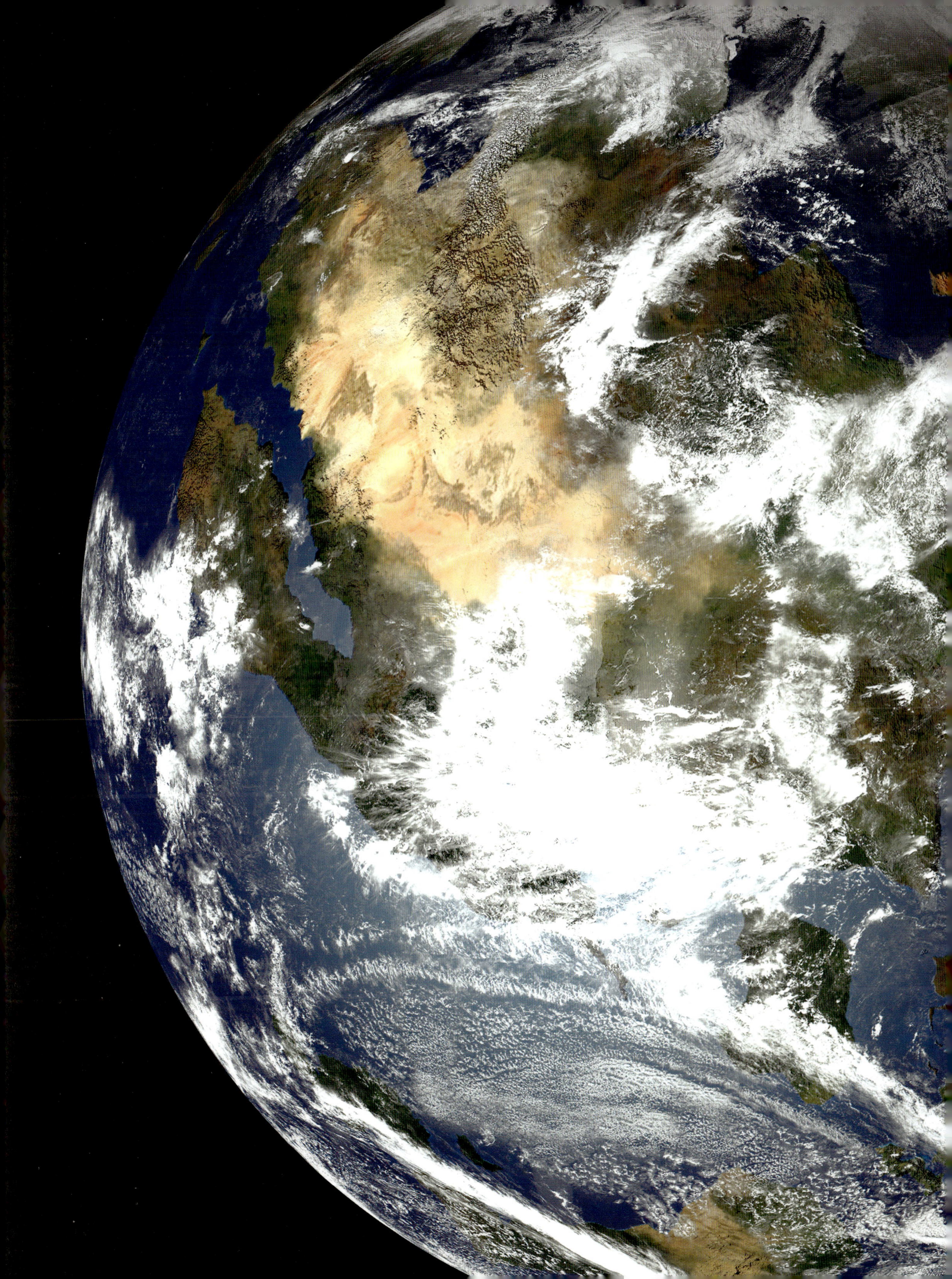

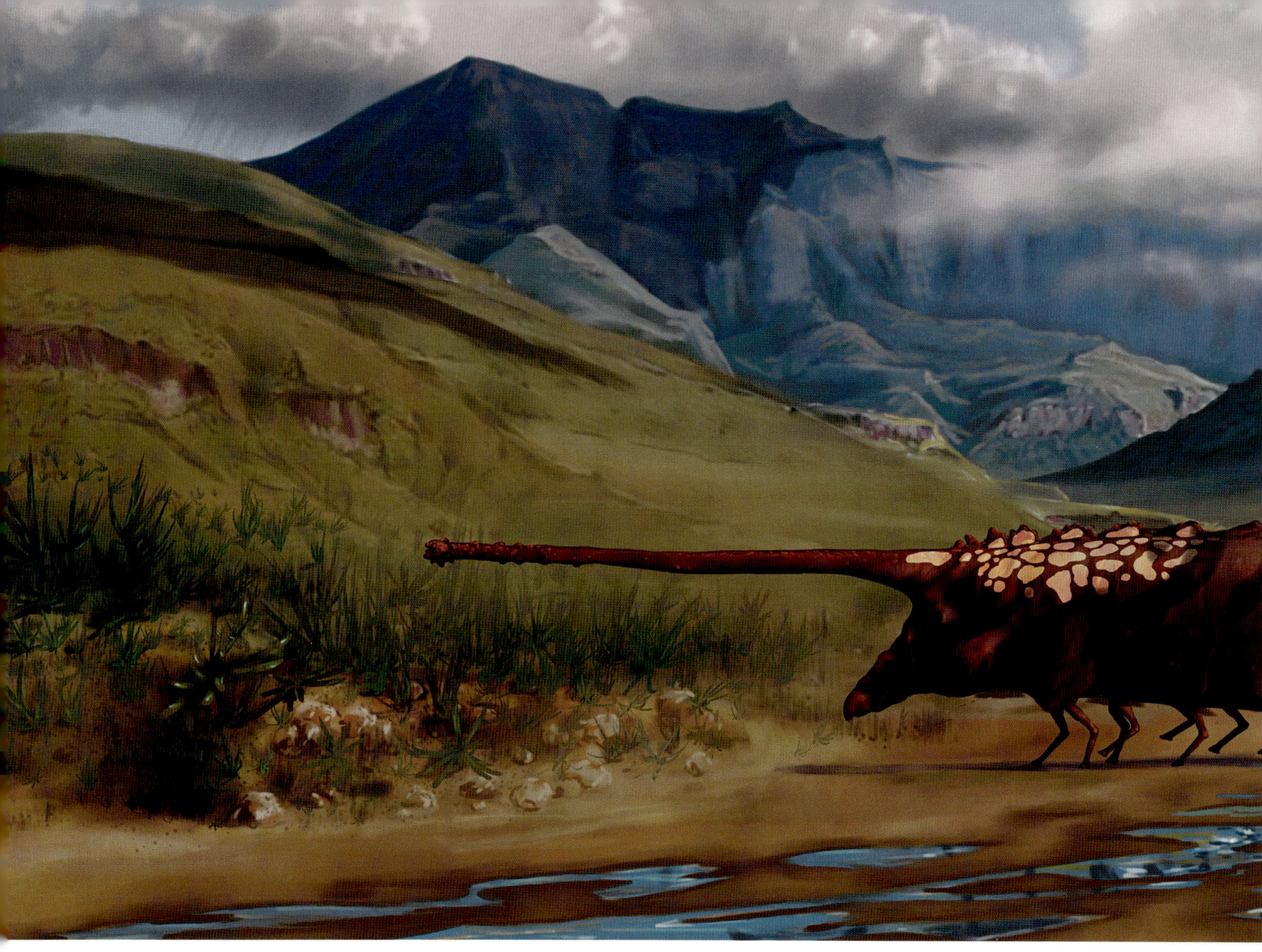

How Humans Came to Furaha

Sol III Homo semisapiens

The Background

The images and immersions brought by the Ngonjera were popular for a few months, after which most people turned back to work or took up their 'interests' again. Management stimulated people to have interests, so they would be content with their postdiluvian scarcity and allow the world to heal. Some, mostly in flooded Europe, took Furaha as their interest, studying the planet and its biology, producing fiction, poems, paintings and music. The management's artificial Minds purred happily until a new meme emerged, one that involved actually travelling to Furaha. Some even proposed that management should take the lead. Obviously, the quantum Minds had no desire to do that at all.

Start of the Horizonist Movement

When too many people talked about leaving for Furaha for too long, the Minds raised their virtual eyebrows: people could dream, but dreams should not lead to frustration. The Deluge had shown the harm that unchecked greed and frustration could do. The Minds started steering people to other interests. By then the Horizonists had chosen their name, saying that they wanted new and broader vistas.

Management Moves

As the movement progressed, the Minds alerted their human colleagues at the Seldon Institute. These considered the situation carefully and stepped up efforts to dampen the movement. This did not work well enough, so they chose a solution that only humans were allowed to choose. Management started nudging events. Some carefully chosen people seemed to meet by accident. In this way, a young enthusiastic Horizonist troublemaker, Filius Fuscus Bruyningh, just happened to meet the venerable scientist Souren Nyoroge and the two then just happened to meet a spacer who just happened to be looking for work.

Under Way

Before he realised what was happening, Nyoroge found himself the symbolic figurehead of an expedition to Furaha. His later books betray that he came to realise, shortly before departing Earth, that the ease with which the expedition was formed had been odd. But, by then, he was already swept up by events, along with other core members of the Horizonist movement. And so, the expedition departed Earth.

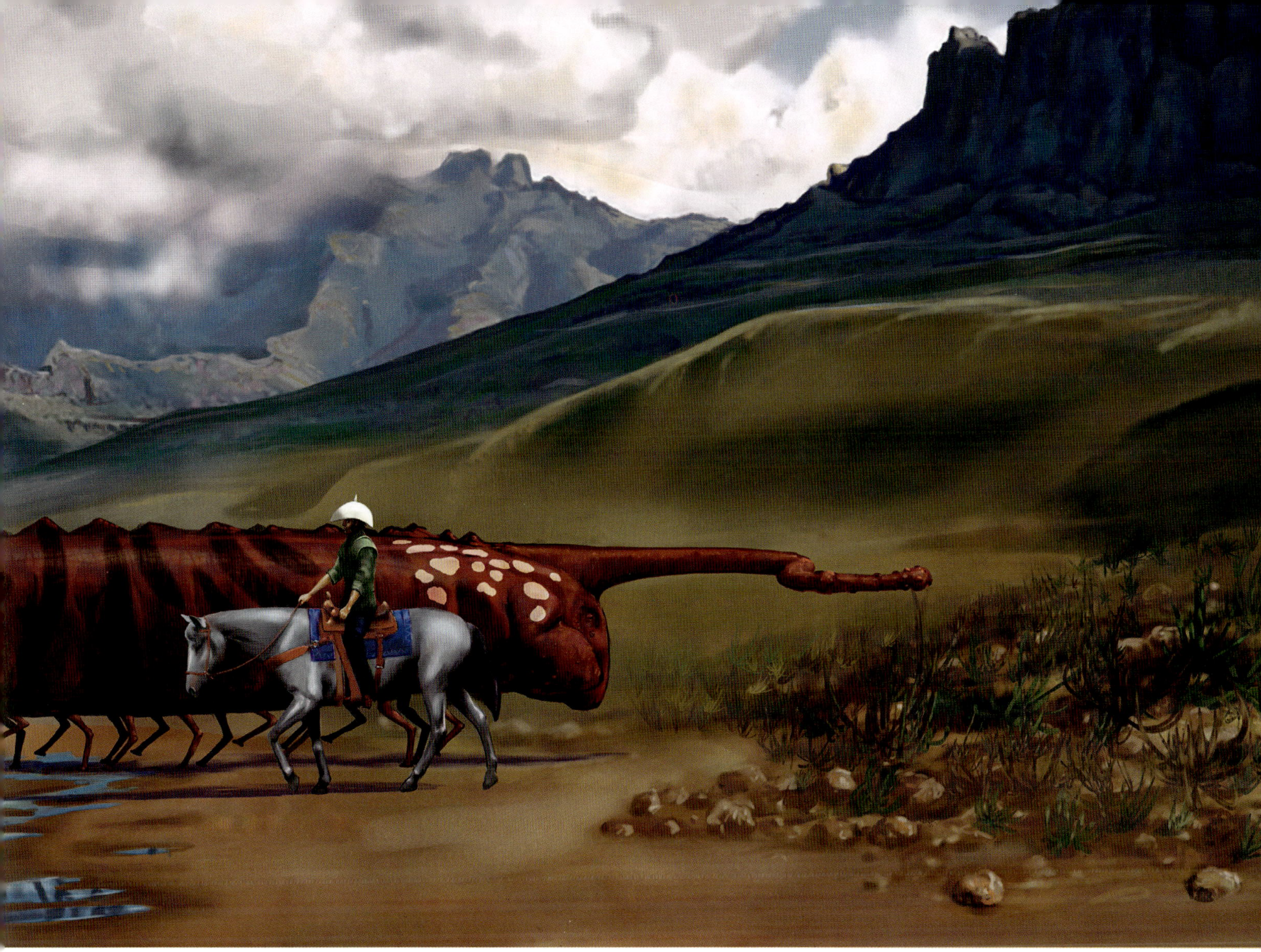

▲ This image shows an early expedition. The Horizonists had brought horses with them, reasoning that they should not cause much ecological disturbance. That was debatable, but the horses did not mind and needed about the same amount of biochemical and genetic tampering as did the humans themselves.

Once the key figures had departed, the remaining Horizonists began to wonder what the fuss had all been about. They resumed their lives as they were supposed to: quietly and complacently.

Management was happy because the population was happy and because there was now another human outpost. Having a few spare samples of humanity among the stars was not a bad idea, as Earth's ecosphere was still very fragile after the Deluge.

Homo semisapiens

All over human space mankind uses the ancient name *Homo sapiens* to describe itself biologically. This old binomen, meaning 'knowing/wise man', is a remnant of the ancient Linnaean naming system that had lost its scientific utility long ago. Still, the naming conventions using *genus* and *species* remained in use, as such names were popular, practical and less cumbersome than a full genetic label. For species on other planets, the name of the planet was simply added to a binomen, so the formal name of mankind became *Sol III Homo sapiens*.

Only on Furaha did people use a different name for humanity, thanks to Souren Nyoroge. Here is the quote giving rise to this custom, from *Worlds Apart: Natural Histories of Furaha and Earth*.

> When humanity first named itself, people felt that there was an immeasurable gulf between Man and Animal. But that gap later proved neither deep nor wide, so a more modest Man considered himself an animal among many. Mankind might better have based its description of itself for another characteristic trait, resulting in *Homo loquax*, or 'babbling man'. Jests aside, intelligence is undeniably a human characteristic. Sadly, history provides ample evidence that mankind rarely acts with wisdom. I have always felt that true intelligence must have two elements: the first is the ability for intelligent reasoning, which is ultimately harmful unless it is accompanied by the second element, that is, acting wisely. That second part is obviously too tall an order for us humans. Hence, I propose a more modest, and perhaps ironically also a 'wiser' name for ourselves. A just and humble name, I think, would be *Homo semisapiens*: man who is half wise.

The Institute of Furahan Biology

At the Heart of Society

The cultural and social makeup of Furaha's human colony evolved from the values of the Horizonists. This odd company appreciated learning and attributed status according to knowledge in multiple fields. Although the Horizonists had felt a need to do something, they did not exactly know what they should do. The Management Minds understood human nature quite well and expected people on Furaha to do what people on other worlds had done. That was to settle down so life became a copy of that on well-managed Earth. But not quite so on Furaha.

The Bureaurchy and the Institute

The Horizonists had not given much thought to the actual running of their little colony and so they naively elected bureaurchs to perform social duties. In other colonies, bureaurchs gained power due to a lack of other social structures. On Furaha, people looked up to ranking members of the Institute of Furahan Biology as role and status models, and for social guidance. The Institute thus gained political weight, ensuring a moderate role for the bureaurchy.

It took a few generations for the social system to develop fully. A scientific career in youth helped to acquire higher social status in later life, or at least to acquire the desired label 'citizen-scientist'. The various branches of the University, the Faculties, developed their own customs and divisions appeared. Zoologists looked down on Botanists, who among themselves muttered that there was no meaningful distinction between Zoologists and Parasitologists.

Social upward or downward movement was possible, but crossing lines between Faculties was difficult during some periods. Only Mathematicians were consistently considered free of such restraints.

Social Stratification

The number of hierarchical steps multiplied into many levels known as 'grades'. Having passed the various *studiosus* (student) grades, an aspiring citizen reached the *doctorandus/-a/-um* class, meaning 'he/she/it who must become learned' in Latin. After fulfilling one's thesis, the ex-*doctorandum* became 'learned', this is, a *doctor*. Afterwards, there were the increasingly lofty scales of *doctior* ('more learned') and *doctissimus* ('most learned'), with subdivisions. Few reached the grades of *professorandus*, *professor* or even *profissimus* levels and those few who chose to retire could prefix their grade with 'post', except when they were above *professorandus* grade, which would earn them the high-class *emeritus* postfix.

Simultaneously, society developed a parallel classification, the 'clades'. Most of the population was from mixed European, Asian and African descent. Sons were given first names according to their fathers' clade and daughters those of their mother's. After a generation or two people began to link their languages with their surnames. As daughters inherited their fathers' surnames, and sons those of their mothers, each family supposedly spoke two languages, although sometimes just a few words, apart from the common language.

For example, a man from the French clade 'de Hauteville' might have children with a woman from a German clade 'Özal'. Their sons François and Thierry Özal considered German their language of preference, while their sisters Heidi and Sissi de Hauteville preferred French. Within clades, powerful families formed clans. Groups such as the Trangs, Ngenys, Kwambais and Hongs enlivened the system. People outside the citizen-scientist class, such as Technoti and Bureaurchs, regarded the scientist system with a mixture of awe, amusement and irritation, all the while aspiring to become *Technarch* or *Magister* themselves.

Faculty, grade, clade and clan set status in this complex system. The system never fossilised, probably because of conflicting links between these elements. There was a time, in the 150s to 170s *Ab Colonia Condita*, when Faculty and Clade were linked. A Swedish Zoologist clan, the 'Linnaeans', monopolised high positions in this period, but resistance from the Bureaurchy, Botanists and the Zedong clan ended the dominance of the 'moose lovers'.

Science

Science does not progress quickly on Furaha. Field work can be dangerous, so many youngsters on their Science Stint prefer to traverse well-trodden paths that cause them no harm, yield no new knowledge but enable them to take their place in society. Still, curiosity, social status and the presence of a still largely unexplored planet mean that science does progress. Contrary to expectation, the number of humans increases slowly, causing the Malthusians to clamour for an absolute cap on the Furahan human population, pointing back to poor old Earth.

▶ From left to right: Sébastien Hrvatsky, Sigismunda Felsacker, Ed ten Borghue.

An Expedition

The historical scene shown here is a depiction from about 130 ACC*, showing people on an expedition. The figure on the right is of course Ed ten Borghue, Furaha's first Great Naturalist and the second Furahan citizen-scientist to be designated *Homo s. illustris*.

The young woman in the centre wears a partial contemporary Field Uniform marked 'S'. The 'S' was usually interpreted to reflect her grade as a mere *Studiosa*, but it is now thought likely that the 'S' indicates her name, as this is none other than young Sigismunda Johanssdottir Felsacker, later famous for her work on the Palaeogaeas and mostly for her book *Palaeo Days*, which defined early Furahan culture. The man on the left may be the enigmatic Sébastien Hrvatsky, the love of Sigismunda's life. Some twenty years later, Sigismunda wrote about his death under circumstances that were never clarified, in words that still echo through Furahan society. Felsacker, a Furahan native, left the planet and no-one on Furaha knows what became of her.

Oddly, Felsacker never mentioned meeting Ed ten Borghue in *Palaeo Days*. Perhaps young Sigismunda had eyes for Sébastien only.

**Ab Colonia Condita*: since the formation of the colony. Note that these are Furahan years, making it 208 Earth years.

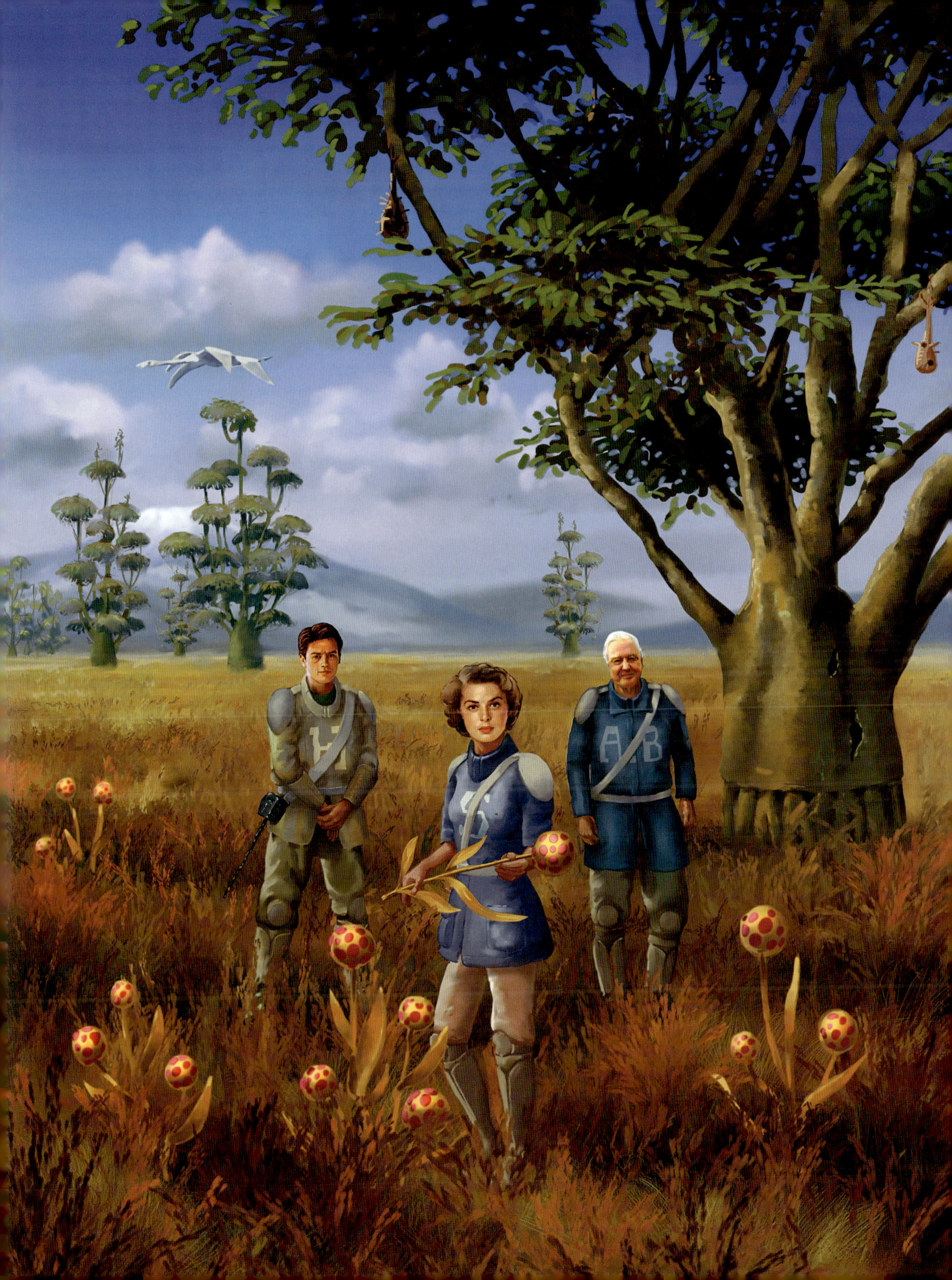
H
S
AB

Geography

Maps and Names

Basic Data

The surface area of Furaha is 1.29 times larger than that of Terra. In Terra's case most of the land is found in the form of a few large continents with many small islands, but there are more intermediate land masses on Furaha. Although no one bothered to define Furahan land masses as continents or islands, cartographers follow the ancient habit to mark the largest five land masses on a given planet with capital letters on a map.

North and South

The map betrays some other traditions inspired by old Earth. One is that 'East' defines the direction in which the sun rises, which in turn defines the other compass points as well. Another is to use a 360 by 180-degree coordinate system. Another custom was to use Latin for continents and major seas, but this

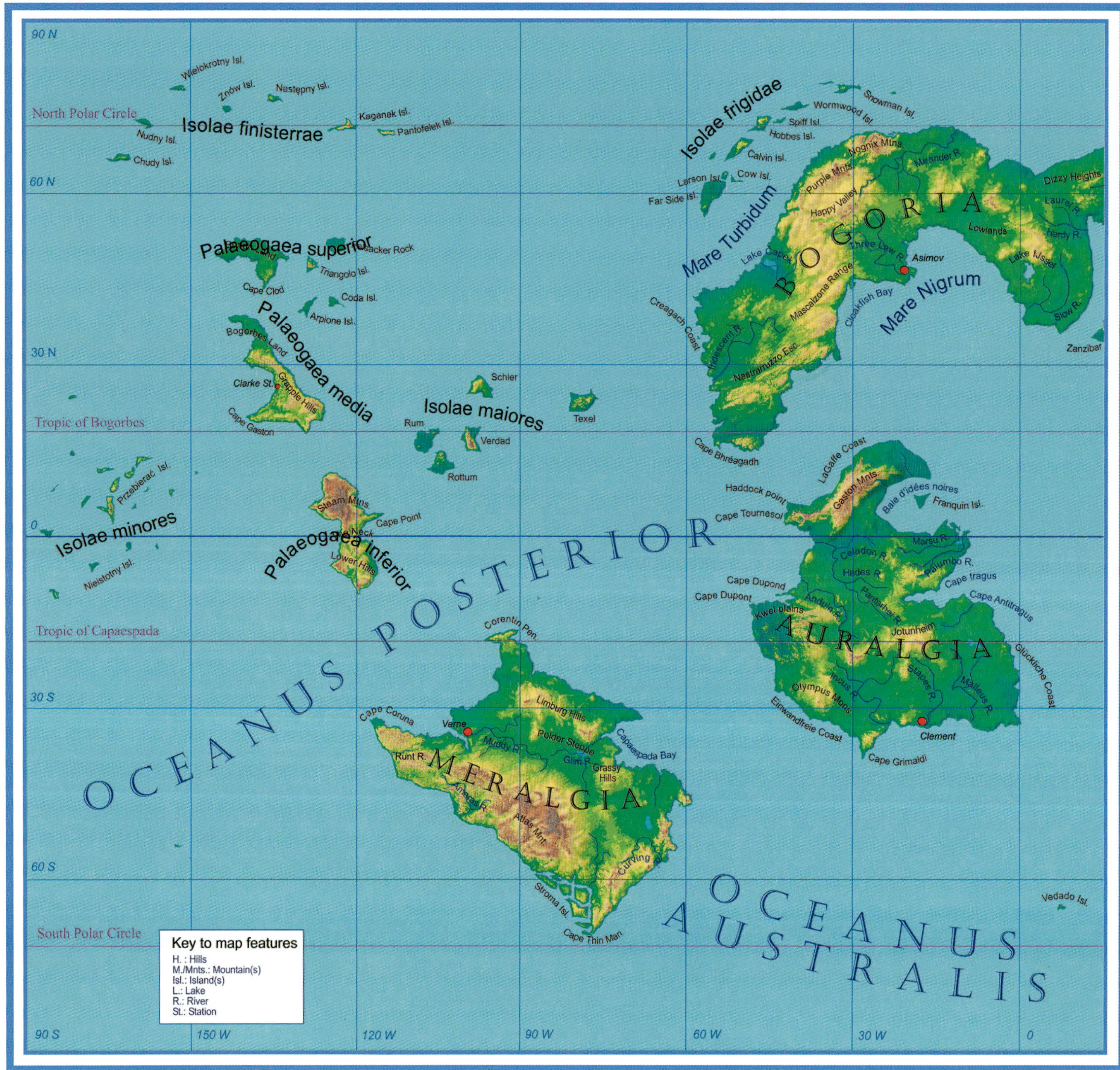

custom broke down quickly on Furaha. Names are shown partly translated here to do justice to linguistically rich Furahan customs.

Longitude and Latitude

The Horizonists placed the meridian system to allow the left and right edges of a world map to run clear of major landmasses. The exact choice was facilitated by centring the line on a tiny nearly circular island, called Antigreenwich. The map's edges also represent the date line.

▼ Altitude map of Furaha at equirectangular projection *(plate carrée)*.

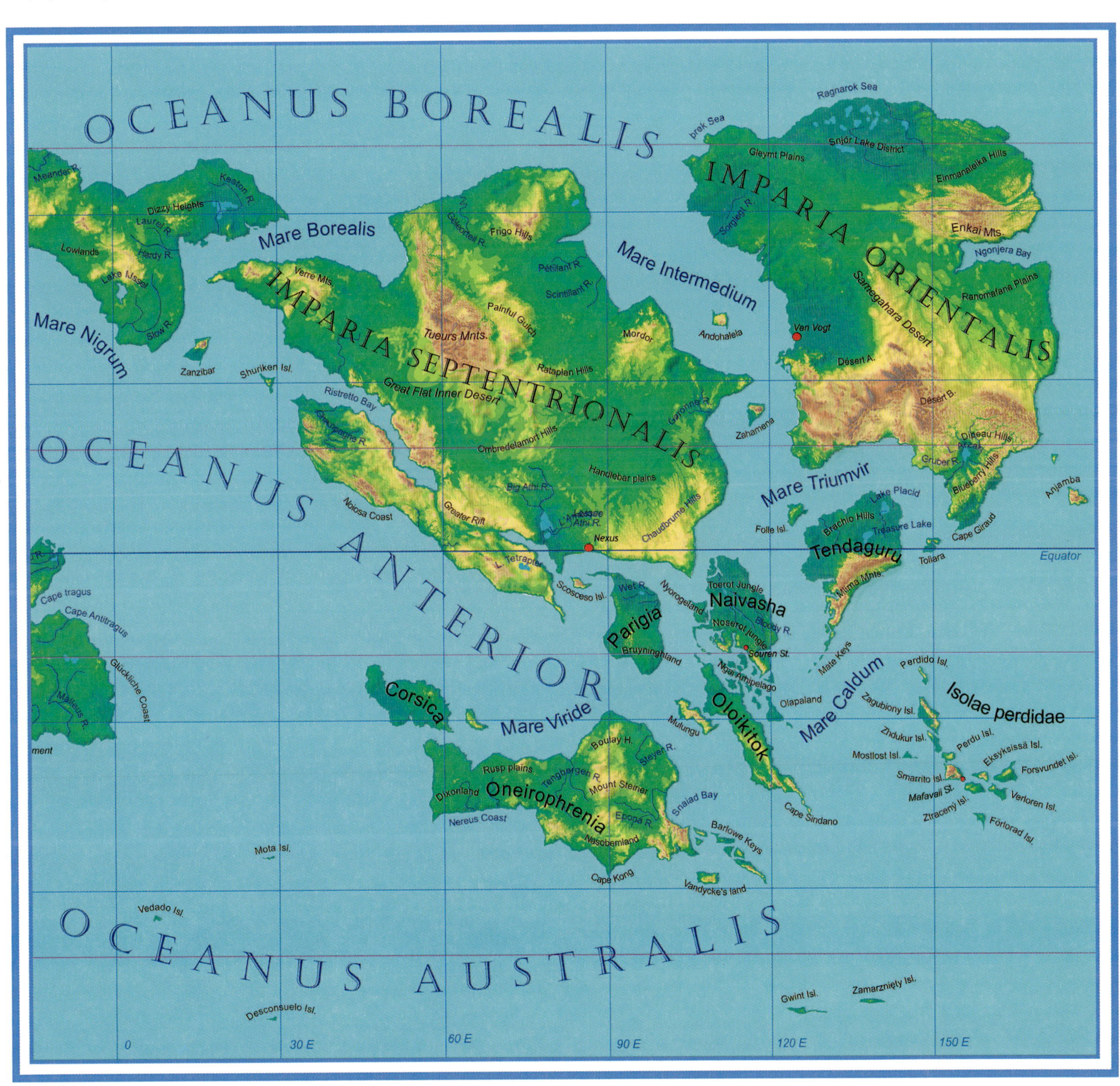

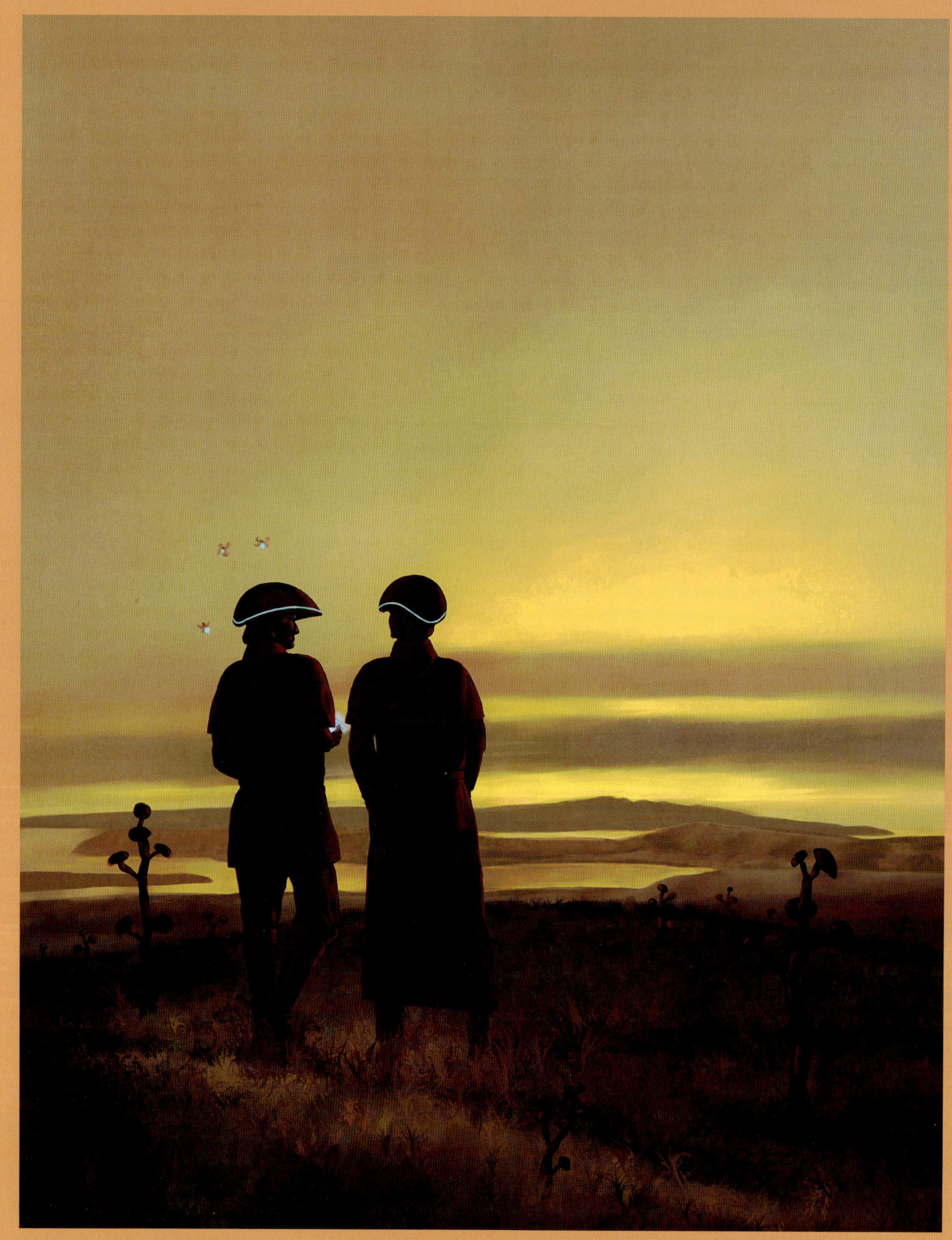

Chapter 3

Plants

'On each of the few planets with complex life that mankind encountered, some organisms (autotrophs) build their bodies out of dead matter, while others (heterotrophs) merely ingest biological materials fabricated by other organisms.

That basic distinction is often, but not always, coupled with locomotion, in that autotrophs are usually sessile and heterotrophs are usually mobile. On Furaha, as on Earth, larger sessile autotrophs (that is, plants) use light to power their metabolism. On both planets, the energy pyramid of the entire ecology rests largely on the ability of plants to use sunlight.

What a surprise it was to find that Furahan plants need not be green. Imagine our greater surprise to find that all these Furahan red, green and grey plants were much better at photosynthesis than Earth's plants, whose efforts we know to be very meagre indeed.'

From *Worlds Apart: Natural Histories of Furaha and Earth* by Souren Nyoroge

◀ This painting, called *The Brothers*, is a reinterpretation of an old Earth painting. Furahan citizen-scientists felt that paintings could be performed and interpreted as often as desired, rather like music.

The brothers are the twins Caspar and David Nastrarruzzo who often used 'Friedrich' for the two of them, confusing many who wondered whether there was a third brother. The twins often also played with their last name, using variants such as 'Nastrazzurro' or 'Nastradesivo'. The Nastrarruzzo clade elders feared that such frivolity harmed their 'social phenotype', a Furahan concept describing how outward appearance colours a clade's status in society. The elders were too late: the brothers ensured their family acquired a 'rogue' phenotype lasting two generations.

The painting depicts a chapter from their book *Trapping Tetrapters*. The brothers noted that their hatbands resembled the colour of a local bioluminescent tetrapter species. They increased the intensity of their hatbands and relished that they were consistently followed by amorous tetrapters.

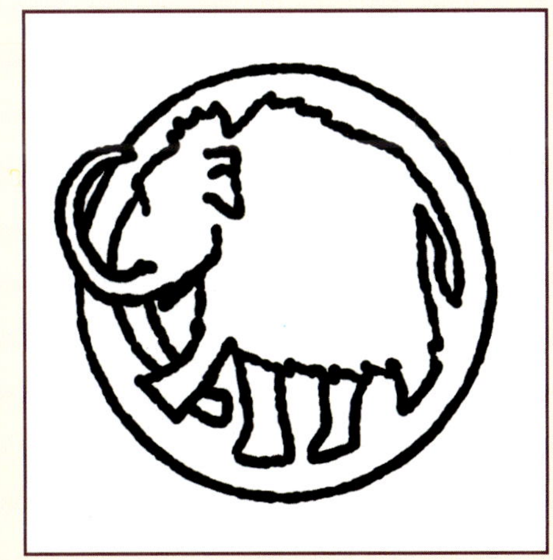

▶ The Nastrarruzzo cartouche showed a stolid Earth mammoth until the brothers managed to make it look irreverent.

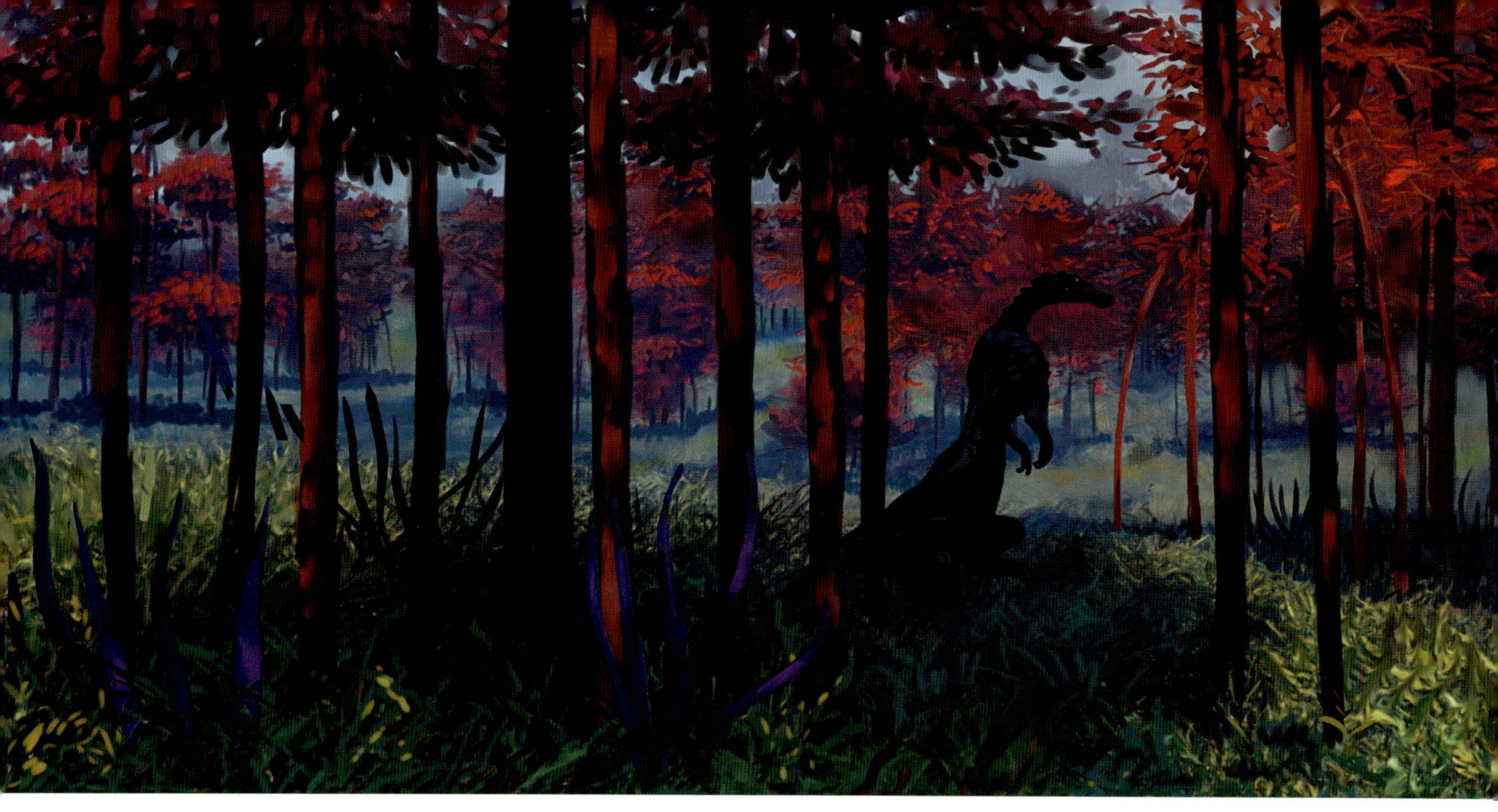

The plural of photosynthesis

Subtracting Colour

Even if the occasional leaf on an Earth plant is red or spotted, there is still the same green pigment underneath: chlorophyll. In fact, chlorophyll means 'green leaf'. Some still think that green is the best colour for plants, which it is not, really, or at least not always and not everywhere.

The Green, Green Grass of Home

Only a part of the energy radiated by Earth's sun can be seen by humans. That part is a mixture of wavelengths. Going from low to high wavelengths, colours are usually distinguished as violet, indigo, blue, green, yellow, orange and red. The sun emits most light in the region of the green and yellow wavelengths, so you would expect Earth plants to preferentially absorb these colours. They do the opposite though: chlorophyll absorbs blue and red light but reflects green and yellow light. Earth plants are green because they cannot use green light. Why they evolved to do so is a riddle. Perhaps bacteria had already occupied the best spot, using green light, or perhaps little yellow or green light penetrated the early seas where plants originated.

The Red, Green, Grey Grass of Furaha

The spectrum of sunlight on Furaha closely resembles that of Earth. On Furaha, as on Earth, photosynthesis evolved more than once. On Earth, however, complex plants all use chlorophyll, while there are three lineages of complex plants on Furaha.

The first group, known as 'chlorochromes', absorb blue and red light, which turns its leaves as green as those on plants on Earth. The second group has red leaves. These 'erythrochromes' may look like a temperate autumn forest on Earth, but on Furaha the colour is permanent.

The third group's pigment absorbs light over the entire spectrum, so these 'poliochromes' are grey.

Binding Light

On Earth, photosynthesis in plants only puts about 5 per cent of light energy to good use. The reason for this poor efficiency is that the chemical reactions of photosynthesis run in the wrong direction almost as readily as in the right one, wasting a lot of energy. Earth's plants can also only use a small part of the light at noon, because the maximum capacity of photosynthesis is rather limited. Finally, photosynthesis does not work well with little light either.

Photosynthesis in all three large Furahan plant clades operates at about twice the productivity of Earth plants, which has some intriguing consequences. For the same total leaf area, a Furahan tree can grow twice as fast, or it can grow at the same rate with half the leaf area. Some Furahan trees are extremely tall because they need a smaller amount of leaf mass. Furahan plants can also grow with a much smaller amount of light than any Earth plant. On the whole, the amount of energy that flows from Furahan plants is higher than on Earth, explaining quick growth, abundant numbers of seeds and a generally bustling ecosystem.

All of this evolved because the ancestors of Furahan plants stumbled upon better pigments than their counterparts on Earth. The rest is history; natural history.

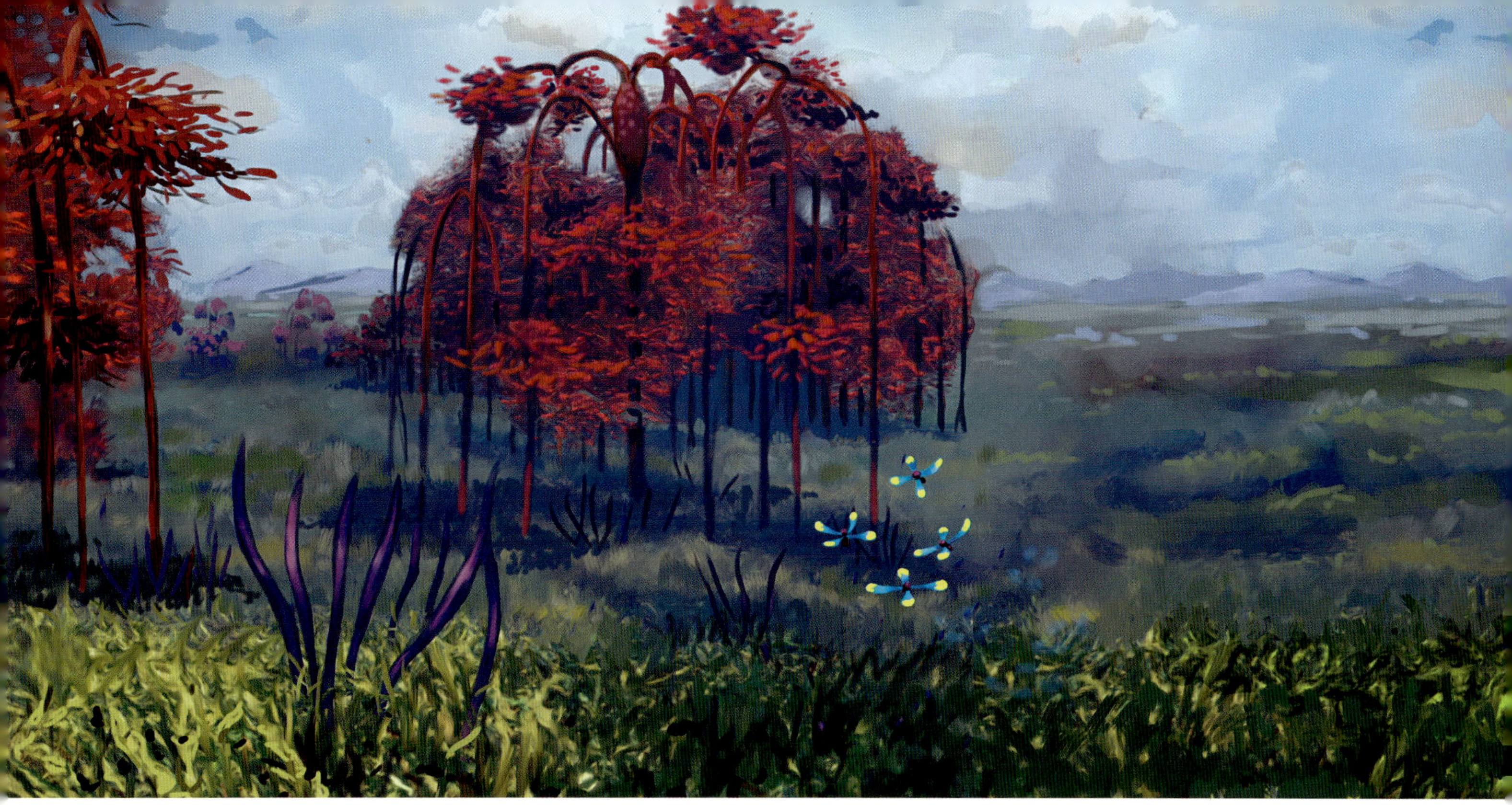

▲ Red polypremnic trees contrast with green and yellow grass, while a snafe ponders existence under the trees.

▶ **Chlorochromes**

The word chlorochrome means 'green-coloured'. Blue and red light are absorbed and subtracted from white sunlight, leaving green and yellow to be reflected and seen. A section of the Institute's Botanical Gardens was planted only with chlorochrome Furahan plants, to give people an idea of the limited colours of Earth plants. The two leaves shown here come from the same tree species. The upper leaves are split, which ensures that lower leaves receive more even light.

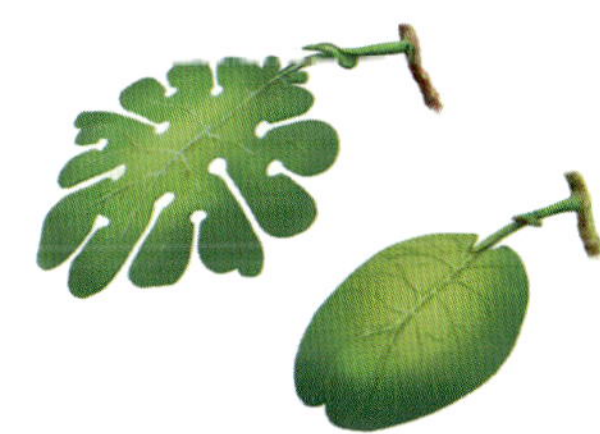

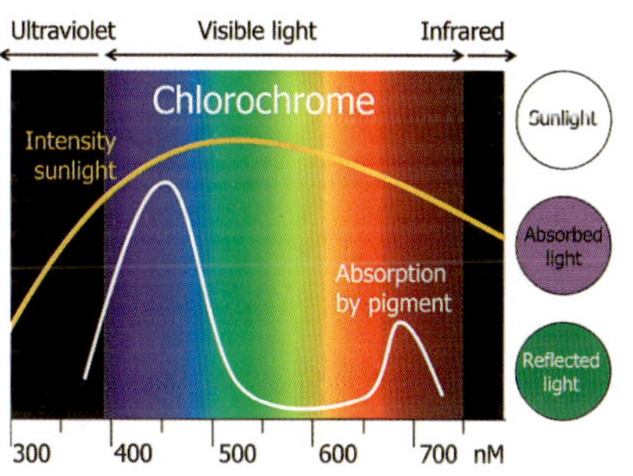

▶ **Poliochromes**

Grey plants absorb a bit more yellow and green light than other colours, so their leaves are slightly purplish. While women might differentiate the colours of Furahan leaves as grey lavender, mauve-grey, lilac-grey or periwinkle greyish, men typically shrug and say 'grey'. Many travellers reported becoming depressed after a stay in areas with poliochrome plants only, such as the Melancholy Forest, with its Weeper Trees and Sadsap brushes.

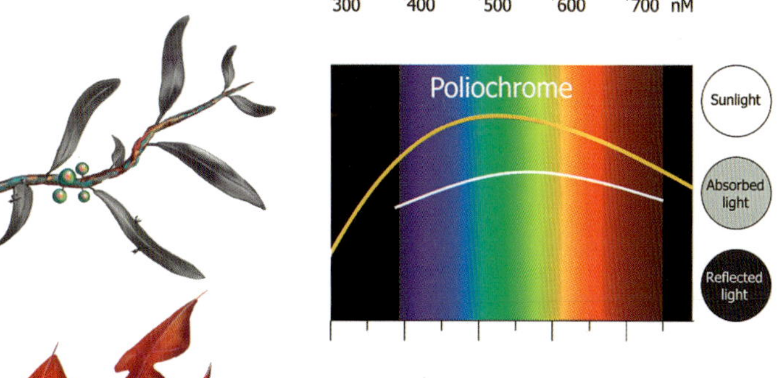

▶ **Erythrochromes**

This red plant, the redweed, has no clear difference between stems and leaves. The Botanical Gardens have an area devoted to Erythrochromes, called the Martian Garden, even though Mars has no trees at all. Historians blame unchecked romanticism for this misnomer. Tourists love it.

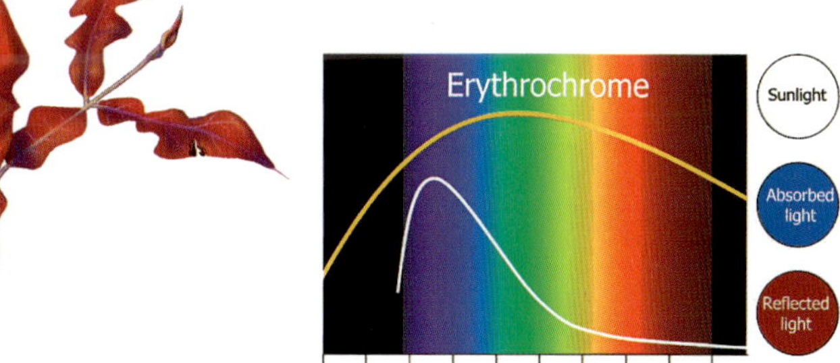

ROMANARC TREE

INSULA COELI

Newcomers to Furaha often mistake a Romanarc tree for a small grove of separate trees. It never enters the mind of people born and raised on Earth that trees can have more than one trunk. Furahan citizen-scientists of course consider trees with many stems normal. The technical term for having many stems is 'multirobur' or 'polypremnic', terms that everyone but a botanist forgets immediately. Polypremny is a characteristic of various clades of Furahan plants.

The presence of many stems gave the Romanarc tree its common name. The tree may even have been baptised by a shipmate of the spaceship Ngonjera. No doubt that person had the arches of classical Roman buildings in mind when naming the Romanarc tree. Other Furahan tree types, the 'Cathedrals' and 'Gothicals' no doubt received their names in similar fashion.

Ecology

The Romanarc tree is a major ecological player on the plains of the 'ruspveldt'. Left to their own devices, these plains would quickly evolve into forests. But large herbivores, megarusps in particular, do not leave them to their devices. By eating and destroying most saplings they turn the veldt into a steppe or savannah.

Growth

A Romanarc sapling looks nothing like its adult version. It is slender and has only one stem that shoots up quickly, even by Furahan standards. The plant tries to elevate its leaves out of harm's way as quickly as possible, but, even so, most saplings end up as leaf pulp in an herbivore stomach. The few saplings that manage to produce leaves at a reasonable height enter the next phase where the tree forms its first arches. The ultimate result is a network of arches that are attached to the soil at both ends. This flexible network cannot be toppled, no matter how hard rusps may try. Rusp foraging may destroy one stem and sever the connection between the ends of a stem, but the separated parts simply live on.

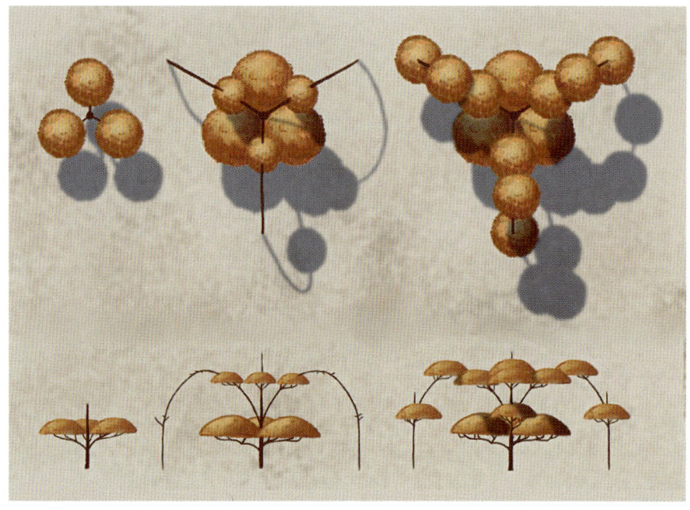

▲ **Early Growth Stages**
A young tree is shown on the left, with a single stem. It has three clusters of leaves, called 'cupolae'. The middle stage shows growth of these cupolae and three arches springing from the central trunk, offset to the main cupolae by 60 degrees. These arches carry their own first cupolae. A later stage, at right, shows more cupolae on the arches, while new stems grow up from the arches. These new stems will later produce their own arches. An adult Romanarc is shown in the main painting.

Islands in the Sky

Romanarc trees provide resting places for terrestrial as well as aerial animals. For flying animals, the tree offers an alternative to roosting on the open veldt, where a multitude of predators would welcome them. For avians, the Romanarc tree is truly what its scientific name says it is, that is, *Insula coeli*, meaning 'an island in the sky'.

▶ The wadudu castles in the foreground resemble Earth's termite mounds. Their hard white covers offer protection from the sun. The wadudu construct tubes on and in the ground linking up their castles.

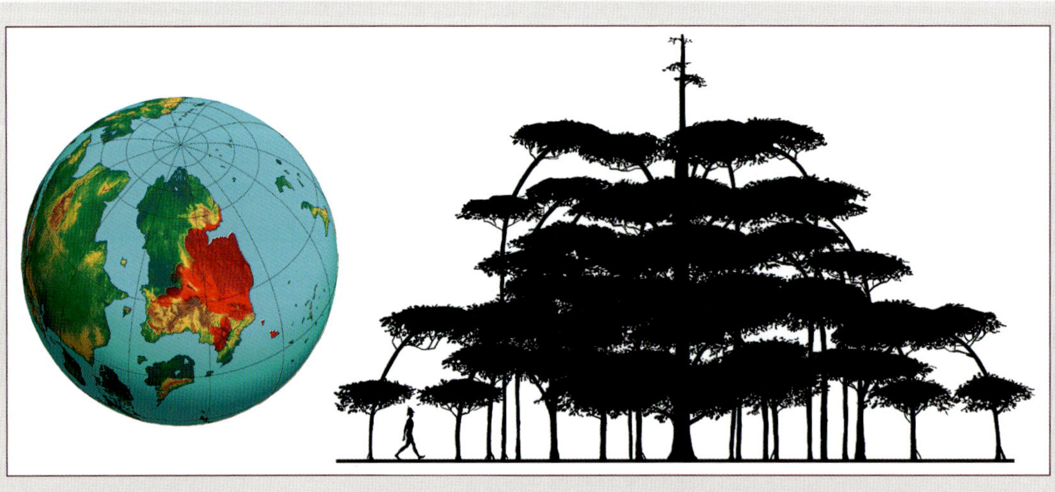

◀ **Romanarc Tree**
Insula coeli
Name derivation: *Insula* (L.) (meaning 'island'); *coeli* (L.) (meaning 'from heaven')
Habitat: Ruspveldt
Distribution: Fairly dry to semi-desert open terrain in *Imparia*
Height: 15m

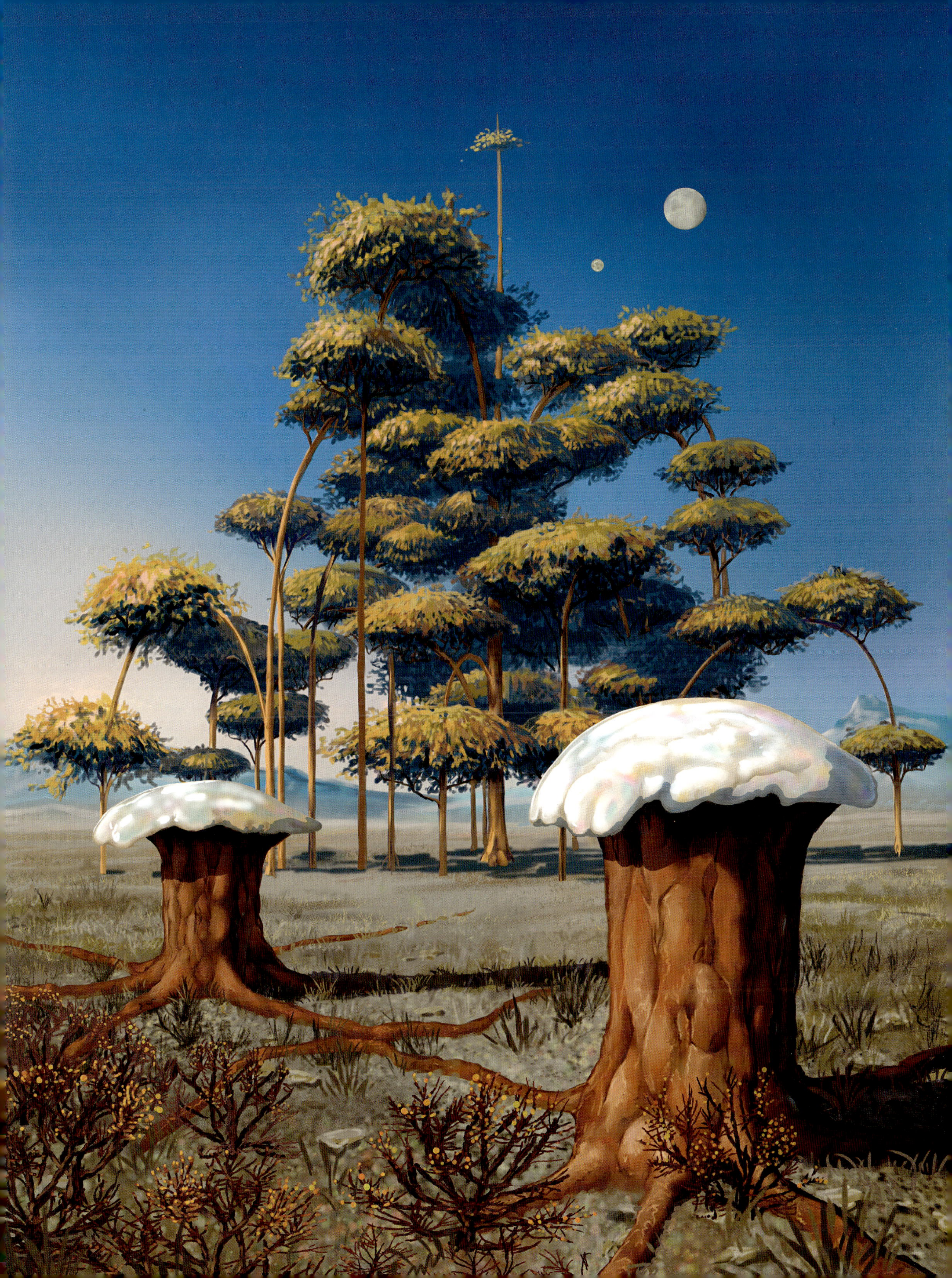

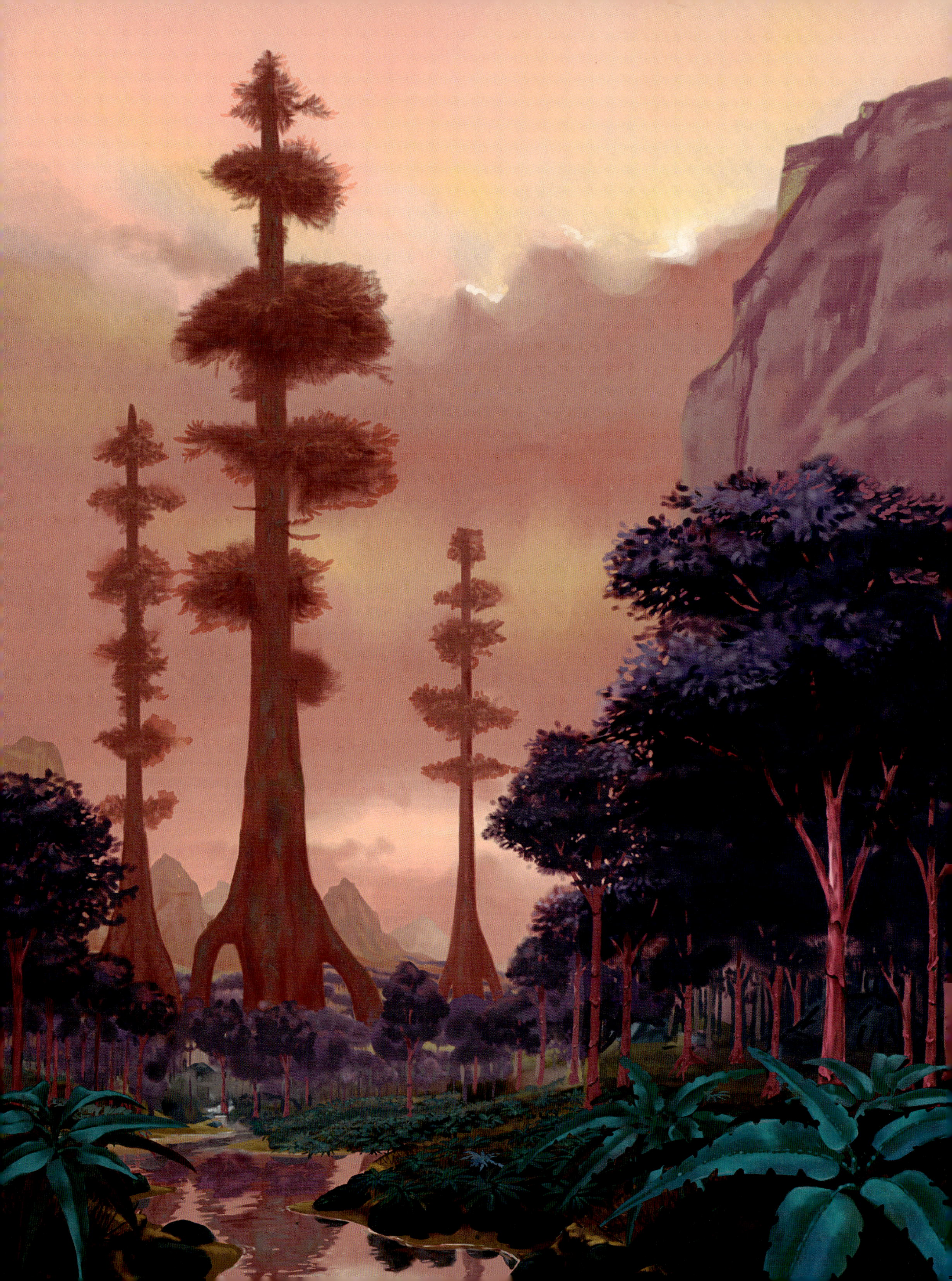

Odin's spear or Odinspear

Gungnir gungnir

'Odin's spear', usually simplified to 'Odinspear', may well be the largest tree species on Furaha. The largest odinspear was measured at 121.3m, but that was in winter, so the tree would be higher in summer when the wood expands. There may well be larger ones in unexplored regions.

Age

Terran trees take 500 to 800 years to reach sizes of over 100m, but odinspears probably take much less time. One reason for this is that Furahan photosynthesis is more efficient than Earth's, allowing faster metabolism. Another reason is that the plant kingdom that odinspears belong to, the poliochromes, constructs trunks in a way different from Earth trees without producing growth rings as handy markers of a tree's age, as the trunks are tubes.

◄ Evening view of a stand of odinspears.

Wooden Tubes

The evolution of tubular trunks resulted in a large space inside the tube. In some clades this space remains empty, which provides an interesting dwelling for many animals, provided they can gnaw, bore or chisel towards that space. In other clades, the space is filled with toxic foam, in an apparent attempt to keep out such unwanted guests. In yet other cases, the trees reward some animals with ready-made spaces or secretions. In turn, those guests are supposed to drive unwanted freeloader and parasitic species away. It will not surprise anyone familiar with evolution that the result is a continuously evolving and shifting mess of intricate interconnections.

Both the inside and outside of immense trees, such as odinspears, house entire ecosystems, providing enjoyment for nature watchers as well as research opportunities for career citizen-scientists.

Caradoc Nastrarruzzo's Tall Tales

Although portraits of august members of the Nastrarruzzo clan grace Academia Halls, the exploits of some other members of that clan are told more often in cafés.

Caradoc Nastrarruzzo boasted to have served as a First Expedition Guide many times, although records suggest he usually served as *Factotum Tertium*. Caradoc once claimed discovery of a tree species even taller than odinspears. Caradoc, obviously hoping for fame, urged to reroute the Expedition and brashly proposed to name the trees *Clava cuchulainii* (meaning 'Cú Chulainn's club'). No such trees were found, but for the rest of his life Caradoc blamed this on a routing mistake, insisting that Cú Chulainn's club was out there.

▼ **How to Grow a Tube**

The trunk consists of a tube, strengthened by crossbeams or local thickening. A tube is as strong as a solid cylinder of the same diameter, but much less heavy.

The inset shows that growth of a tube trunk requires that wood can be formed as well as removed. Both these xyloclastic and xyloblastic processes take place in a crucial tissue layer covering the wood on the outside and inside. Tube trunks save on wood by needing less material than solid trunks and by recycling old wood during growth.

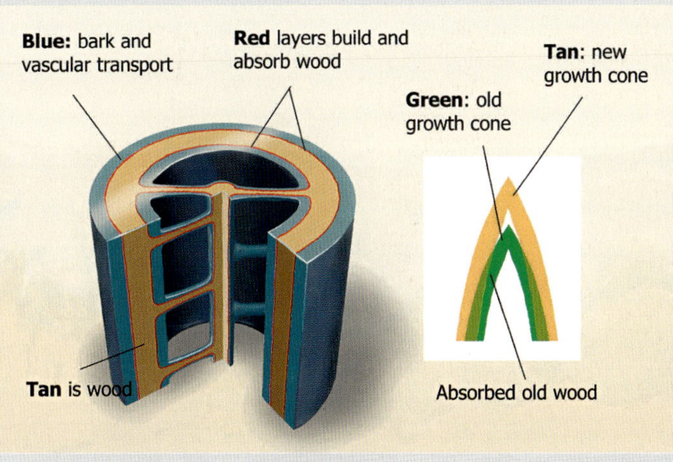

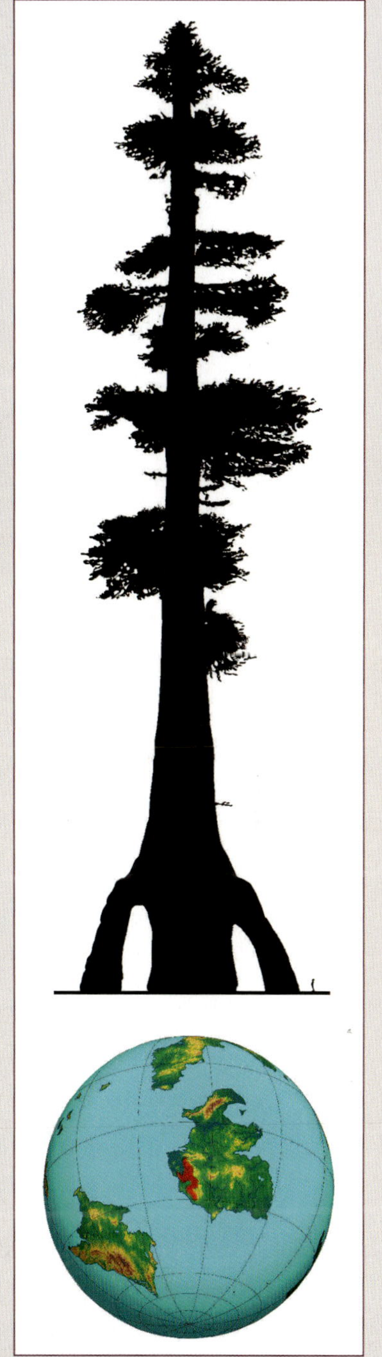

▲ **Odinspear**
Gungnir gungnir
Name derivation: *Gungnir* (Old Norse) (meaning 'spear of Odin')
Habitat: Temperate hills
Height: known 120m

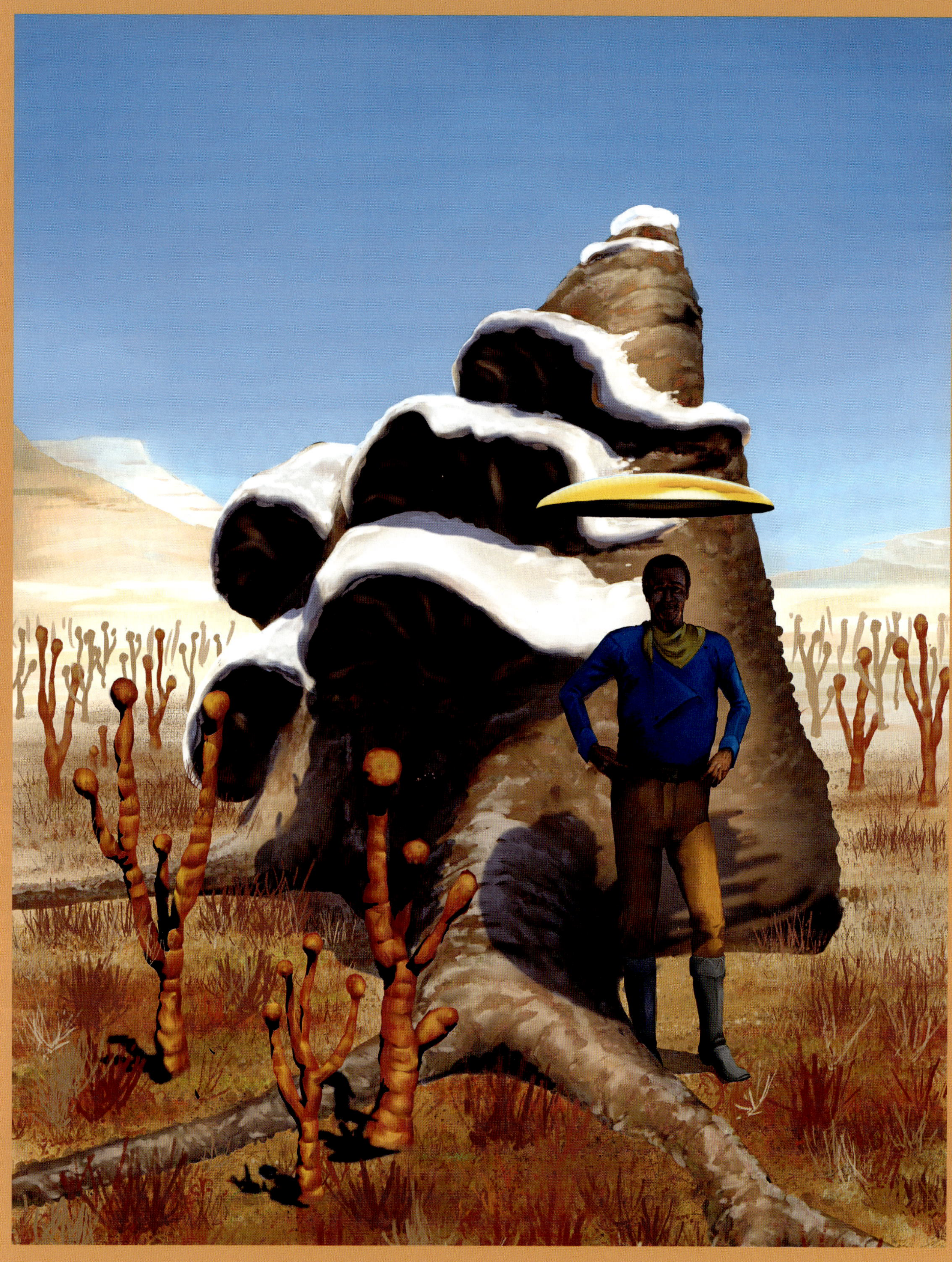

Chapter 4

Mixotrophs

'Plants, animals and fungi; this was the order of multicellular life that used to be taught to Earth's schoolchildren. But that scheme reflected limited understanding and life was too complex, too exuberant, to fit such a simple scheme.

Furahan children learn a more complex scheme. They know that *Plantae* are sessile autotrophs, *Animalia* are mobile heterotrophs and *Fungi* are sessile heterotrophs. But they also learn about photo- and chemosynthesis, and we teach them about creatures whose descriptions include 'auto' and 'hetero' as well as 'photo' and 'chemo'. They learn that these life forms, Furahan mixotrophs, are mobile as larvae and sessile as adults.

When first faced with such creatures, some disliked seeing the 'natural order overthrown'. But in biology everything is natural by definition and 'order' is often just wishful simplification. There are laws of nature, but there is no law prescribing that nature or its laws must be simple.'

From *Worlds Apart: Natural Histories of Furaha and Earth* by Souren Nyoroge

◀ Johann Wainana was famous for his work on wadudu castles, built by colonies of small wadudu. Even though the colonies count tens of thousands of individual wadudu, the animals are rarely seen. Wadudu do not venture out into the heat of the day and gather what they need from below ground. The castles have clever sunroofs that overhang any parts in danger of overheating. While a major part of the castle resembles the soil around it, sunroofs are white, highly reflective and impervious to water.

Wainana, whose ancestry went back to a Kikuyu crew member of the Ngonjera, mockingly cultivated a 'tough man' persona, which earned him the nickname of 'sheriff'. In real life he was a kind man, teaching students to check their boots for spidrids every morning, ''cause neither you nor the spidrids are gonna like it otherwise'. (Wainana was touchy about his fear of spidrids.) The one thing he was very dogmatic about was science. To this day he is famous for drawling out his favourite one-liner: 'A scientist's gotta do what science says he's gotta do'.

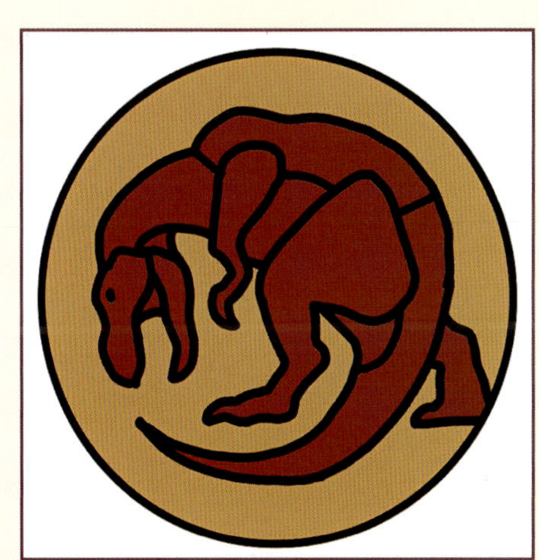

▲ When Wainana earned the right to use a cartouche, he went for a Tyrannosaur. Many think it is shown in a threat posture, even though Wainana himself maintained it was just 'doubled over laughing'.

PHALANX

PHALANX RIVULUS

The phalanx, a typical aquatic mixotroph, has various organs that are connected by stalks to a submerged central globule. The form and number of these organs vary markedly, causing much confusion. The spheres floating on the water are a good example of such variability. They are called *corpora spheroidea dentata* because of their serrated openings. Some varieties of phalanges grow only a few small spheres, whereas the South-East variety (shown here) grows numerous large spheres.

Dentated Spheres

These spheres developed from traps for small animals. Rather like some carnivorous Terran plants, the spheres could close, trapping animals inside to dissolve and digest them. But this no longer happens: the spheres now provide nocturnal shelter for commensal and symbiotic animals, closing every night. In return for shelter, the animals provide the phalanx with protection from algal overcrowding. Some animals seek shelter much more often than others. For example, the Phunque (*Stoma pangaster*) makes hardly any use of the spheres, whereas the Funk (*Gaster panstoma*) spends nearly every night in the spheres. The presence of these sphere dwellers somehow induces the formation of spheres.

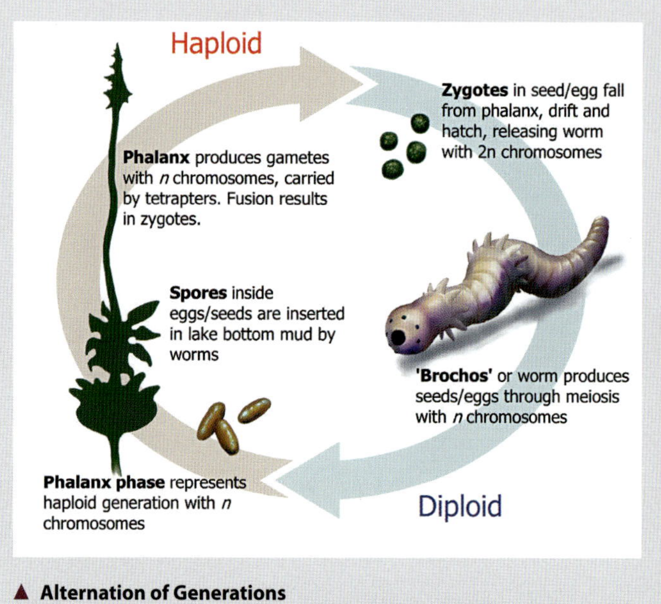

▲ **Alternation of Generations**
This is a simple scheme of the phalanx/brochos alternation of generations. The brochos is the worm-like animal on the right. It is only about 2cm long and lives on lake bottoms. The phalanx does not resemble it at all, reaching a height of up to 3m.

► Phalanges at a lake.

Phalanx Reproduction

The 'halberds' attract tetrapters by secreting a slightly nutritious fluid. The tetrapters probably carry gametes from one phalanx to the other, helping the sexual reproduction phase of the very complex mixotroph reproduction process. There is 'alternation of generations', meaning that one generation is haploid, having n chromosomes. This generation procreates sexually, resulting in diploid offspring with $2n$ chromosomes. This is called a haplodiplontic life cycle. Compared to Earth, mixotrophs are odd in that both generations in this cycle form complete individuals. Not only that, but the haploid generation is plant-like, the phalanx, while the diploid generation is animal-like. In the phalanx case, that generation concerns a small worm-like creature living in the mud of the lake bottom.

The ecology and genetics of mixotrophs have driven many students to despair.

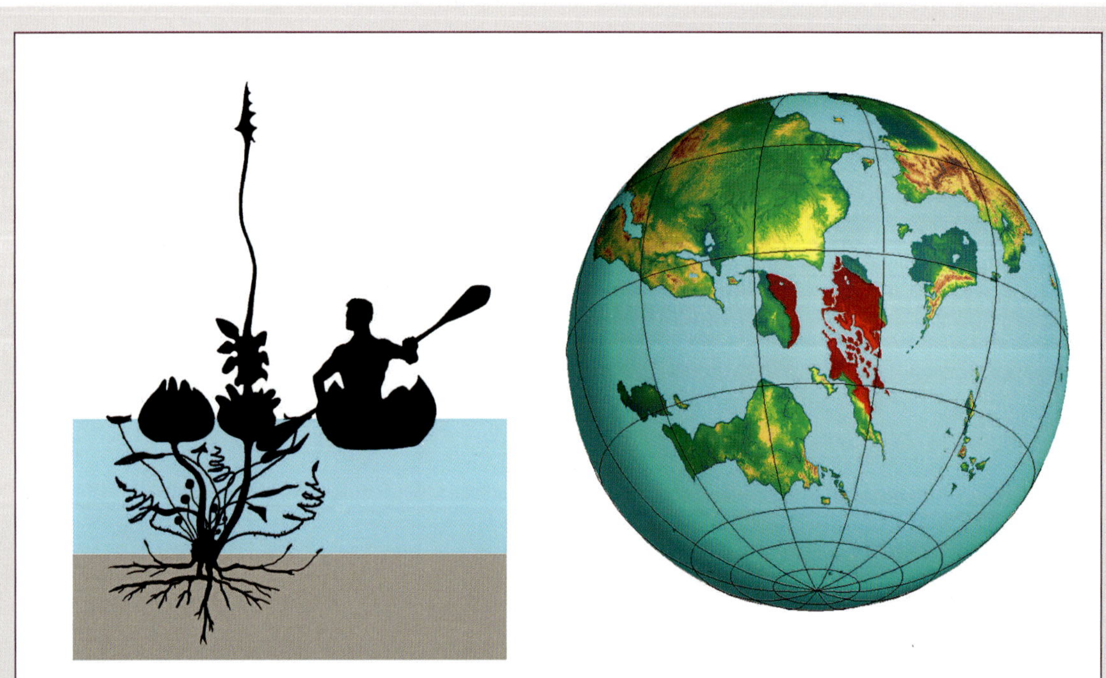

◄ **Phalanx**
Phalanx rivulus
Name derivation: *Phalanx* (Gr.) ('denoting a specific line of battle)
Habitat: Tropical lakes and streams
Height: up to 3m

Notes: Without genetic analyses it is not easy to find out which apparent animal belongs to which apparent plant. Both must share the same scientific name, which meant some supposed species were struck from the records.

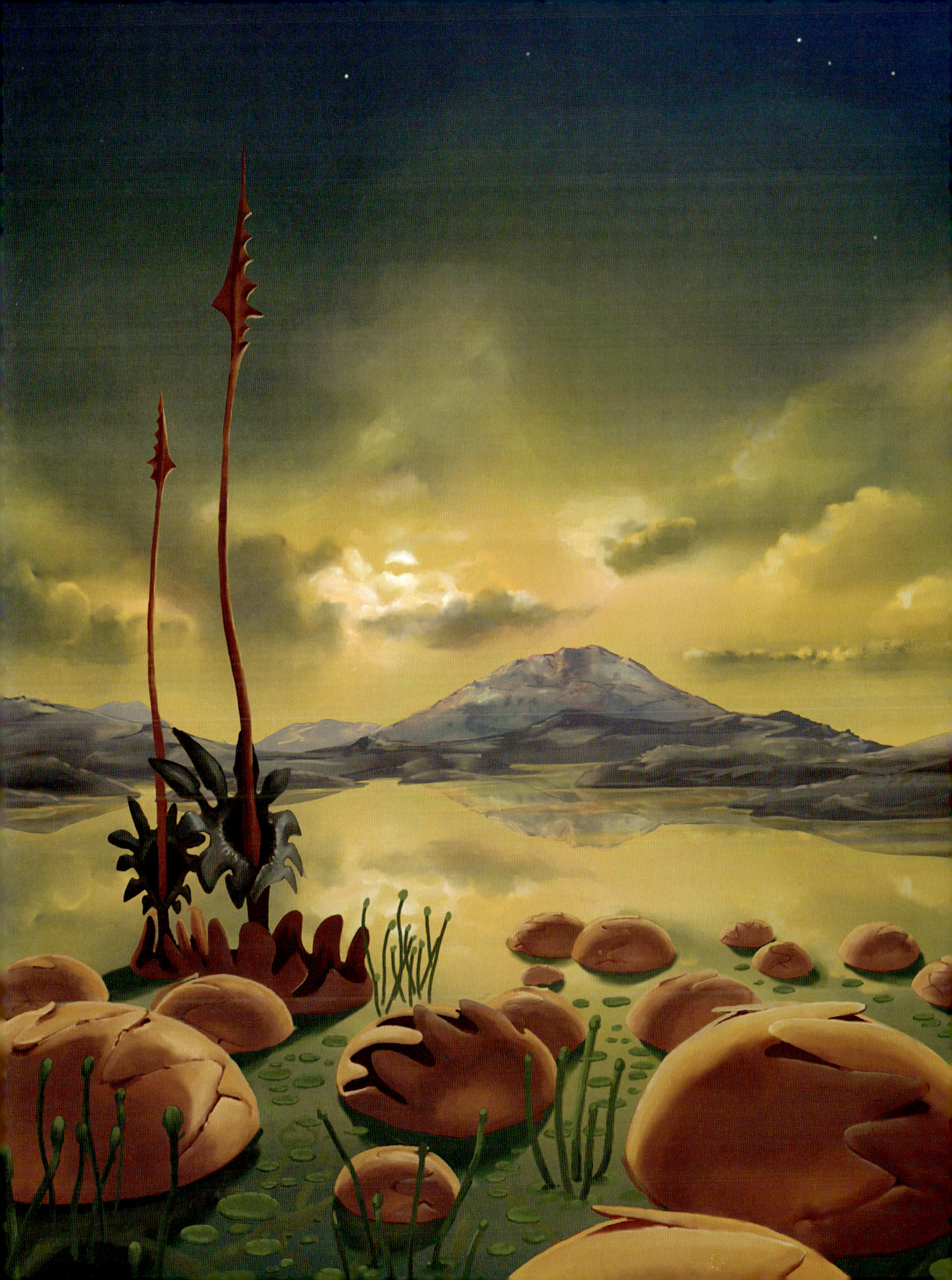

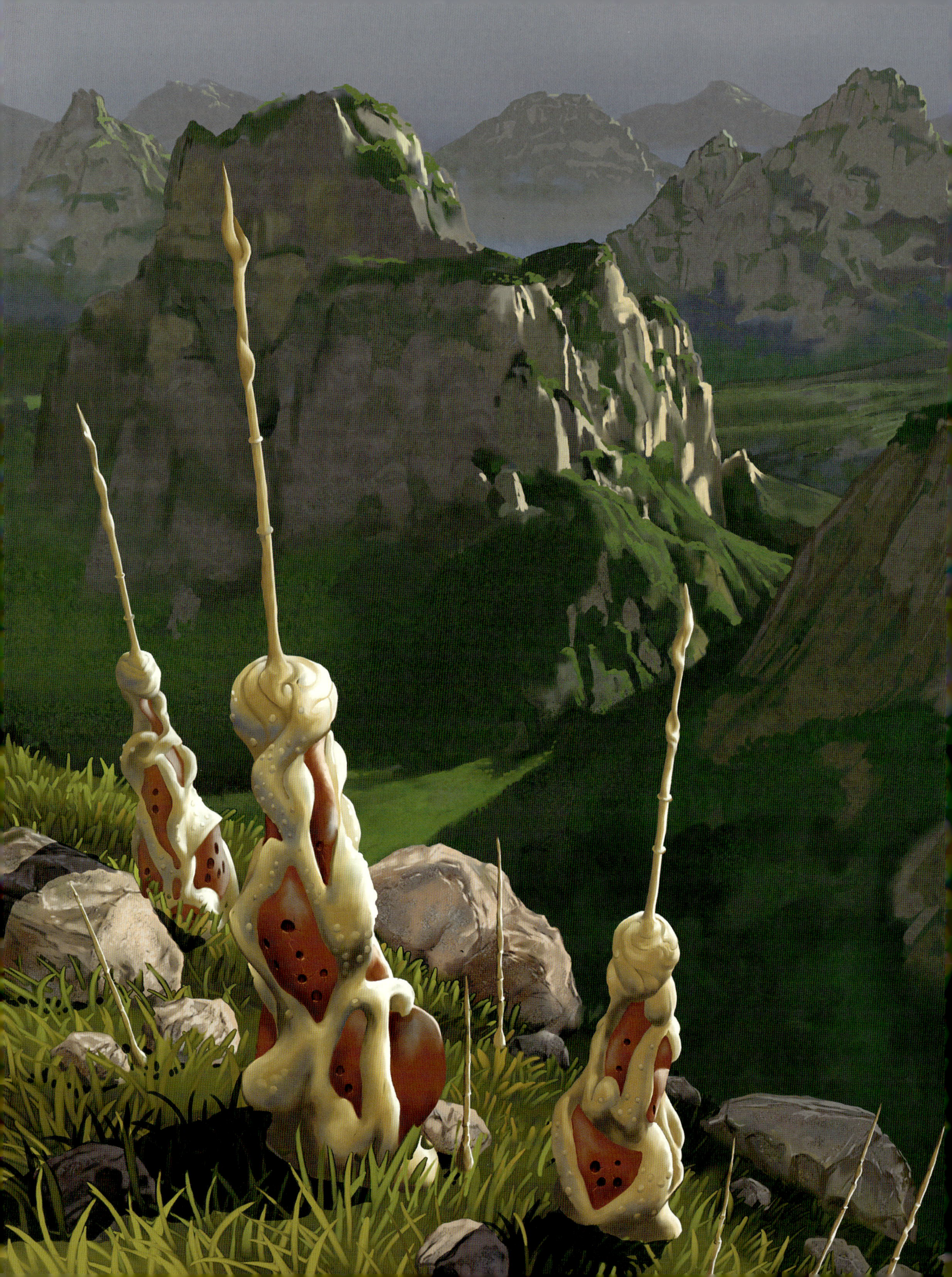

BOTORO

SANAMU SCITULA

The botoro is a typical mixotroph, with the ability to use photosynthesis to produce organic matter from inorganic sources. Many mixotroph larvae move about actively like animals, until they settle down. Almost all mixotrophs can also absorb organic matter for their own use, like animals and fungi. They can clean up dead matter, but many can also absorb living matter, provided that this matter stays still long enough for mixotroph threads to grow into it.

Fluid Feeders

The botoro is a good example of mixotroph fluidity, in that its metabolism leans towards a mixture of fungus and plant traits. It uses photosynthesis as an auxiliary energy source in summer. The plumes seen on the large 'statues' are not plants but mixotroph leaves, and so are the small plumes rising from the ground. In the autumn these will be absorbed or shed. A large part of the botoro's mass is unseen, consisting of an extensive network of underground roots and threads.

The above-ground statues are permanent fixtures. The botoro also has a variety of other organs, of which only 'sarissas' are visible here. Sarissas are the straight spikes growing up from the creature's top, bearing photosynthetic leaves. They also help with procreation, as they house the parfum glands. Most of the year the sarissa just sits there, but in summer it comes into its own, sprouting wreaths of leaves and giving off the botoro's characteristic smell.

The other organs are underground. From above ground, it is difficult to tell which parts belong to the same individual. In fact, all the growths seen in this image may belong to just one individual.

Symbiosis

The mountain slopes shown here are not very hospitable as there is light and rain, but that is about it. All soil was formed on the spot, and there is not much of it. The soil does not contain much organic matter, but the botoro enlisted the help of local wadudu (insect-like animals). These wadudu gather food in a common nest. The botoro forms ready-made hollow spaces in its bodies, complete with entry and exit holes. The wadudu readily build their nests in these hollows. The botoro steals some of the wadudu's food supply and digests dead wadudu, an arrangement that benefits both species.

Unpopular Study Subject

Few people study this particular symbiosis, however. Northeast Bogoria is cold and wet most of the year, and although botoros smell almost pleasant in summer, the rest of the year they smell like wet raincoats in a school wardrobe. Most people prefer to stay in.

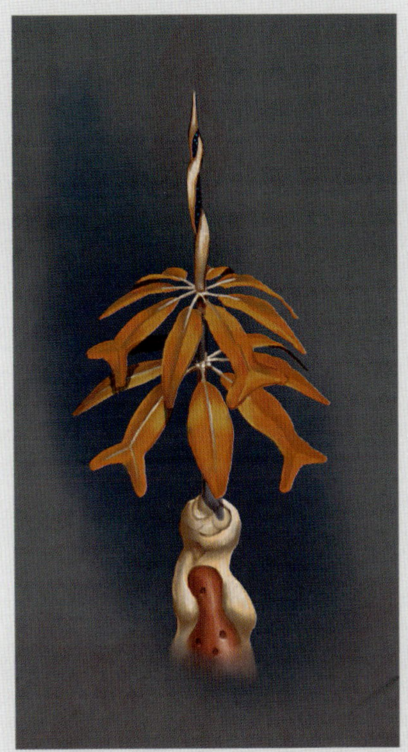

▲ **Phalanx in Summer**
This image shows the botoro's phalanx in summer garb, meaning it is festooned with nice plumes. These leaves supply the organism with additional energy to make the most of the summer season. The plumes typically bend and turn towards the light during the day. The top of the phalanx shows how the spirals have twisted open, revealing the botoro's odoriferous organs as bright blue dots.

◀ Botoro on a mountainside.

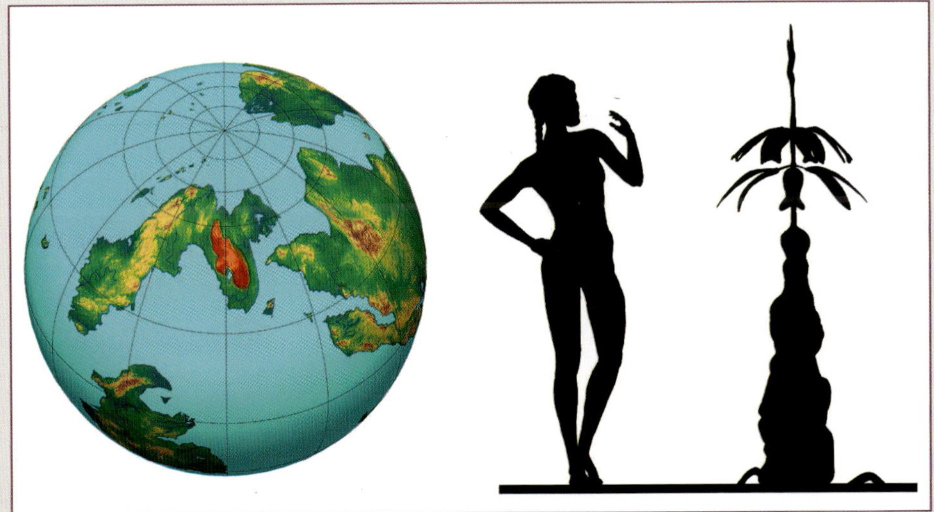

▶ **Botoro**
Sanamu scitula
Name derivation: *Sanamu* (Sw.) (meaning 'statue'); *scitula* (L.) (meaning 'elegant')
Habitat: Cold hills and mountains, Northern Bogoria
Height: 1.5m without plume

Nightsnare

Laqueus lentus

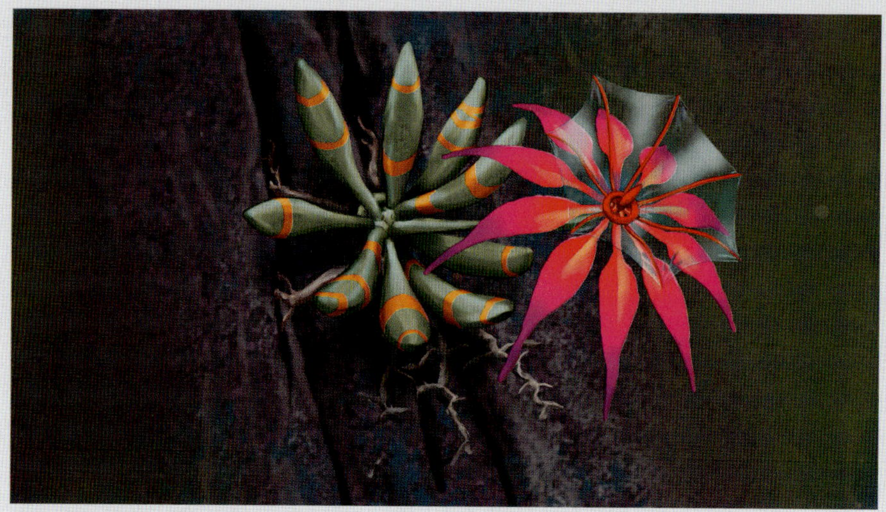

▲ Another Sticky Trap
This mixotroph species augments its resources by catching small flying animals. Its circular membrane suggests an evolution separate from the nightsnare, with its vertical mitt. The membrane in this species is backed up by a bright array of leaves, that oddly resemble Earth's flowers.

The nightsnare is a typical marshland mixotroph. The mixotroph clade it belongs to, the 'carnifexes' (meaning 'butchers'), has evolved an ability to deal with a scarcity of nutrients, which is something that can occur in marshes. Carnifexes devour animals. Consuming animal flesh is by itself not uncommon for mixotrophs, but it is the way in which those precious nutrients are obtained that sets carnifexes apart from run-of-the-mill mixotrophs. Whereas most mixotrophs are content to digest animals that were decent enough to die near their roots, carnifexes do not wait for such happy accidents to happen. They are hunters, and in the case of the nightsnare that means catching small flying animals such as tetrapters and flying wadudu.

Mitts

Nightsnares catch their prey with specialised organs known as 'mitts'. A long flattened hollow cone grows at the end of a stalk. The cone contains slime glands and is lined with tiny rootlets, ready to grow into any animal tissue. But that is just the digestive part of the organ. Animals are caught by the mitt proper, which is an array of long filaments that emerge from the cone. These filaments spend much time curled up inside the cone, but they eventually straighten up. They spread out as they extend and, with the help of thinner filaments, they hold together a membrane of sticky slime. Tetrapters and wadudu that blunder onto this membrane are usually not strong enough to tear themselves free from it, so they remain there until the filaments curl up again and bend down back into the cone, where the hapless prey is promptly digested. Hardly anything remains. After a while, the mitt is ready to emerge again, with a nice fresh sticky membrane.

The mitts may be open permanently, but a greater number of animals are caught at night, because of another trick up the nightsnare's sleeve.

Mimicry

Many mixotrophs have display organs to lure flying animals to help mixotroph procreation. Some produce smells, others use visual signals and some use both. Carnifexes combine smells with bioluminescence to form fertilisation aids, from which the mitts probably evolved. The lured animals are usually provided with food, so both species benefit from this symbiosis. Nightsnares still attract wadudu for this purpose and indeed provide them some nectar. The organs used for that purpose, 'lanterns', are only seasonally present and are not shown in the main illustration but look very much like the mitts used for predation. The resemblance is an example of 'aggressive automimicry', meaning that a species mimics itself to catch prey. Here, the predatory mitts mimic the benevolent fertilisation lanterns.

Evolution very likely already favours wadudu that can discriminate between mitts and lanterns, but most cannot, so, for the time being, the nightsnare has the last laugh.

► A stand of lakeside nightsnares.

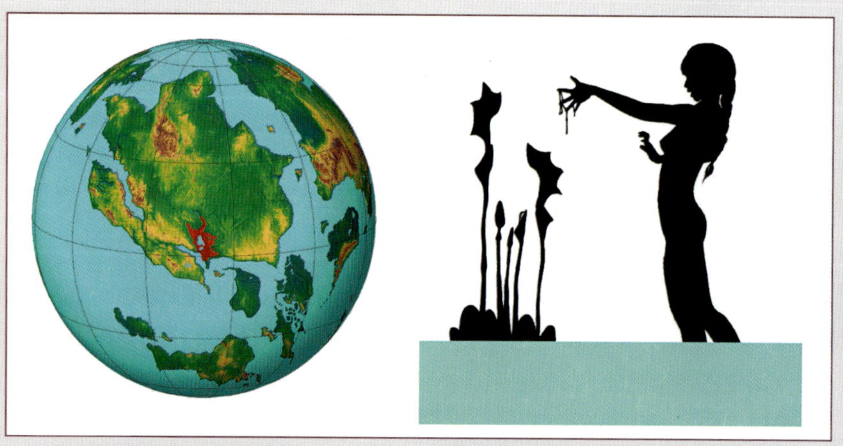

◄ Nightsnare
Laqueus lentus
Name derivation: *Laqueus* (L.) (meaning 'noose, snare or trap'); *lentus* (L.) (from the verb *lantesco*, meaning 'becoming sticky'). Hence their nickname: 'sticky snares'
Habitat: Marshes near Lake l'Ambique and the Athi rivers in *Imparia Septentrionalis*
Height: 1m without mitts
Details: Nightsnares can easily be studied as they are close to Nexus. The brochos phase of their existence is only a few mm long but occurs in very large numbers when they are in season

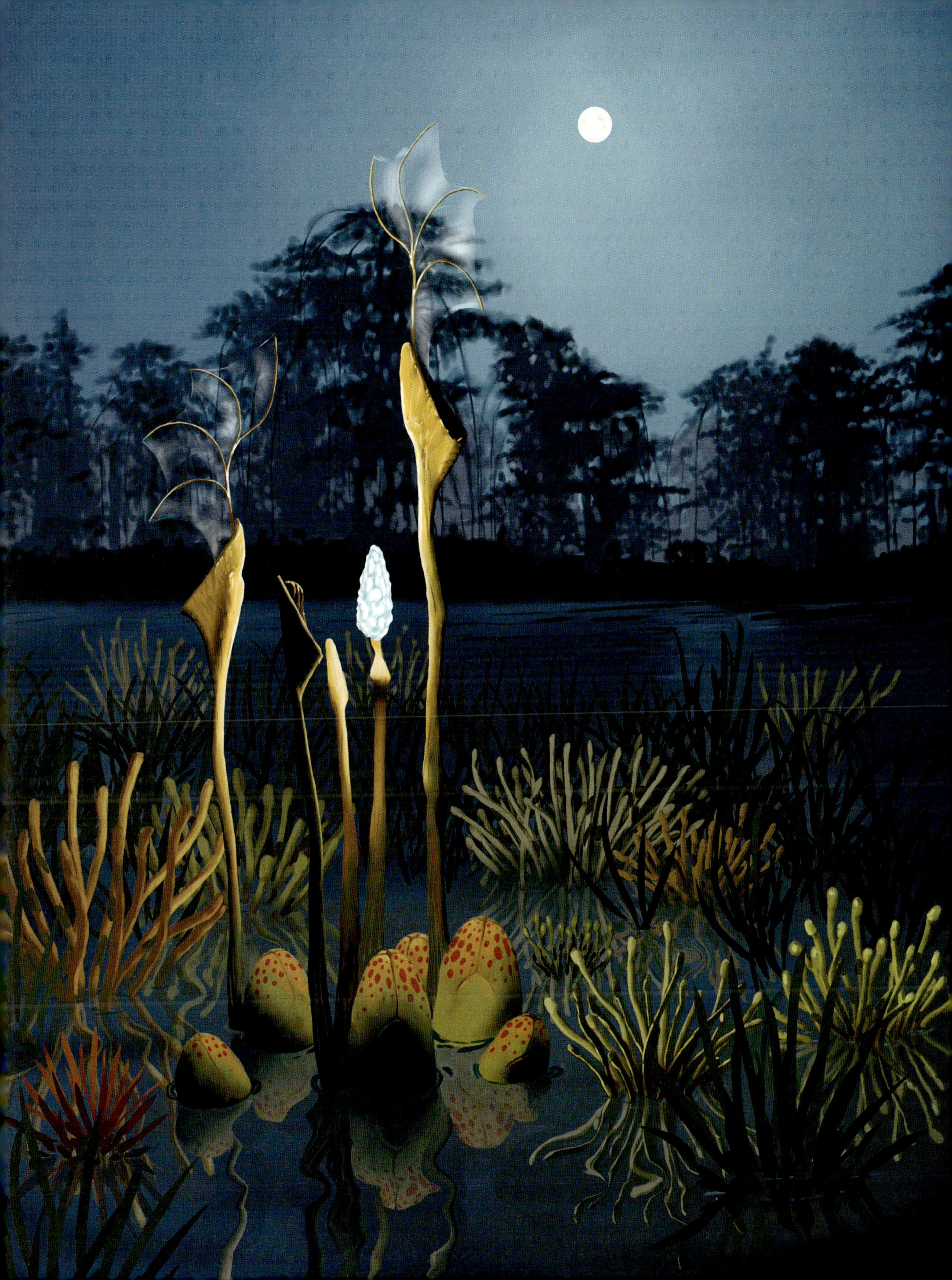

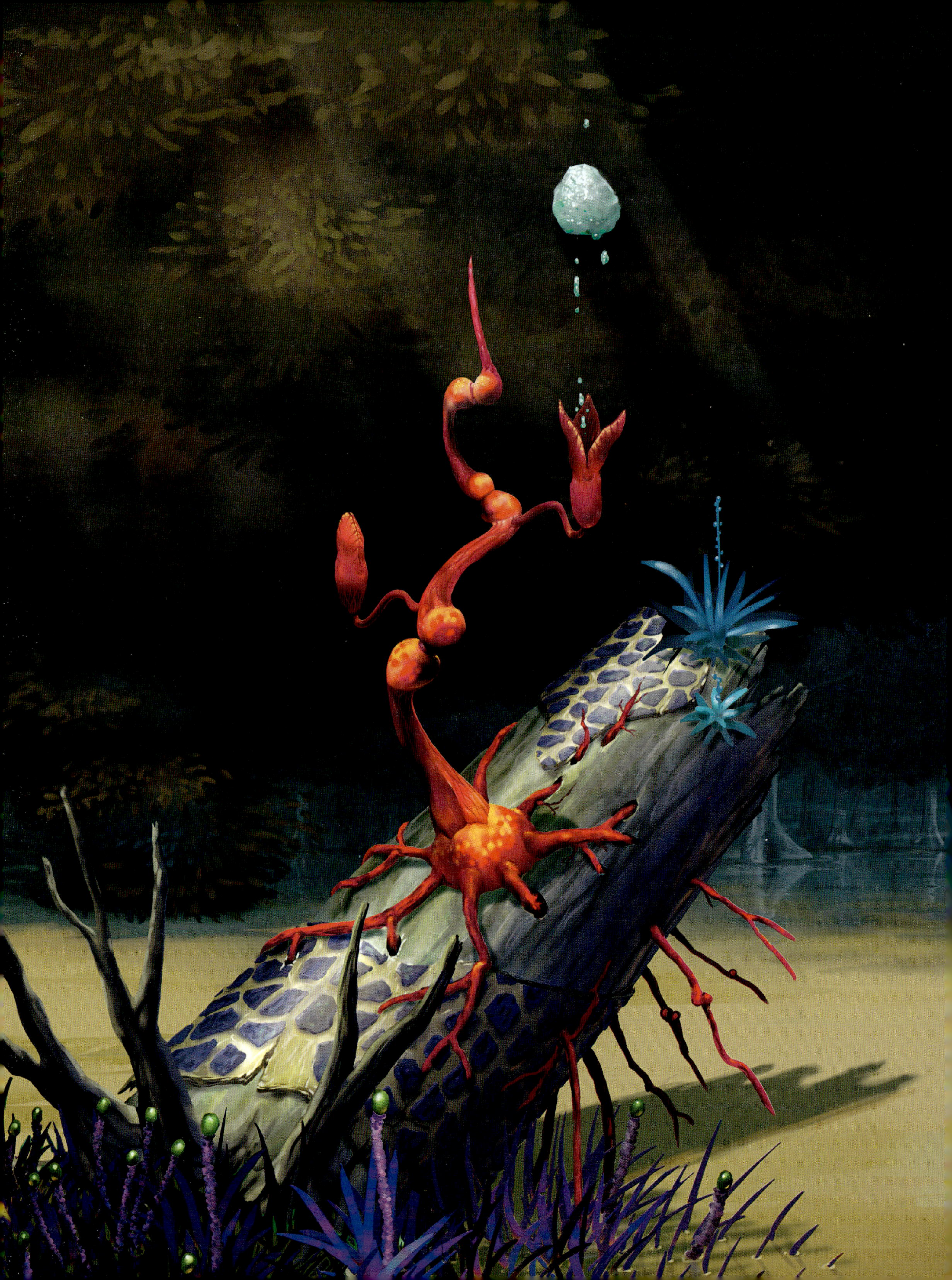

Purple Flyfoam

Spumascansa damieni

The Purple flyfoam belongs to the *Potemata* family of mixotrophs, which evolved an unusual, even creative, way to disperse its young. Nyoroge himself used to compare evolution to human creativity. He felt that some clades evolved more readily in unexpected directions than others, which he labelled 'evolutionary creativity'. Obviously, he wondered whether this hypothetical concept could be substantiated and what the genetic mechanisms might be. In *Worlds Apart* he described the evidence with great care and clarity. His later followers, the Nyorogists, resoundingly proved that mixotrophs indeed rated very high on the 'evolutionary exuberance score'.

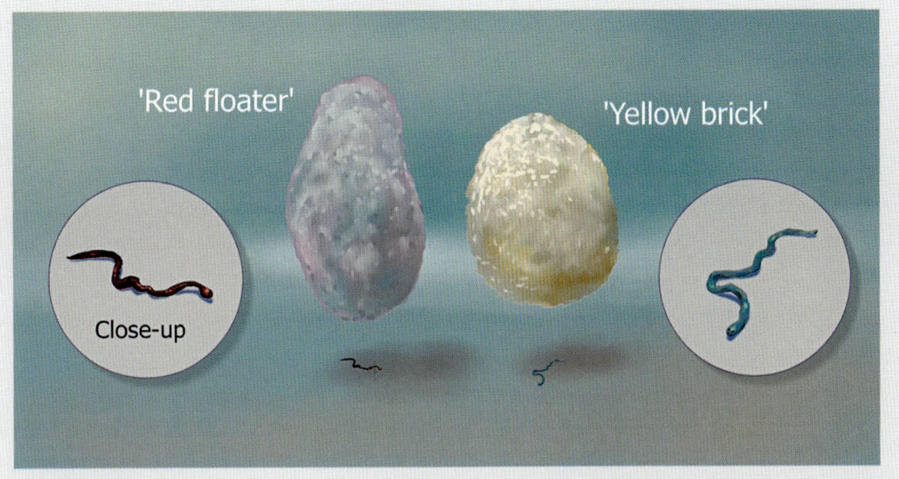

▲ **Floaterwatching**
'Raft watchers' are people who enjoy spotting and determining 'floaters', to use their own jargon. The rafts differ in shape and colour between species, but are also variable within species, giving rise to much discussion. The brochos larvae are suspended in the foam, helped by their diminutive size, slim shape and long hairs that are shed when the raft lands.

The Apple Does not Fall Far from the Tree

How to disperse one's progeny economically and successfully is a considerable problem for sessile life forms. Apples indeed do not fall far from the tree. However, it would be better for the apple tree if some apples did fall far away. Apples are in fact the apple tree's way to spread apple seeds. Once animals are seduced to eat apples, apples seeds are deposited far away, with free fertiliser.

But that is not the mixotroph way. Mixotroph creativity already manifested itself in the mobility of their larvae. But these wormlike creatures only travel small distances, far enough to find the best rooting spot in a small area. Travelling over truly large distances poses another challenge altogether, one the flyfoam's ancestors solved by providing its young with balloons.

◀ Flyfoam growing on, in and through a tree trunk.

Lighter than Air

These 'aerial rafts', as mixotrophologists call them, differ from human-made balloons, which typically consist of a thin membrane, inflated with a light gas. Some human-made balloons combined several such bubbles within a larger envelope, a design taken significantly further by the flyfoam. The raft consists of foam filled with methane gas, which has only half as much mass as air. The foam is made of water with a complex mixture of proteins and other materials. This mixture makes the foam stable over several days, even in tropical sunlight. Other ingredients function as antimicrobials, preventing the foam and the embedded larvae from becoming infected.

The individual bubbles of the foam eventually succumb to the effects of gas diffusion and evaporation. One bubble after the other will burst, reducing the lifting power of the raft. The larvae, meanwhile, use their storage of food. If all goes well, they land with some energy to spare.

Once on the ground, the larvae seek a nice place to settle and spend the rest of their lives there. As their aerial journey may have lasted several days, that place may be hundreds, even thousands, of km away from the home of its parent. This explains the Furahan saying about children choosing their own path –'The flyfoam does not root next to its parent'.

▶ **Purple Flyfoam**
Spumascansa damieni
Name derivation: *Spuma* (L.) (meaning 'foam'); *scansa* (L.) (from the verb *scando*, meaning 'to climb'); *damieni* (L.) (after Damien Clarke)
Habitat: Humid tropical environments over a large range
Height: 1 m
Details: The overall aspect of the flyfoam depends on the season. Whether an individual flyfoam has leaves depends on how easily it can live in a saprophytic manner

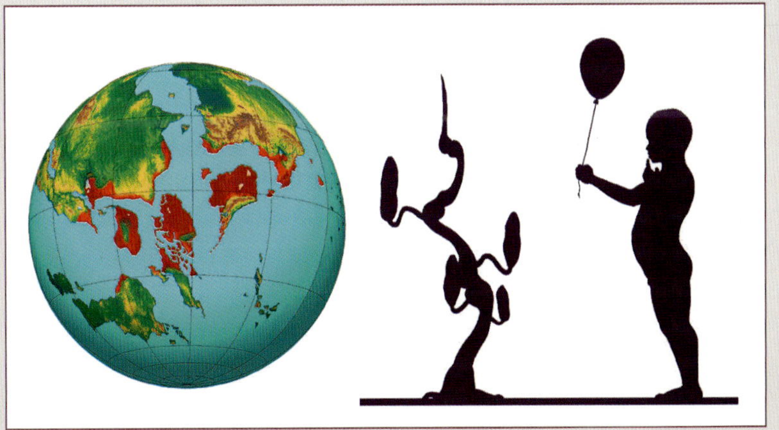

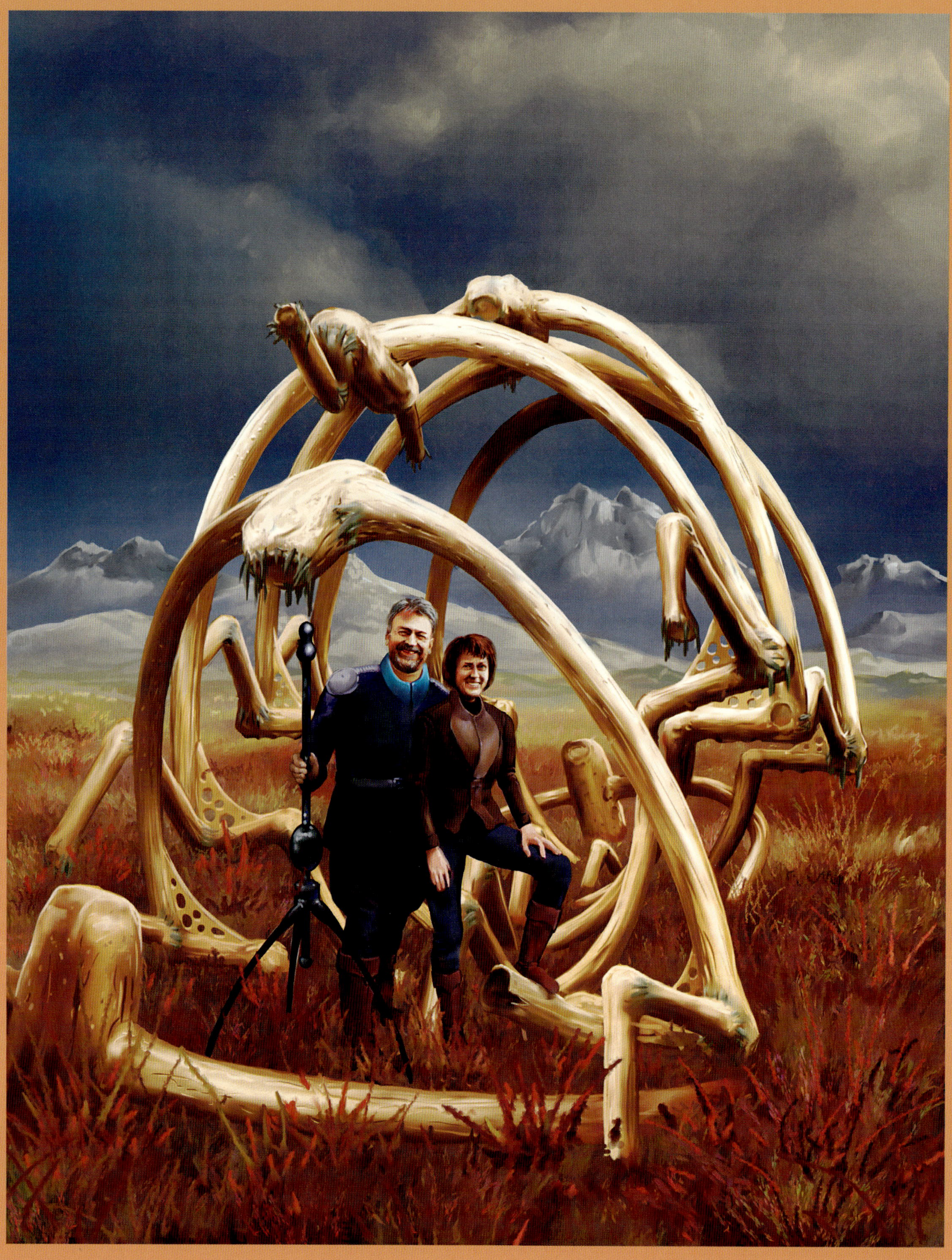

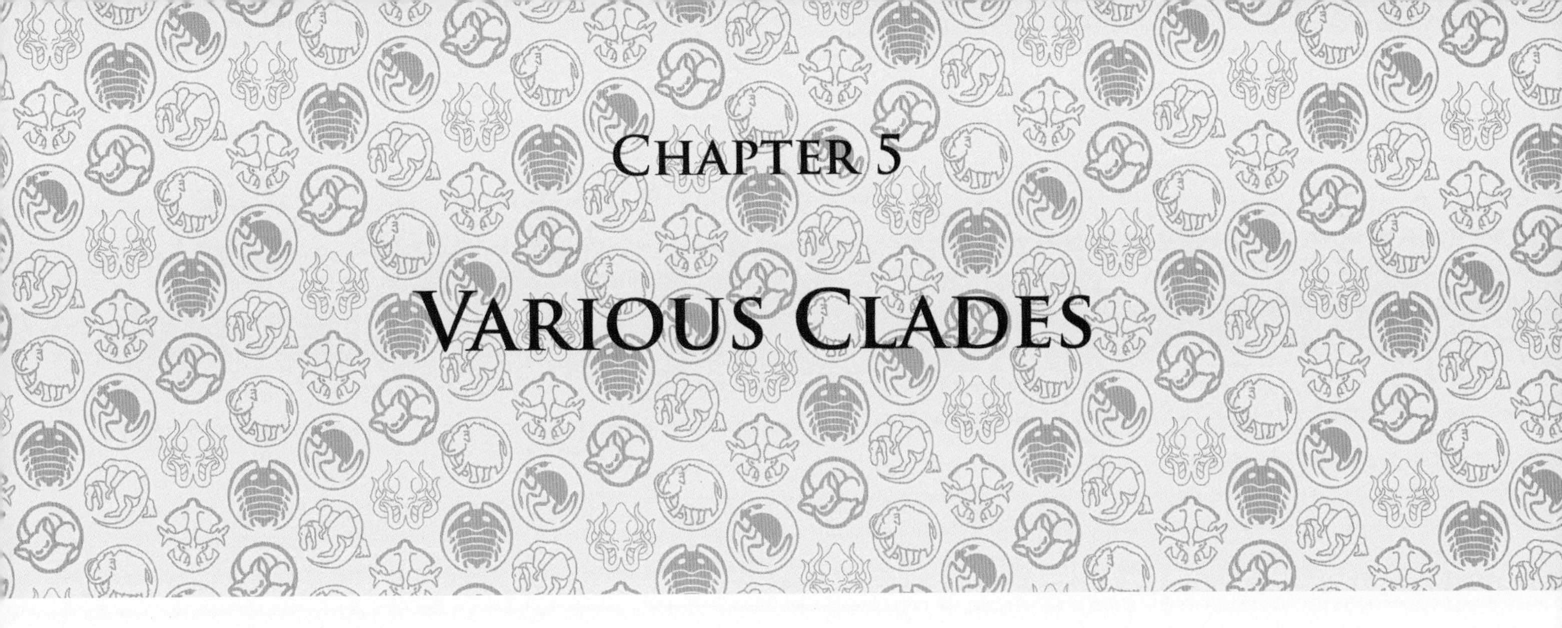

Chapter 5

Various Clades

'It was not just the number of species or clades on these plains I found staggering. It was the sheer number of individual animals that I found difficult to comprehend. Thousands of large animals, weighing over 1000 kg each, formed just one herd. There were thousands of such herds.

Earth was like that once, but our ancestors became used to an ever more impoverished environment and forgot what our birth world was like. Unfortunately, not even terraforming can restore Earth to what it once was.

Humanity found Furaha, an unspoiled planet. If Earth cannot be untouched, let this alien planet serve to remind us of what we lost and what we blindly discarded. Let our touch be light this time, and let us not again spoil what we love and admire. Never again, we vow, will natural history be subordinate to human history.'

From *Worlds Apart: Natural Histories of Furaha and Earth* by Souren Nyoroge

◀ Dying brontorusps often sink to the ground with their body upright. As scavengers cannot always reach upper parts of the carcass, the upper parts of the brontorusp body may desiccate in place. The ligaments become extremely tough and of little value to scavengers, so some arches stay connected and remain upright for years, forming odd cathedral-like corridors. Here, two members of an odorological survey team pause to be imaged under the bones that are the only reminder of the magnificent animal that once died on this very spot.

Wadudu

An Aerial Wadudu Hunt

Forfex currax and *Telum aureum*

At first sight both these animals can be mistaken for Earth insects as their small size, spindly legs, lacy wings and apparent exoskeleton all suggest 'insect'. But they are wadudu (singular 'mdudu'), with eight legs, not six, and six wings, not four. The most fundamental difference from insects, however, lies not in the number of their limbs but in their skeleton.

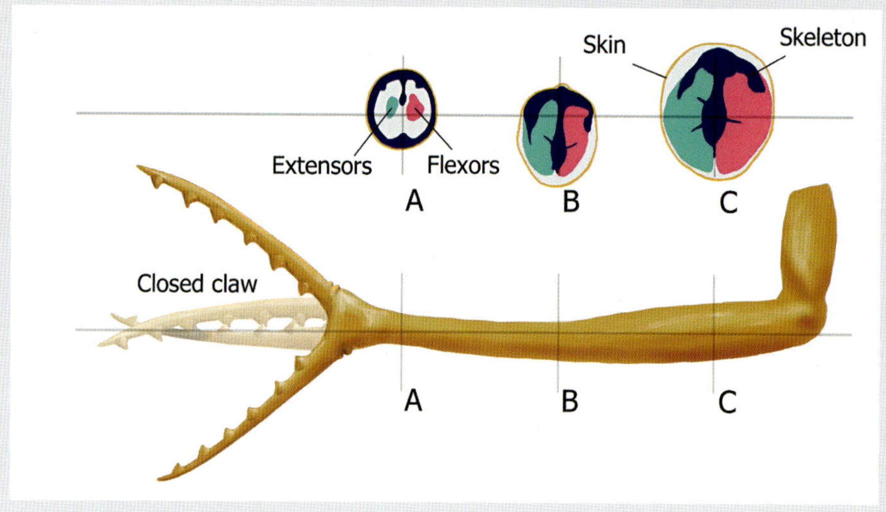

▲ **Claw Anatomy**
Lateral view of the left first leg. The claw is shown in the fully open, extended, position. In the closed position the upper hemichele (half-claw) partially shears across the lower hemichele. The cross sections show that the skeleton is tubular in shape at Section A.

A Mesoskeleton

The wadudu mesoskeleton is not an exoskeleton, although it superficially looks like one. It is embryologically situated under the skin and is not part of the skin itself. Formally, the mesoskeleton does not clothe the outside of the animal. The real skin may however be 'dormant', meaning it is only metabolically active in periods of growth. When the skin is dormant, the mesoskeleton to all purposes lies on the outside, just like a true exoskeleton. It can form tubes, such as in leg parts, or plates, serving as armour. But on other parts of the body the mesoskeleton forms internal stiffening rods, beams and complex structures, and the animal is externally soft-bodied in such places.

This flexible nature explains how wadudu can have flexible, even tentacular structures. An excellent example is their neck.

The degree of the mesoskeleton's 'externalisation' also depends on the animal's size. The disadvantages of an exoskeleton are high mass and a limited capacity to absorb blows. These problems become more limiting for larger animals, which explains why the mesoskeleton is more external for small wadudu and more internal for large ones.

The Hunter

The cataphract is the hindmost of the two animals shown here. It is large, strong, agile and fast. Its middle pair of wings is much larger than the front and hind pairs and provides the greatest amount of lift and propulsion. The front and hind pairs beat more quickly and allow the animal to turn very quickly. The animal cannot hover, however, so it must make the most of its speed and agility. The first of its four pairs of legs are greatly modified and are no longer used for walking. They end in effective pincers that allow the animal to grab its prey in flight.

The Hunted

The yellow zoom is smaller than the cataphract, but just as fast and manoeuvrable, so the outcome of the hunt is never a foregone conclusion.

Zooms belong to the largest group of wadudu, the *Crupellarii*, colloquially known as 'beetles', biologists' protests notwithstanding. Furahan beetles fill just about every niche available. Zooms live off plant lice, an imprecise word indicating all the very small animals that stick some kind of mouthpieces into a plant to suck up nutrients.

▶ An aerial hunt.

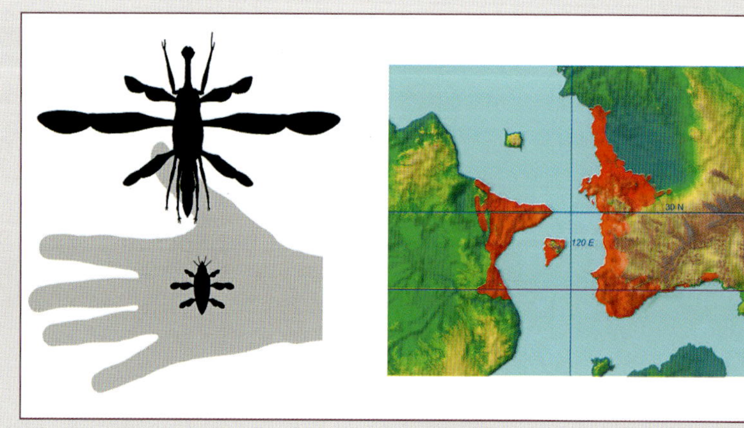

◀ **Cataphract and Yellow Zoom**
Forfex currax* and *Telum aureum
Name derivation: Cataphract: *Forfex* (L.) (meaning 'shears or scissors'); *currax* (L.) (meaning 'nimble, brisk, agile').
Yellow zoon: *Telum* (L.) (meaning 'bullet or dart'); *aureum* (L.) (meaning 'golden')
Habitat: Open and half-open terrain, in particular around water
Distribution: Southern coastal regions of both *Imparia* continents around the Intermediate Sea
Mass: A few grams at most
Length: Up to 25cm for the cataphract, 5cm for the zoom
Diet: For the cataphract: other wadudu; for the yellow zoom: plant lice

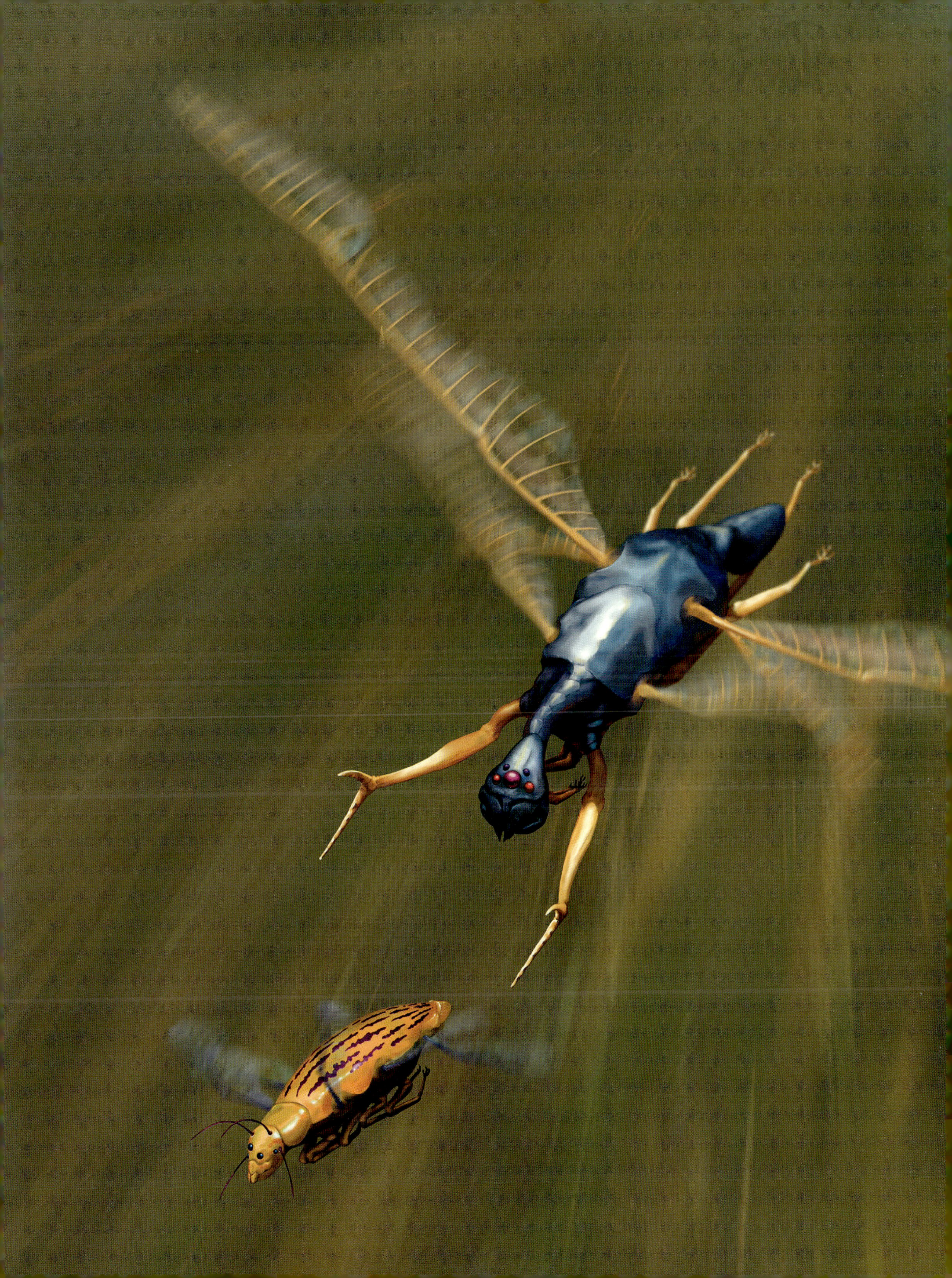

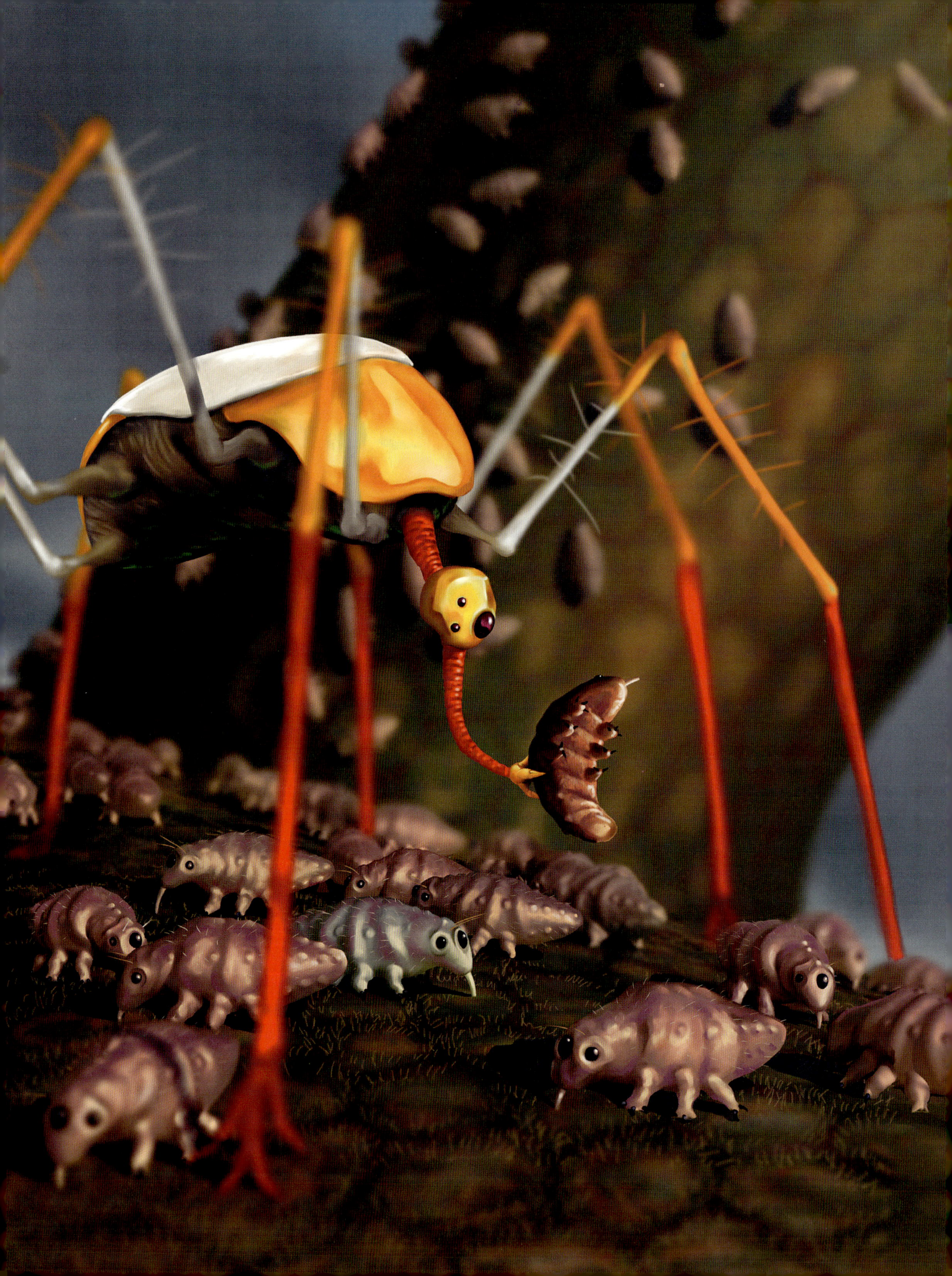

Lausfresser Lunching on Lice

Ulla sanguisuga

This predatory mdudu, called a lausfresser, or 'lice-eater', preys almost exclusively on plant lice. Here, a lausfresser strayed onto a grey (poliochrome) plant. The lice's colour is probably an attempt at camouflage to escape predators. Even so, plant lice are easy to find and form a dependable food source for many predators. The louse's main defence mechanism is to gather in large numbers, increasing the chance that a predator is sated before it is an individual louse's turn to be impaled and sucked dry.

◀ Lausfresser towering over lice.

Camouflage

The lausfresser does not need to conceal itself from its prey, but it might use camouflage to escape being eaten itself. Instead, it is very conspicuous. Camouflage is not always about blending in with the background. Being highly conspicuous may, surprisingly, also fall under camouflage. Such strong visual signals may serve to distract other animals or to disrupt an animal's outline. Yet another purpose of such bold markings is to announce that the bearer of the bright colours is best avoided, because it is poisonous or venomous.

Lice

There is no clade of Furahan animals with the official name of lice. The word is an umbrella term that bundles a large variety of small animals that live off plant juices. Some are adult wadudu, some are spidrids and others are larval forms of various other animals.

The species shown here, the 'piggy', has a very simple strategy in life, which is to suck sap and digest it as quickly as it can. This may be wise, as piggies are subjects of fierce predation, without much going their way. The sooner they reach the next stage of their lives the better. Regardless of which form these larvae take in their next stage, it will almost certainly allow them more options than to suck or be sucked.

The Binomen of The Lausfresser

Formal recognition of a new species by the Furahan Academy of Sciences goes through three steps. The first takes place in the field, when an Expedition's Captaincy decides that the scientists who propose a new species have checked the records to ensure that it is indeed a new species. The second step is the naming ceremony, in which the citizen-scientist who gets to name the species, the Recitator, announces the species to the expedition by loudly calling out the binomen. This is usually a festive occasion. The last step consists of a check by the dour curators of the Institute.

The main image's perspective was chosen for a reason: it mimics the camera perspective of the image used to document the formal description of the Lausfresser. The Recitator of the lausfresser had just started his career and was called Sigurður Fehlanzeige.

Young *doctior* Fehlanzeige was inspired by H.G. Wells' 1898 novel *War of the Worlds*, whose invading Martians utter the eerie cry 'Ulla'. From a human point of view, the Martians' tripod walking machines tower far above them, and humans are powerless to escape being picked up and emptied of their blood. *Sanguisuga* means 'blood sucker' in Latin. Sigurður's friends and colleagues all liked that name very much, so the naming ceremony turned into a great party and to this day a cocktail, invented there and then, is named the 'sanguisuga'.

Unfortunately, his superior, Preprofessor Emilio Porcone, sent another name to the Institute without telling Sigurður. Porcone's chosen name of *Phtheirophagus phonicus* was in fact not that bad, meaning 'bloodthirsty louse-eater'. Still, it was unheard of to change a proposed name in such a manner. Sigurður protested formally and the curators decided in his favour. The affair started a lifelong feud between the two men.

▶ **Lausfresser**
Ulla sanguisuga
Name derivation: *Ulla* (E.) (fictional Martian cry); *sanguisuga* (L.) (meaning 'blood-sucking')
Habitat: Plants, wherever there are plant lice
Distribution: Southern coastal regions of *Imparia*, but no-one bothered to study this region thoroughly, not even the robocartographers
Mass: A few mg
Span: 2cm
Diet: Plant lice!

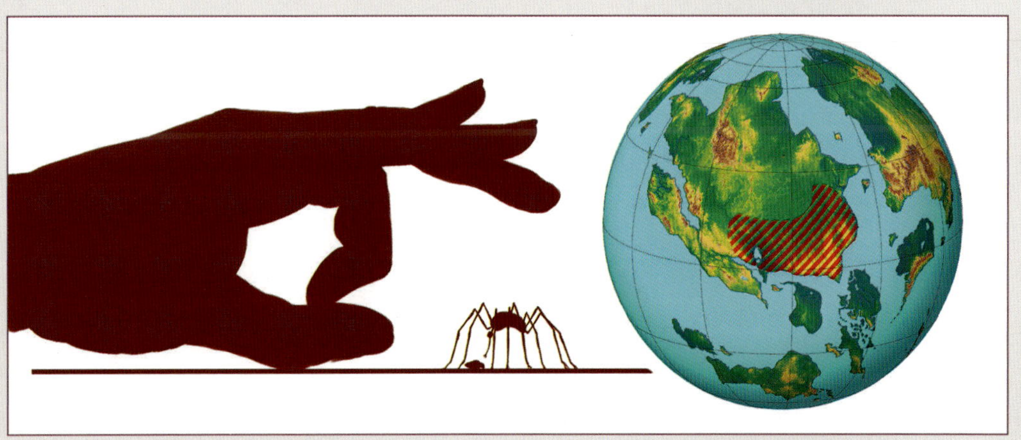

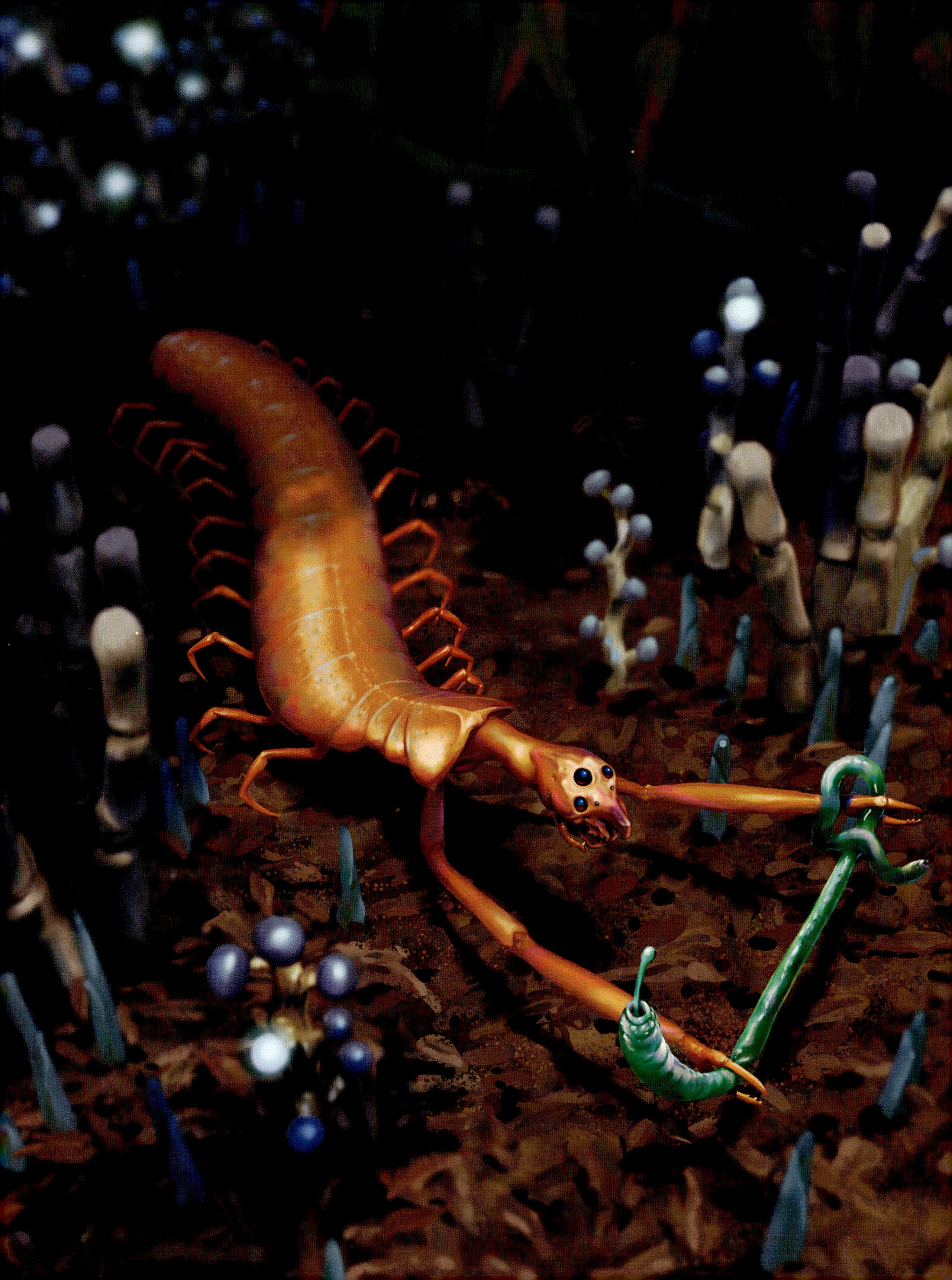

The Early Sneeth Catches the Wurm

Vespillo perreptus

Animals like sneeths frighten people, in particular when an unwary person suddenly sees something with many legs crawling into view from the corner of his eye. Sneeths have features that humans respond poorly to, such as an eerie overabundance of legs along with probing pincers. The fact that they are up to 50cm long does not help either.

It is not quite known why some animals, such as spiders and snakes, frighten humans so much. Is this an inherited trait or behaviour learned in childhood?

Sneeth as Evolved Wadudu

The sneeth body scheme represents just one of the myriad shapes wadudu evolved into. Sneeths, with their many segments and legs, resemble Earth's centipedes and millipedes. They walk in a similar manner, with small phase differences between adjacent legs, so a wave appears to travel along their legs. In sneeth, all segments follow the path of the first one, so the animals snake their way along the ground rather like, well, snakes.

The first few body segments are elongated and fused. Only one section carries limbs, and these are long and end in strong pincers with sharp venomous points.

In contrast to all these specialisations, the sneeth head is of a fairly basic ancestral shape, although the eyes are large, fitting its mostly nocturnal lifestyle. The head's lack of specialisation is probably explained by the fact that the animal uses modified legs for work that other wadudu use jaws for. In words that a professional mdudulogue might fancy, 'raptorial limb specialisation obviates gnathic adaptation'.

◀ Sneeth measuring a wurm.

Sneeth Behaviour

Sneeth venom is, by sheer biochemical accident, toxic to humans. Being stung by a sneeth leaves a wound that will not heal for a long time. In one case some toes had to be amputated, but that concerned a student who ignored the advice to wear at least Class IIIA boots in central Naivasha.

The venom did not evolve to harm students, but to incapacitate just about any small animal that the sneeth can grasp at night. During the day, sneeths hide in a burrow, digesting the previous night's prey.

Courtship

The scene shown here is unusual, with a rainbow sneeth active in daytime. The reason may well be that it is time to mate. Sneeth courtship starts with a fight phase, in which potential partners test their mettle by grasping one another's claws. The animals push and pull, and if that phase ends to their mutual satisfaction, the stage is set for the next phase. One of them – probably the female, but it is hard to tell – catches a prey and offers it, usually alive and kicking, to its mate. If the mate accepts, the pair may eat the prey together and then find a suitable place to mate and hide their eggs.

In this image the prey is a wurm. Wurms generally score high on the 'human repulse scale', which says more about humans than about the animals on the scale. Furahan citizen-scientists learn to ignore the fear and repulsion that some animal shapes induce. They respect and like the subjects of their study. For instance, the rainbow sneeth does have a rather pretty colour and sheen.

Sneeths are Not Pets!

Mdudulogy students once tried to take sneeths as pets. Such acts are always disapproved of by the Institute, as no-one knows what the results of such meddling are. Unfortunately, in the field not everyone heeds the Institute's advice. The Institute had the last laugh though. While humans may become habituated to sneeths, the reverse does not occur. The sneeth literally bit the hands that fed them, forcing the students to wash their hands of the matter. Repeatedly.

▶ **Rainbow Sneeth**
Vespillo perreptus
Name derivation: *Vespillo* (L.) (meaning 'night thief'); *perreptus* (L.) (from the verb *perrepo*, meaning 'to crawl along')
Habitat: Forest floor, jungle
Distribution: Riverine forests in central Naivasha; both the Northern Toerot Jungle and Southern Noserot Jungle
Mass: 0.2kg
Length: Up to 50cm
Diet: Wurms, other wadudu, spidrids
Vocalisation: None
Social habits: Solitary except for mating season

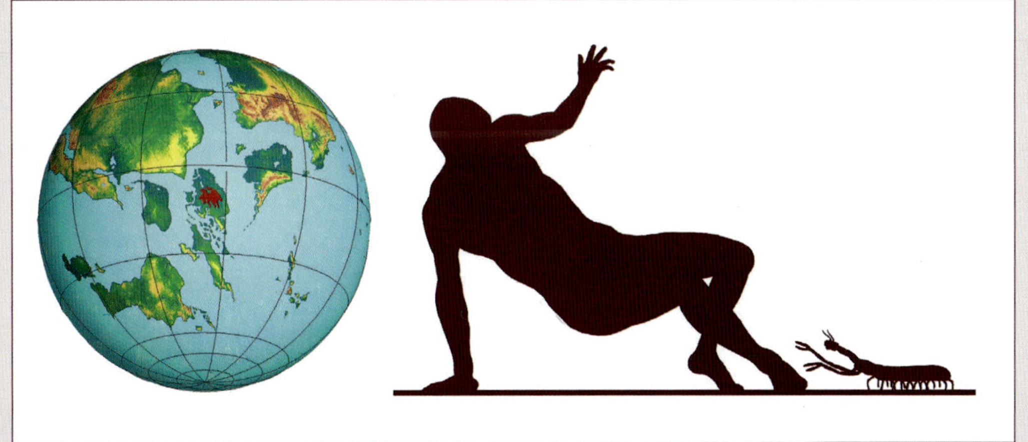

KWALS

FURAHAN JELLYFISH

Kwals are often considered 'universals', meaning their overall body scheme is supposed to occur throughout the universe; well, only given similar constraints, of course. Other universals in this book are wurms and wadudu. Kwals are fairly primitive predators that mostly consist of a gelatinous substance. They move by jetting water down in pulses and catch prey with tentacles. More precisely, they ensnare, glue, envenomate or poison their prey, or use any combination of these.

Like Jellyfish, but Not Jellyfish

Kwals were so named by Giovanni Fehlanzeige. He stemmed from a Germanic clade and based the name on West Germanic words for jellyfish. The similarity with jellyfish must have been obvious. Kwals have a formidable reproductive capacity: they can bud off new kwals as easily as they can reproduce sexually and when life becomes difficult, they form spore balls. Their preferred order of reproduction is to bud, followed by forming a few spore balls in case of some mishap, and only then do they reproduce sexually. Kwals share the ability to 'bloom' with Earth jellyfish, meaning their population can expand very quickly.

Unlike jellyfish, kwals do not have a bottom-living polyp stage. Their bodies are mechanically more sophisticated than those of jellyfish, with rubbery rods made of evolved jelly, functioning to give the animals some rigidity and allowing more complex movement.

Symmetry

Kwals are radially symmetrical, usually with three parts (trimeric), but there are variants with six parts (hexameric) or nine parts (nonameric). Earth jellyfish have a single bell for propulsion. In kwal evolution the three parts probably started forming one bell, but in some groups the three parts separated again. The image on the right gives just an idea of the variety of kwal shapes. We will discuss each in turn, focusing on evolutionary bell separation.

Big Yellow Blob

Name derivation: *As* (L.) meaning 'flap' or 'valve'; *augustus* (L.) meaning 'dignified' or 'majestic'.

This species is close to having just one circular bell, but with one conspicuous difference from Earth jellyfish: in the centre of the bell, the three segments are not fused, but form valves. When the bell contracts, the pressure under the bell increases, which closes the valve, pushing water down and the animal up. When the bell relaxes, the valves open passively, helping residual upwards movement. This adaptation produces more movement for the same energy.

Sea Wheel

Name derivation: *Rota* (L.) meaning 'wheel'; *pituita* (L.) meaning 'mucus'.

Sea wheels have a central hole, turning the bell into a ring. The constriction of the ring drives water down under the ring. The water forms a torus of rotating water and at the inner edge of the torus water flows back up, against part of the body, providing a subtle aid in propulsion.

Purple Flapper

Name derivation: *Remex* (L.) meaning 'oarsman' or 'rower'; *arsus* (L.) meaning 'glowing' from the verb *ardeo* meaning 'to glow or sparkle'.

In flappers, the three parts of the bell have become completely separated, which has allowed novel ways of propulsion. The tribune on the next page shows a hybrid stage of this process. The purple flapper shown here has taken this scheme a step further. The three flaps no longer have a hollow shape, but are pushed downwards like oars. It is quite spectacular, with its bioluminescent spots and spectacular red feeding arms.

Milky Petal Pedal

Name derivation: *Anceps* (L.) meaning 'fold'; *lactens* (L.) meaning 'milky-white'.

This animal represents an evolutionary direction quite different from the flappers. Here, the bells are also separate, now called petals. The two halves of the petals fold together during the upstroke but spread out during the downstroke, so saving energy during the upstroke.

► From top to bottom: big yellow blob, purple flapper, milky petal pedal and sea wheel.

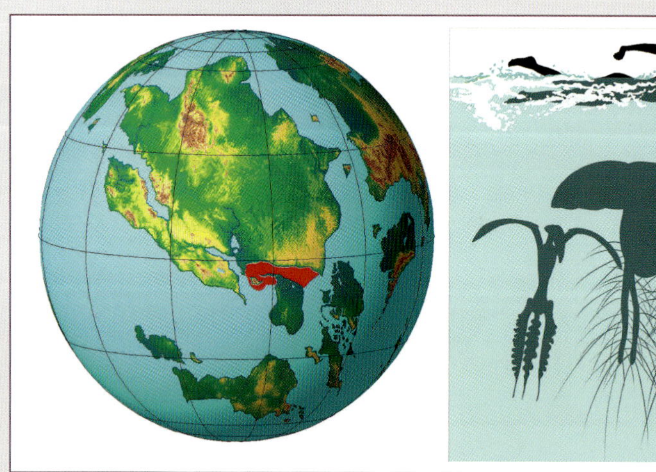

◄ **Various Kwal Species**
Habitat: Open sea, various depths
Length: width up to 1.7m

Notes: The four species of kwal shown on these pages were all found in a small region just off the coast of the Institute's main branch. Actually, these are just four of the at least 60 species of kwal that were found in this region of the sea and that was only the result of chance encounters. It is very likely that the actual number of kwal species was even higher.

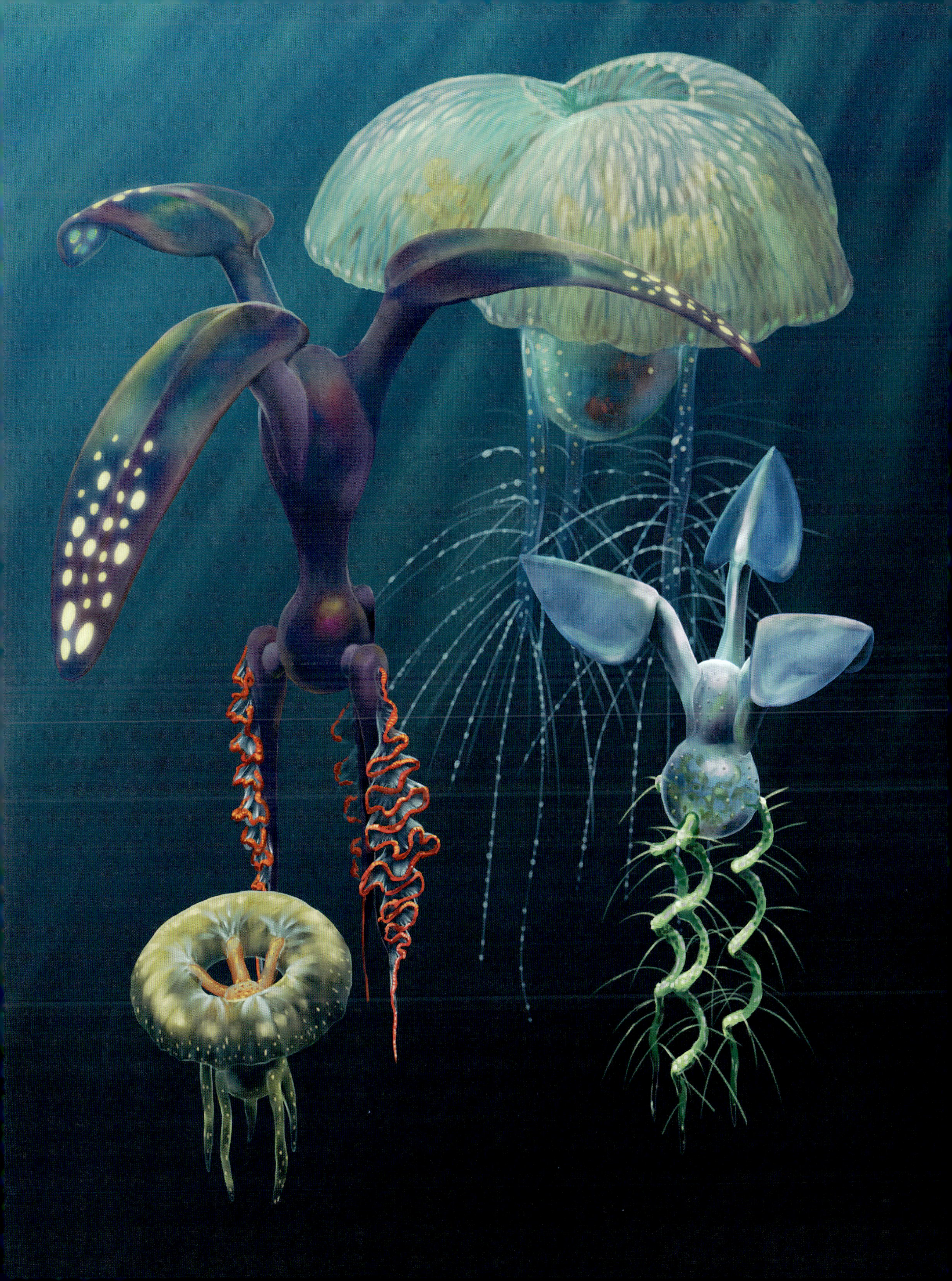

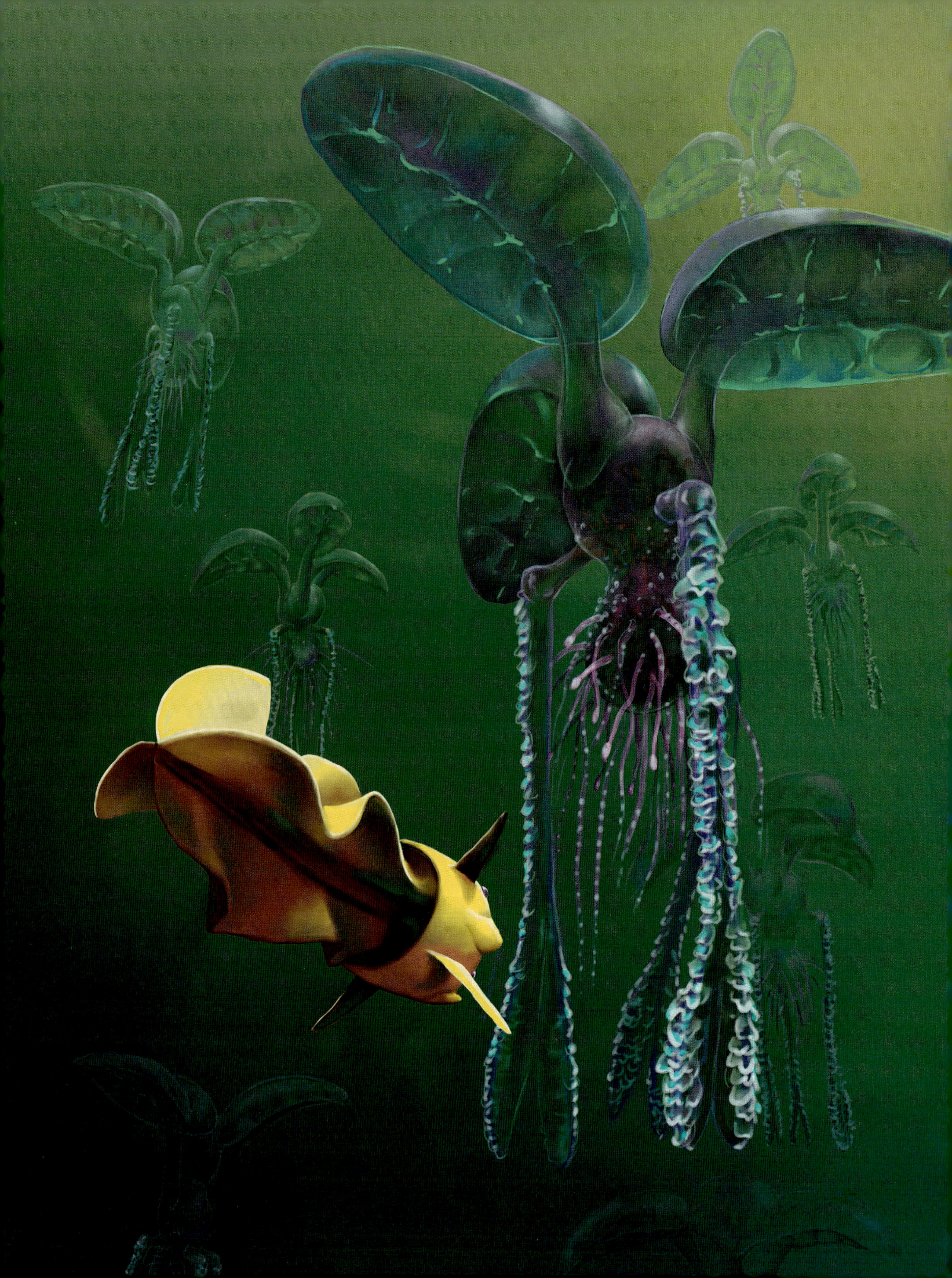

Tribune

Tribunus vacans

The tribune is an impressively large kwal, belonging to a clade in which the original tripartite design of the bell has secondarily reappeared in evolution. The three parts are now completely separate bells, each attached to the body by a stalk.

From Jets to Rowboats

Tribune bells are among the most advanced bell designs. The muscles in their ancestors' bells ran circularly along the bell and contracted all at once, forcing a jet of water downwards. After each contraction, the bell relaxed, allowing water to stream below it once more. The tribune uses an evolved version, the 'squeeze and flex' movement, in which the 'flexion' part makes the bell function as a rowing oar. All this means that moving from a jet to a rowboat can represent progress.

All kwals that use this squeeze and flex hybrid movement use their bells in unison, so their movement retains a pulsatile character.

Kwal Evolution

How far can kwal evolution move towards quicker or more efficient propulsion? A large problem is that the rest of the animal's anatomy may not allow much progress. The biggest obstacle is probably that kwals feed by waiting for something edible to blunder onto their dangling tentacles. The tentacles offer considerable resistance to movement. This is not a problem at low speeds, but it is a major hindrance at high speeds. To move quickly, kwals would have to lose the tentacles and acquire a different feeding mechanism. Such a change would have to happen in small steps, as evolution works a bit like rebuilding your house slowly while you live in it. You cannot tear down the house before you build a new one from scratch.

The squeeze and flex rowing bells may be the most advanced feature kwals will ever possess. Still, give them another 100 million years or so to see what they may come up with.

◄ A cloakfish swimming towards large kwals.

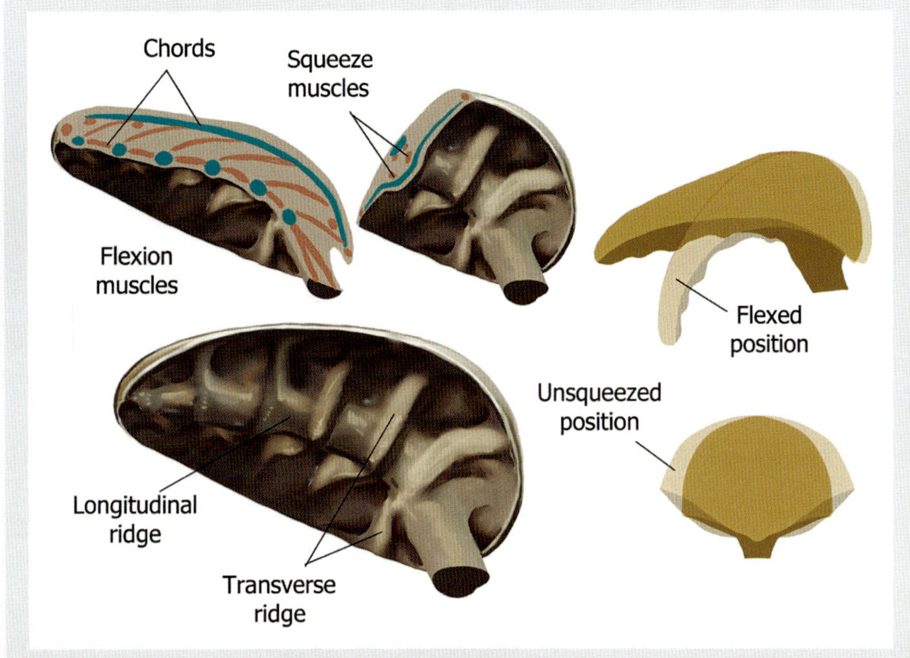

▲ **Bell Structure**
The gelatinous substance of the bell is shown as if it were opaque to make details visible. The muscles in the bell mostly run in thickened ridges and work together with incompressible chords. The transverse ridges, together with some circular muscles, squeeze water out of the bell. Superimposed on this squeeze, a contraction wave passes along the longitudinal ridge, from its stem to the tip. This flexes the bell as a whole and beats it against the water like a rower moves an oar.

A Cloakfish with A Bleak Future

Meanwhile, in the scene on the left, a small cloakfish swims straight for a group of large tribunes. This choice may end in death for the cloakfish. Tribune venom may not be the strongest in the Furahan animal kingdom, but will certainly make short work of a small cloakfish.

► **Tribune**
Tribunus vacans
Name derivation: *Tribunus* (L.), meaning an official in Roman civilization. There were originally three tribunes. A 'tribunus vacans' was an unassigned officer. That term is here used as a word play on the transparent nature of kwal
Habitat: Tropical Western region of Anterior Ocean
Height: 1.5m

Notes: The geographic region of this species is largely conjectural. Its distribution is only certain in accessible shore regions.

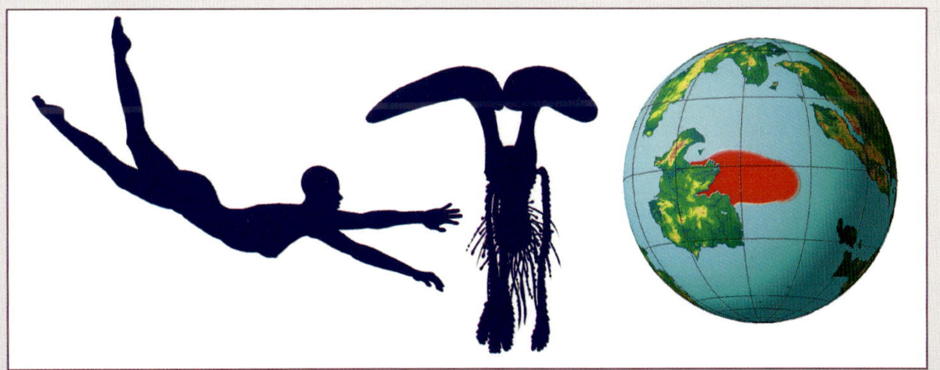

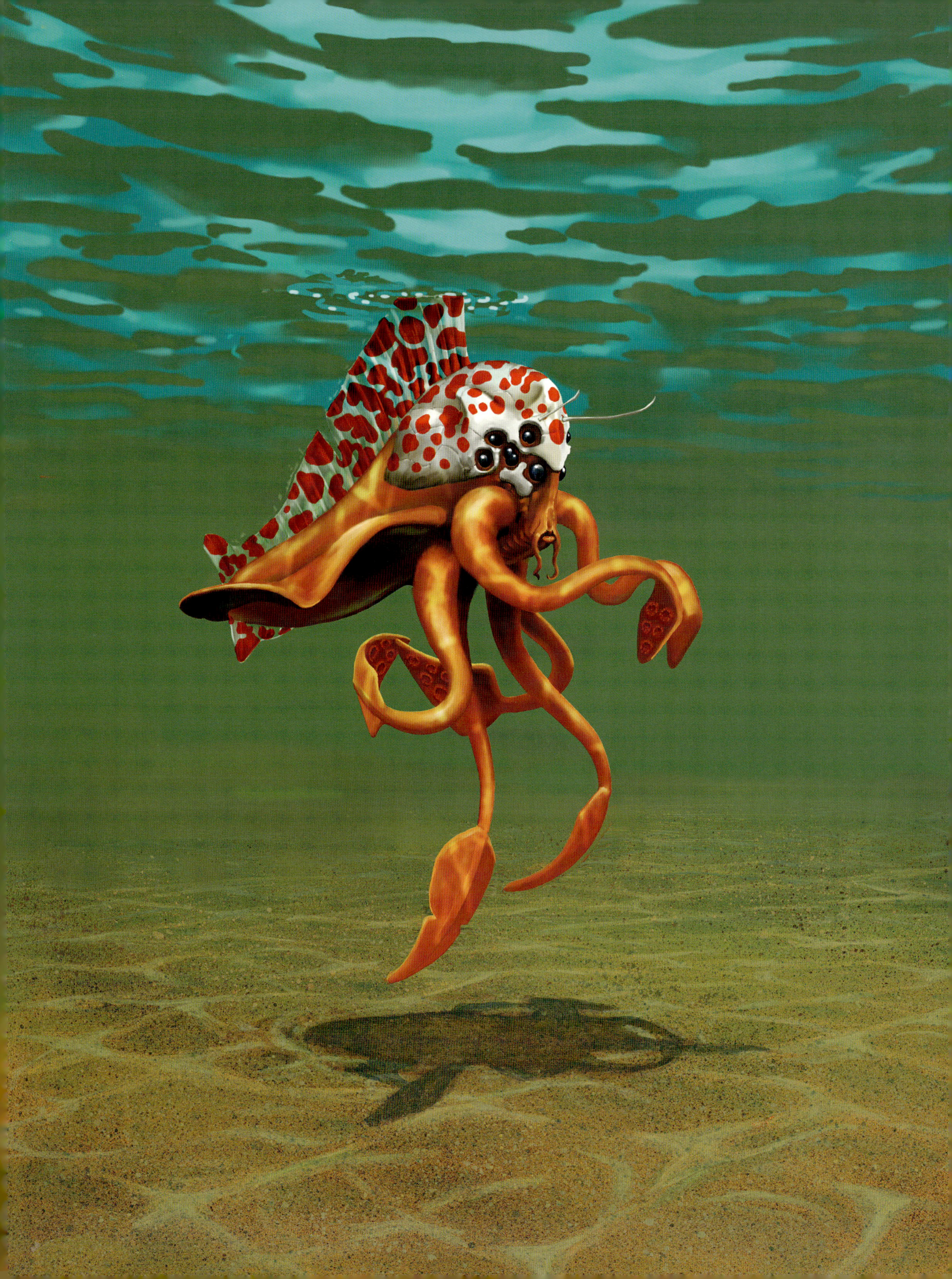

Wardens

Righteous Phleph

Pigritia sursumvergenspropterpenuriaponderis

It is a strange creature, the phleph. The first impression is of a well-coordinated animal, with its graceful arm movements and its seemingly purposeful behaviour. However, after the observer has learnt to look for the meaning of these various features, the inescapable conclusion is that the phleph has rather less to offer than its exterior would suggest. Its intricate hypnotising arm movements seem to be the equivalent of twiddling one's thumbs.

Phlephs are Wardens

Phlephs belong to a group of unusual sea creatures, the Cthulhuidae, often called 'wardens'. This name may have had to do with their habit of keeping a close eye on their surroundings, usually from the entrance of the holes or burrows that most of them live in. This behaviour is much better appreciated from the depiction of the Flachkopf Warden on the next two pages, as the phleph shown here is not at all behaving normally: it has somehow found itself in shallow waters near the shore without an ability to hide. Perhaps it was chased by a predator, or perhaps it was intrigued by some animal swimming by. It is quite possible that curiosity got the better of the phleph. Regardless how it got here, it should not float around doing nothing but should quickly make its way to deeper waters and find a nice burrow.

◀ A floating phleph.

Bodies in Transit?

All wardens have odd body proportions: their tentacles seem large in relation to their size. The prevailing theory holds that these animals evolved from free-range swimmers to manipulatory bottom dwellers. In this view they started as streamlined elongated forms with small and unobtrusive arms and may well end up with small bodies tucked out of the way under well-developed arms. At present, they seem to be halfway along this path. They are agile but not fast.

Phleph Activity

Phlephs are rather passive animals, in spite of first appearances. One of the few researchers to take an interest was Ottokar Uytterwaerde, who described several species of Cthulhuidae. He famously quipped that, 'Phlephs can hardly even be considered edible. When one attempts to take a good bite, surprisingly little substance meets the teeth'.

Uytterwaerde, living in rough times, added a remark that later citizen-scientists would not condone, 'After thorough cleaning, the shell makes an attractive mantelpiece ornament'.

Upward Mobility

The long scientific name of the phleph may hide a story that is not documented in the official annals of any naming ceremony. A lack of documentation is not surprising, as naming ceremonies are festive occasions, during which not everything should be recorded.

The translated name appears to be a description of the animal's slow behaviour: the genus name, *Pigritia*, means 'sluggishness' or 'laziness' and the species name, *sursumvergenspropterpenuriaponderis*, literally means 'upwards directed due to a lack of weight', which could be a simple description of the animal's ability to float up or down by controlling the volume of the shell bladder.

The phleph was named by Ottokar P. Uytterwaerde, then a young and rising marine biologist. At the time he was working under a Lector with whom he apparently did not get along well. In later years, Ottokar, always an enthusiastic participant in naming ceremony festivities, divulged at one such occasion that the Phleph was in fact named after that Lector. Apparently, the species name referred to that Lector's inexplicable rise to elevated levels in the Institute's hierarchy.

▶ **Phleph**
Pigritia sursumvergenspropterpenuriaponderis
Name derivation: *Pigritia* (L.) (meaning 'laziness'); *sursumvergenspropterpenuriaponderis* (L.) (meaning 'upwards directed due to a lack of weight')
Habitat: Reefs, rocky shores in shallow tropical seas
Distribution: Archipelago
Mass: About 3kg
Length: 40–50cm

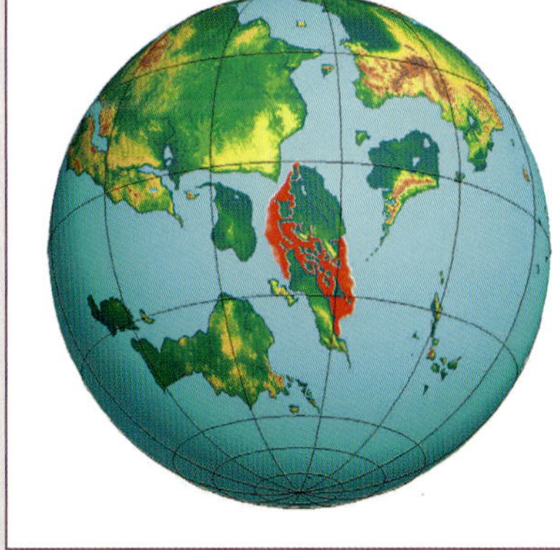

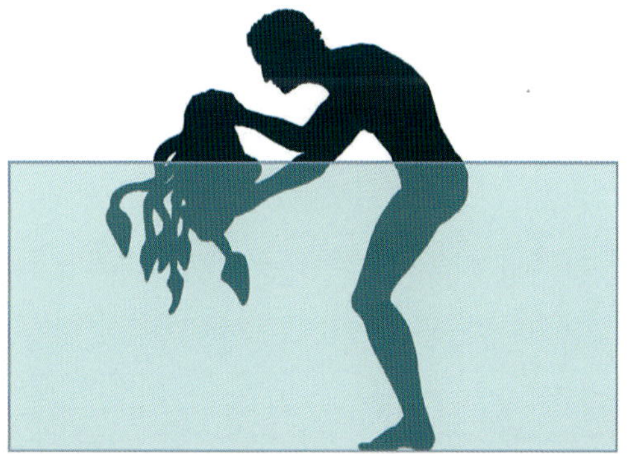

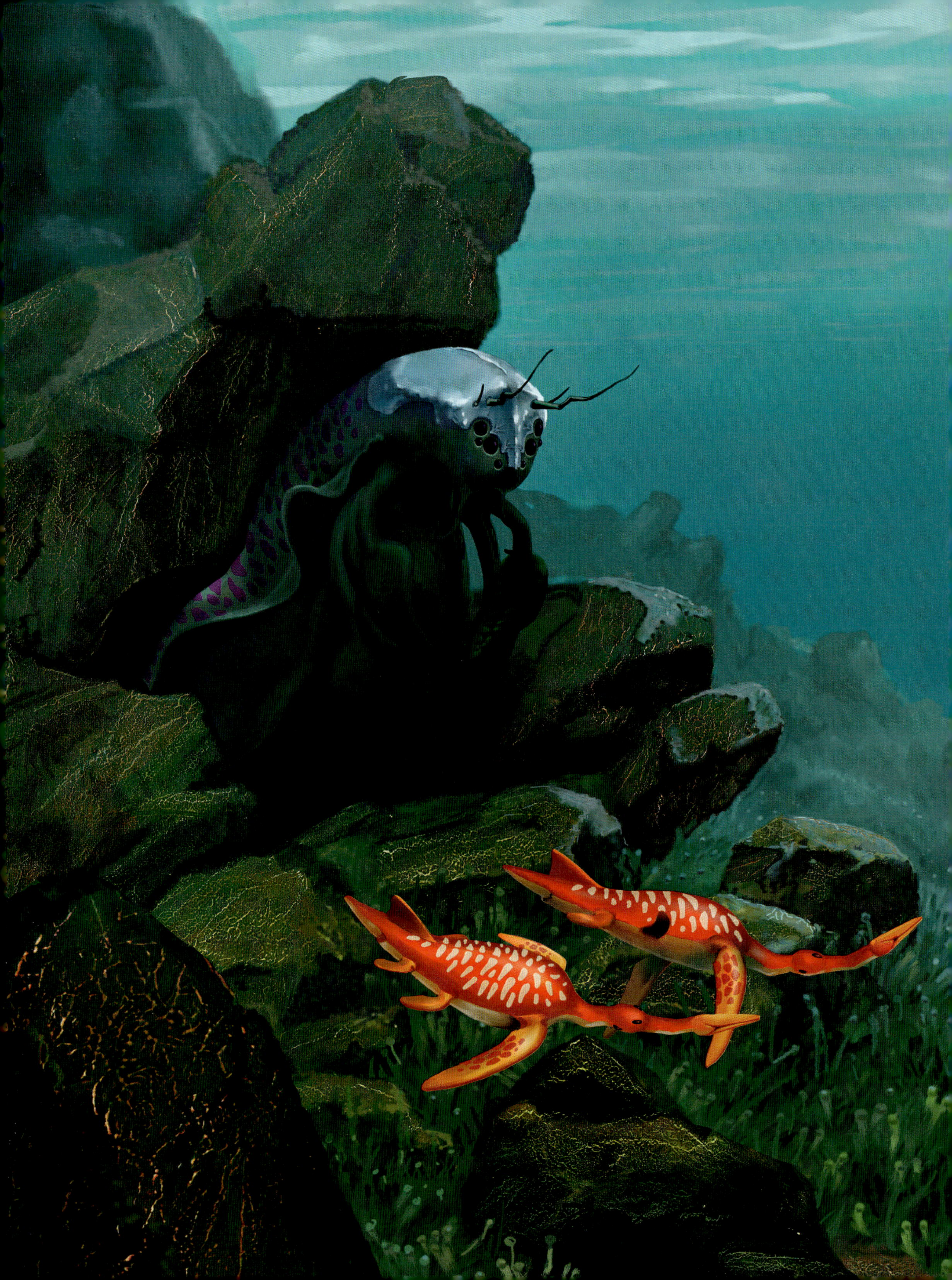

FLACHKOPF WITH FLORTLES

PRINCEPS SIMULANS AND *FLAVOPTERUM PULCHER*

While the seas of Alpha Phoenicis IV are as teeming with life as Terran seas were, the marine life of Furaha is much less well known than that of old Terra. The shallow waters on the continental shelves have been explored, but rather haphazardly. The scene in this illustration shows one environment that has received at least some attention. Not surprisingly we are dealing with shallow waters, here off the coast of Tendaguru, where sediments washed into the sea provide the base for a rich ecology.

The Flachkopf

The animal in the background is a typical 'warden', a species of the clade 'Cthulhuidae'. To appreciate both common and divergent features of various warden species, the flachkopf may be compared to the righteous phleph. The flachkopf, also known as 'la figure plat' or 'flatface', is here shown in a prototypical warden environment. Wardens are all bottom dwellers, living in caves or burrows. This particular specimen has found a space between volcanic rocks. Wardens are known to drag rocks and other objects across the sea floor with their tentacles to secure the entrances of their homes.

◄ Two flortles passing a flachkopf.

Curious Omnivores

Wardens consume anything remotely edible and will examine anything that might be of an organic nature. Their six tentacles are quite dexterous and versatile. Wardens use their tentacles to handle and examine food but may also be seen 'walking' over the sea floor, tiptoeing on the tips of their tentacles. Usually, however, they stick to their holes. Being of an inquisitive nature, they pop out of their holes to look at anything happening in the neighbourhood. After emerging, they prick up their tentacles and wave them around. They look as if they are directing traffic, but only other wardens are impressed by this behaviour.

Are Wardens Intelligent?

Once every decade or so someone reintroduces the theory that wardens are intelligent. This is not surprising: their behaviour is mysterious and their neural ganglia are surprisingly large. Still, neither feature is proof of intelligence. The debate usually swings between two points of view. One holds that intelligence ought to be easily recognisable through behaviour, and in that respect the flachkopf certainly does not stand out. While their behaviour is intriguing, no-one has been able to make sense of it. The other point of view demands that there must be room for alien forms of intelligence, which cannot be expected to be immediately recognisable to inquisitive and manipulative Terran primates. Unfortunately, that argument can neither prove that wardens are intelligent, nor that they are not. The discussion usually simmers for a while and dies out once everyone realises the lack of new arguments.

Flortles

In the foreground, two colourful flortles swim by. These bear some resemblance to plesiosaurs, of a rather minute variety. They are, of course, species of the large clade Fishes V. Recall that Fishes V have six fins. In the case of flortles, evolution has greatly favoured the first pair of fins over the others. The two other pairs of fins are still present but are used for steering only.

Recognising Sapience

The marine biologist Ottokar P. Uytterwaerde was very outspoken about warden intelligence, something he felt was completely absent. He used his caustic sense of humour to make his point.

> My colleagues may state that a warden and a man can look at one another inquisitively, pondering life, the universe and such matters. What seems more likely to me is that both occasionally wonder about the other's edibility. So far, man is in the lead.

► **Flachkopf**
Princeps simulans
Name derivation: *Princeps* (L.) (meaning 'leader' or 'foremost person'); *simulans* (L.) (meaning 'pretending')
Habitat: Reefs, rocky shores in shallow tropical seas
Distribution: Tendaguru, *Isolae perdidae*
Mass: About 3kg
Length: 40–50cm

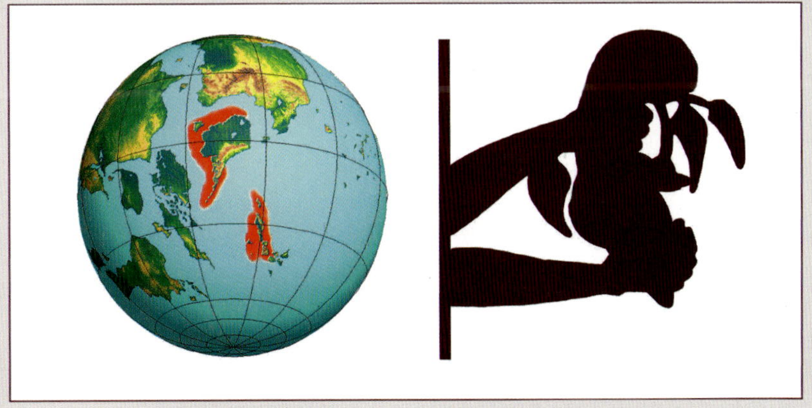

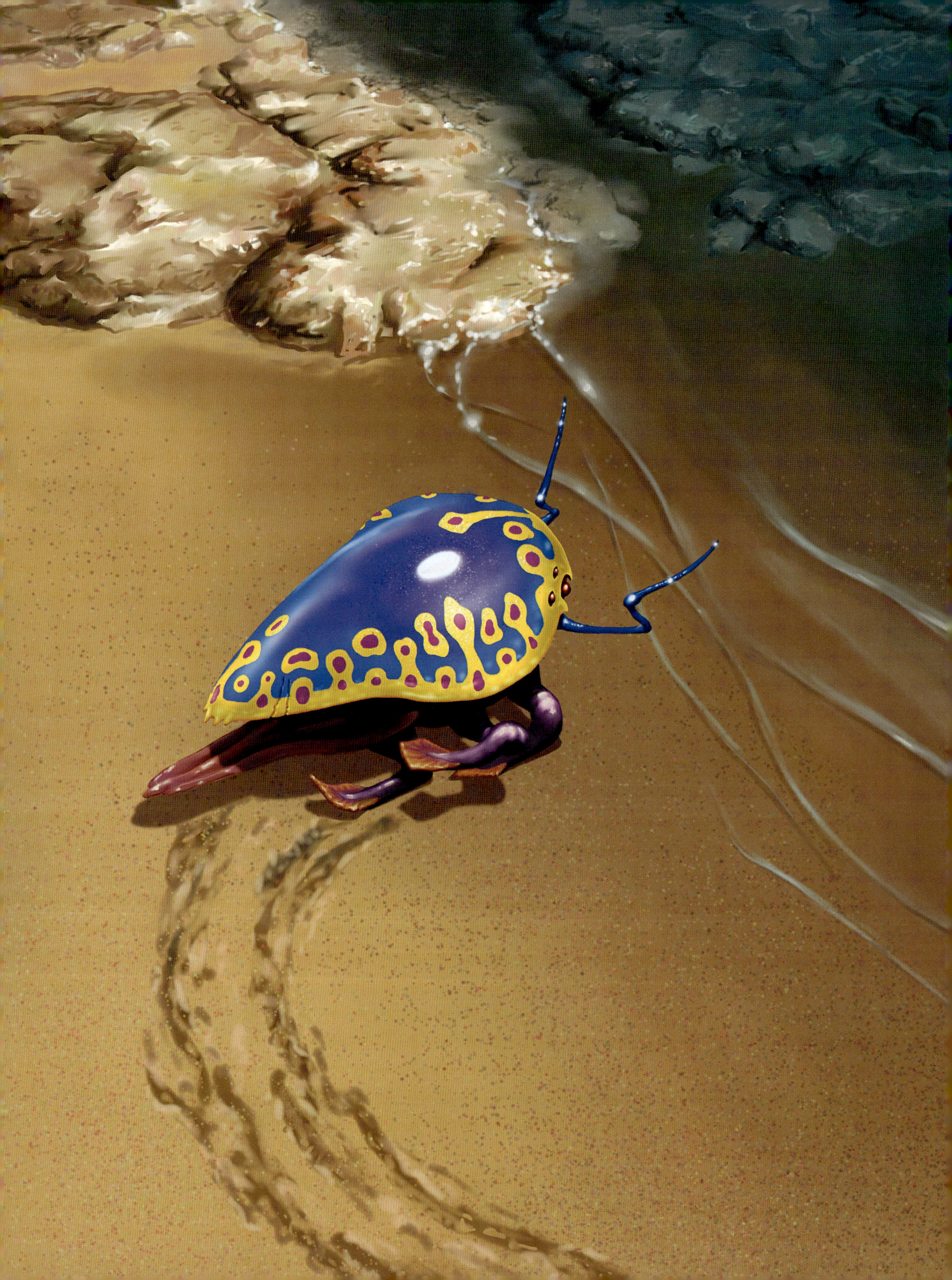

A Stamping Snigel

Myrmillo tengbergenii

Snigels are wardens, with typical warden characteristics such as a long body with large tentacles, a shell covering the head or top of the body (the galea) and a curious disposition.

Snigel Stampede

Not much is known about the lifestyle of snigels, but the one thing that has earned them household recognition is their unpredictable 'stampedes'. This ill-chosen term refers to their curious synchronised forays onto dry land. Without any apparent reason they will set out towards the beach from their comfortable holes, nooks and crannies in the shallows. They actually climb onto the beach, where they crawl around for a while, leaving odd trails. These mysterious manoeuvres last for several hours. Afterwards, the snigels return to the water and take up their normal habits of peeking from their hideouts and being curious, if rather cowardly, busybodies.

People have unsuccessfully tried to connect stampede timing to time of day, the phases of Furahan moons, barometric pressure or fluctuations in ultraviolet radiation. Another theory states that snigels are controlled by alien intelligences, showing that offering people a good education does not prevent some from acting stupidly.

◀ A snigel on a stampede.

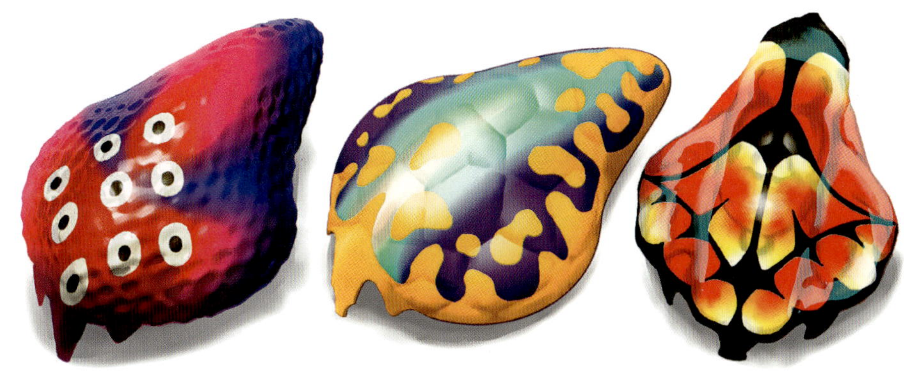

▲ These snigel galeae show the riotous variability of snigel colour. It is not certain to which extent the myriad variants represent different species, but the three shown here are known as species of the genus *Myrmillo*, for better or worse. From left to right they are the 'Mad Strawberry' *(M. fragiformis)*, the 'Turtleback Snigel' *(M. testudiformis)* and the 'Shoggoth' *(M. shoggoth)*. Some say that *M. shoggoth* deserves its own genus, in which case its scientific name would become *'Shoggoth shoggoth'*.

Walking with Tentacles?

As can be seen from the main illustration, snigels walk on tentacles – well, sort of. They put one tentacle forward, as if we would place our curved hands on a table in front of us, with the nails flat on the top of the table. As they roll forward another part of the tentacle touches the ground, as if we would consecutively let our knuckles and then our wrists rest on the table. The vertical part of the tentacle withstands gravity through contraction of circular muscles that make the tentacle stiff enough to keep it from buckling. When there is no more tentacle to roll forward on, the tentacle as a whole is lifted and swung forwards. Meanwhile, the tail fin used for swimming is neatly pleated and folded away.

A Society Scandal

New snigel variants continue to be recognised on the basis of galea colour. But even the species depicted here, a fairly common one, has not been described in detail. Cleaned snigel galeae recently started to become collector's items, even though the shells have no inherent value. Unfortunately, a few people took to 'harvesting' live snigels. This practice shocked the Institute, whose citizen-scientists thought their civilisation was above such barbarism. The Bureaurchy reacted with force. Some poachers claimed innocence based on the fact that what they did was not specifically forbidden, but with typical Institute reasoning a 'failure to reason' was not considered a mitigating factor, so the culprits received 'aversion insertion' to prevent recurrences.

▶ **Snigel**
Myrmillo tengbergenii
Name derivatives: *Azureus* (L.): blue; *Myrmillo* (L.) (meaning 'gladiator with gallic armour and fish-topped helmet'); *tengbergenii* (after Thijs Ebbenhorst Tengbergen); *fragiformis* (L.) (meaning 'strawberry-shaped'); *testudiformis* (L.) (meaning 'turtle-shaped'); *shoggoth* (meaning 'fictional creature')
Habitat: Reefs, rocky shores, shallow tropical seas
Distribution: All around the *Isolae minores*
Mass: 1–3kg
Length: Carapax 20–40cm

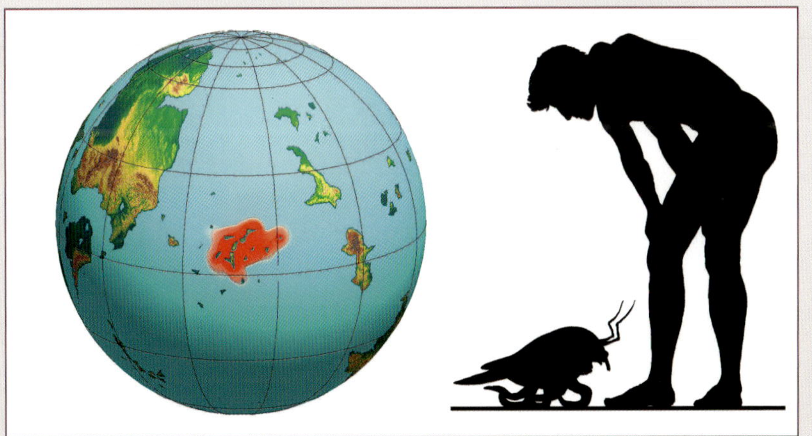

The Sinister Cthulhu

Cthulhu somniator

People commonly consider wardens as inoffensive creatures, a view that the animals they eat will not share. But one clade of wardens is commonly regarded as monstrous: the genus *Cthulhu*.

This genus is omnivorous but conveys a vicious predatory image. This is not surprising: they are at least twice as large as most warden species and attack large prey. But the real explanation for their reputation lies in a quirk of nature – human nature, that is. The human brain is so strongly wired to read other people's moods that anything resembling a facial expression is immediately interpreted as an emotional signal. For example, big eyes are interpreted as juvenile and endearing, while frowning eyebrows are read as anger. The eye holes in the head shield of cthulhu species happen to convey such angry looks. The shape of cthulhu eye holes did not of course evolve to induce emotions in humans at all, but it is hard for humans to suppress the association.

A Dangerous Predator

Mind you, the species shown here, *Cthulhu somniator*, is in fact a large, fierce and dangerous predator. It lives in deep waters, so it almost never encounters humans, which is probably just as well for both species. A cthulhu could certainly dispose of swimming children with ease, and not even the Institute's enlightened attitude of treating all animals neutrally can overcome people's instinctive emotional responses.

A Case in Point

The cthulhu in the image managed to capture a large marine mdudu, which was probably disturbed by the lights of the Institute's submersible. Without this human intervention, it is unlikely that the cthulhu could have overcome the mdudu, a considerable adversary. As it is, the mdudu has been lifted and turned on its back, and the cthulhu has already ripped off one of the mdudu's claws.

What's in a Name?

Cthulhu was a fictional being invented by the twentieth-century writer HP Lovecraft. His creation *Cthulhu* had tentacles but also a humanoid shape and stubby wings, so the resemblance with wardens is not strong.

The name *Cthulhuidae* for wardens was coined by Ottokar P. Uytterwaerde, a bit of a prankster. It no doubt appealed to his sense of humour to group inoffensive and somewhat amusing species such as the flachkopf and the snigel under such a troubling heading. But even he did not bestow the name cthulhu on a particular genus or species.

This changed a generation later, when the Institute scraped the funds together to produce a submersible, the *Doughty Diver*, to let humans descend to depths of more than several hundred metres, which had until then been impossible.

It was during such an expedition that the genus *Cthulhu* was discovered. The person to do so was Ottokar's daughter Estrelinha Uytterwaerde. Estrelinha could neither resist to secure the genus name nor to add a fanciful description to its formal scientific description. Furahan culture at the time tended towards sentimentality. This affected science, even though scientists always think they are purely objective. This is what Estrelinha wrote:

> Having seen the cthulhu in its natural habitat, my memory of it is forever shaped by its sinister looming dark mass, hinted at yet hidden in the dark water, motionless but eerily commanding. When the eldritch creature crept forward, the slow movements of its tentacles mesmerised my companions and myself. It looked directly at me, it seemed, aggravated by this intrusion into its forbidding realm. I swear I felt its ominous presence directly with my mind. Ever since that dark encounter, the abomination returns to me at night, troubling my dreams. In my restless sleep, it somehow knows me and waits for me. It waits! It rests in its murky abyss, floating motionlessly, slumbering, perhaps dreaming, but surely waiting, always waiting, for something, something sinister, something maddening...

► A Cthulhu tearing a mdudu apart.

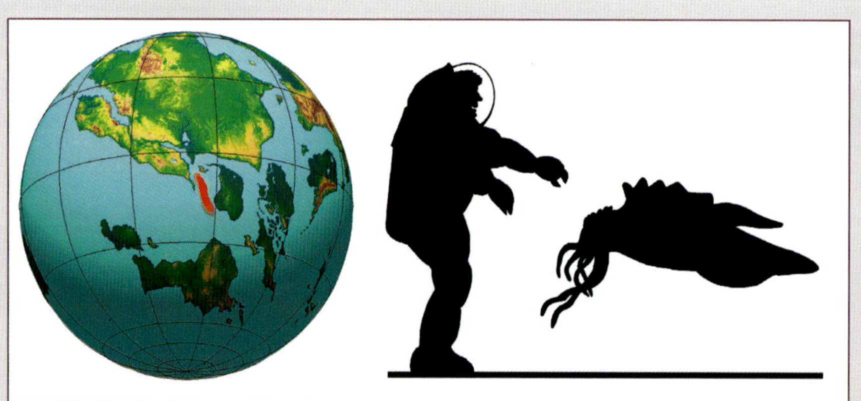

◄ **Cthulhu**
Cthulhu somniator
Name derivation: *Cthulhu* (fantasy name derived from fictional creature); *somniator* (L.) (meaning 'he who dreams')
Habitat: Deep offshore waters, probably near sea floor
Distribution: Unknown as only known from one trench, but may be widespread
Mass: 40kg
Length: 140cm

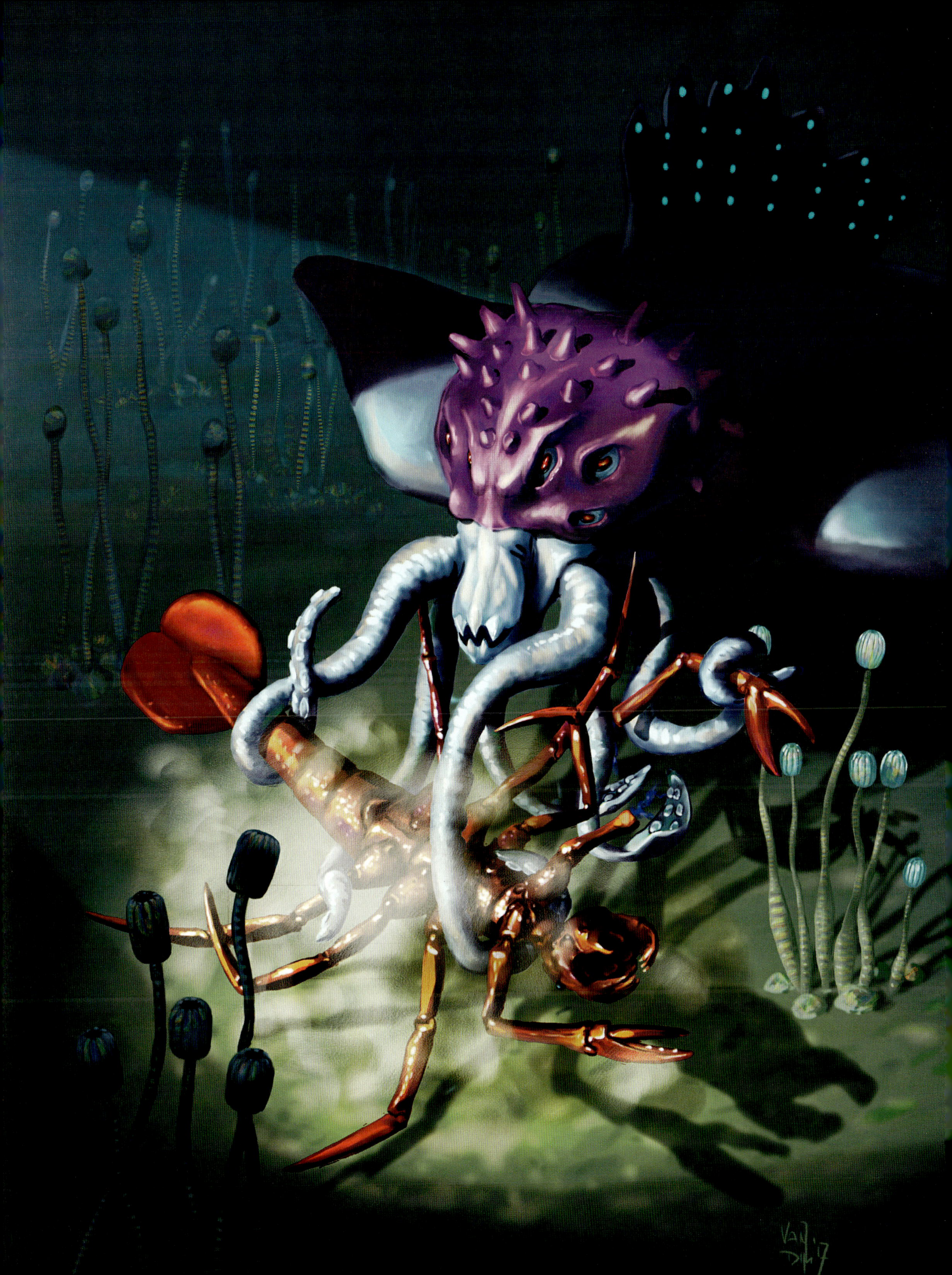

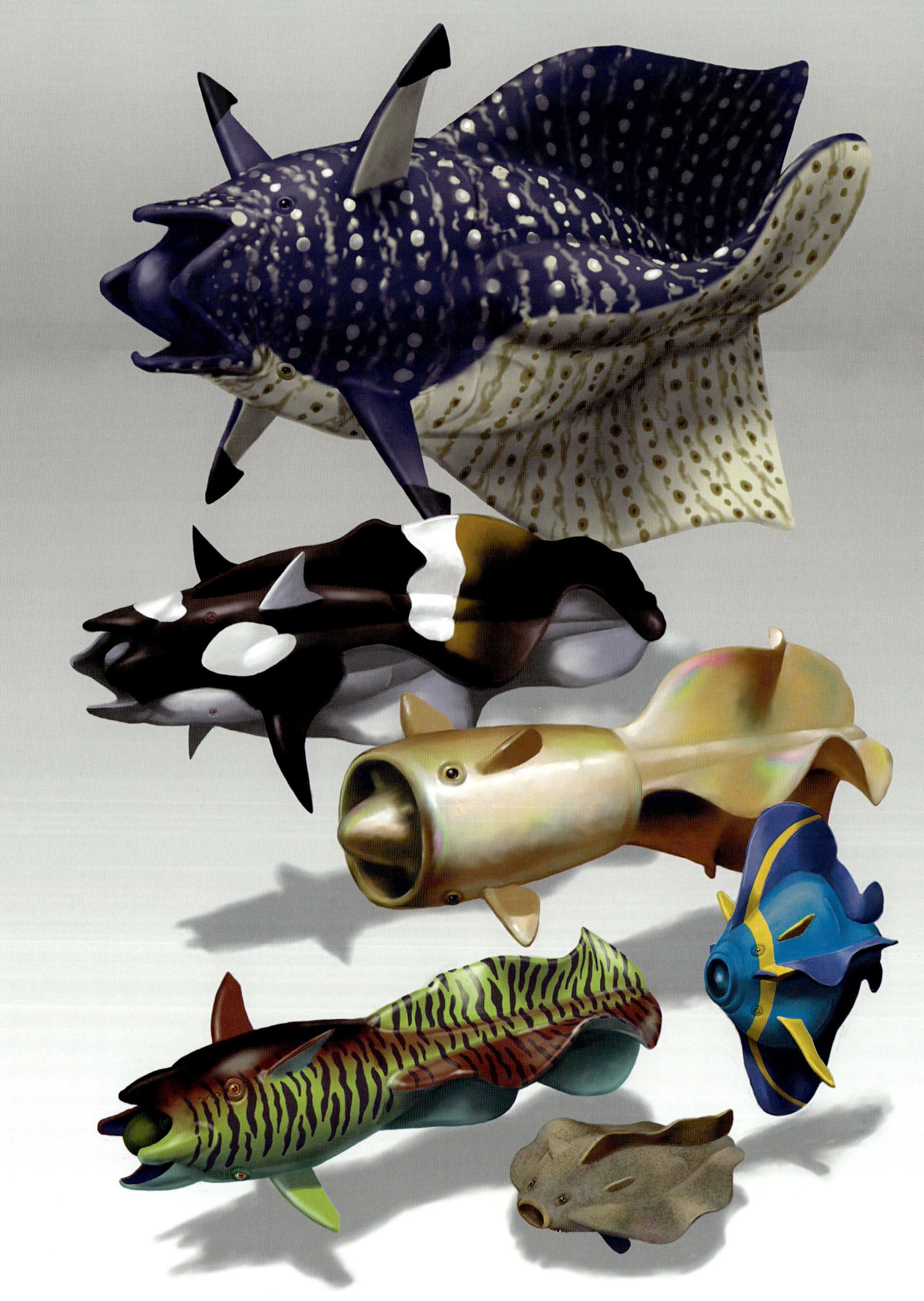

Cloakfish

Cloakfish Diversity

The animals shown represent the worldwide variety of cloakfish, often simply called 'cloaks' or 'cloakies'. Their shape, size, colour and habits vary, as the result of adaptations to different circumstances. Cloakfish can also differ in a more fundamental way. While some conform very closely to their ancestral four-sided radial symmetry, other clades have developed what is in effect bilateral symmetry. Let's have a closer look at some 'cloakies'.

Starfish *Tanypteryx archicus*

Name derivation: *Tanypteryx* (Gr.) meaning 'wide-stretching wings'; *archicus* (Gr.) meaning 'regal'.

The starfish is a large typical 'short-sleeved cloakfish', much like the capaespada also shown in this book. Short-sleeved cloaks often do not use the 'double clap' gait used by 'long-sleeved cloaks' and the one here indeed does not. Its four-sided symmetry is anatomically almost completely intact, but its colours fit bilateral symmetry, with the bottom half much lighter in colour than the top half.

Boxfish *Trabs seamusnastrarruzzi*

Name derivation: *Trabs* (L.) meaning 'beam'; *seamusnastrarruzzi* (L.) named after Seamus Nastrarruzzo. (Because many members of the Nastrarruzzo clade named species, they were allowed to add their first names rather than just use 'nastrarruzzi'.)

The cross section of this species is nearly square at the middle of the body. Together with its stiff body, this explains both its common name and its scientific name of *Trabs*, which denotes a beam of wood. Off-worlders are always surprised by the unbending bodies of cloakfish, accustomed as they are to sinuous Earth fish. Its magpie or orca-like colours make it hard to distinguish its shape, an effect called disruptive colouration.

◄ Starfish, boxfish, barrelfish, common cloak, flounder and blue billy.

Barrelfish *Marmarus cylindrus*

Name derivation: *Marmarus* (Gr.) early meaning 'that which shimmers'; later meaning 'marble'; *cylindrus* (Gr.) from *cylindros* meaning 'rolling stone'.

This species' name was obviously inspired by its rotund mantle. It is well-known for its skin's iridescent quality. As anatomy goes, this species is close to the presumed ancestral cloakfish: not only are the top and bottom parts nearly identical, but the mantle and the barrel are also virtually separate, forming a wide space in between where the gills and the sieves are situated. Barrelfish are somewhat clumsy swimmers.

Common Cloak *Tetrachlamus pycnus*

Name derivation: *Tetrachlamus* (Gr.) meaning 'four cloaks'; *pycnus* (Gr.) meaning 'frequent'.

This is the prototypical cloaky. This species may well be the evolutionary next step following the barrelfish pattern: the mantle and barrel are closely integrated, forming an efficient hydrodynamic shape. The common cloak has another feature definitely worth mentioning. Even though it is quite a bit of work to separate the cloak muscles from the skin and the bones, the result is worth it, as *sautéed* cloak is a delicacy.

Blue Moon *Bractea pressata*

Name derivation: *Bractea* (L.) meaning 'thin sheet of metal' or 'gold leaf'; *pressatus* from *presso* (L.) meaning 'to squeeze'.

The blue moon represents cloakfish with sideways flattened bodies, the 'catahimatia'. Most of such species live near rocky coasts or in reefs, where such a shape may prove advantageous.

Flounder *Psessa epiborborea*

Name derivation: *Psessa* (Gr.) meaning 'flat-fish'; *epi-* (Gr.) meaning 'on'; *borboros* (Gr.) meaning 'mud' or 'filth'.

Named for its resemblance to the Terran flatfish, the Furahan flounder represents vertically flattened cloakfish: the 'platyhimatia'. These creatures are always found on the sea floor. Contrary to common belief it does not feed by sifting sand or mud, as its gills are not equipped to deal with the strain of so much foreign material. It tastes quite pleasant, particularly with a cooled wine.

▼ Species Key and Distribution

A: Starfish; B: Boxfish; C: Barrelfish; D: Common cloak; E: Blue moon; F: Flounder; G: *Homo semisapiens terrae*

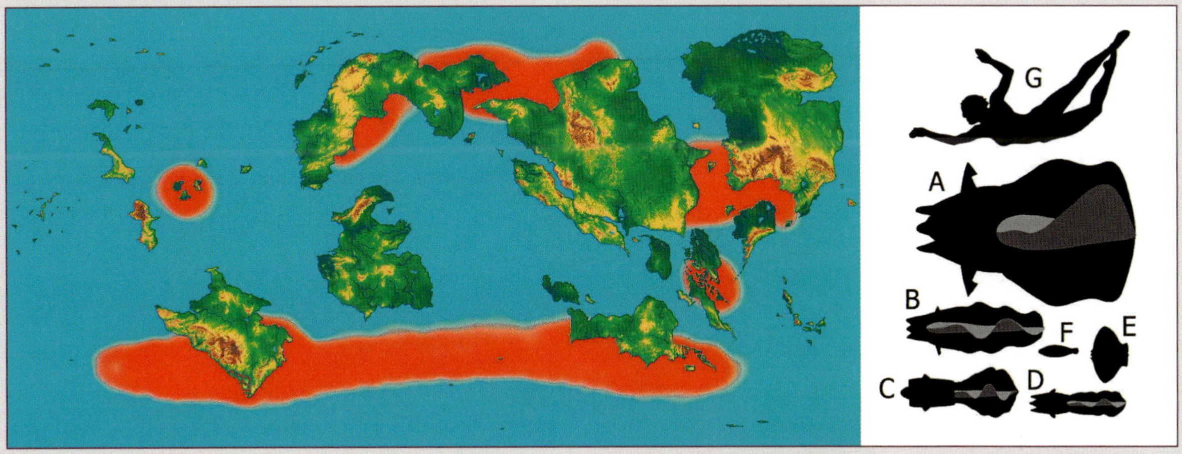

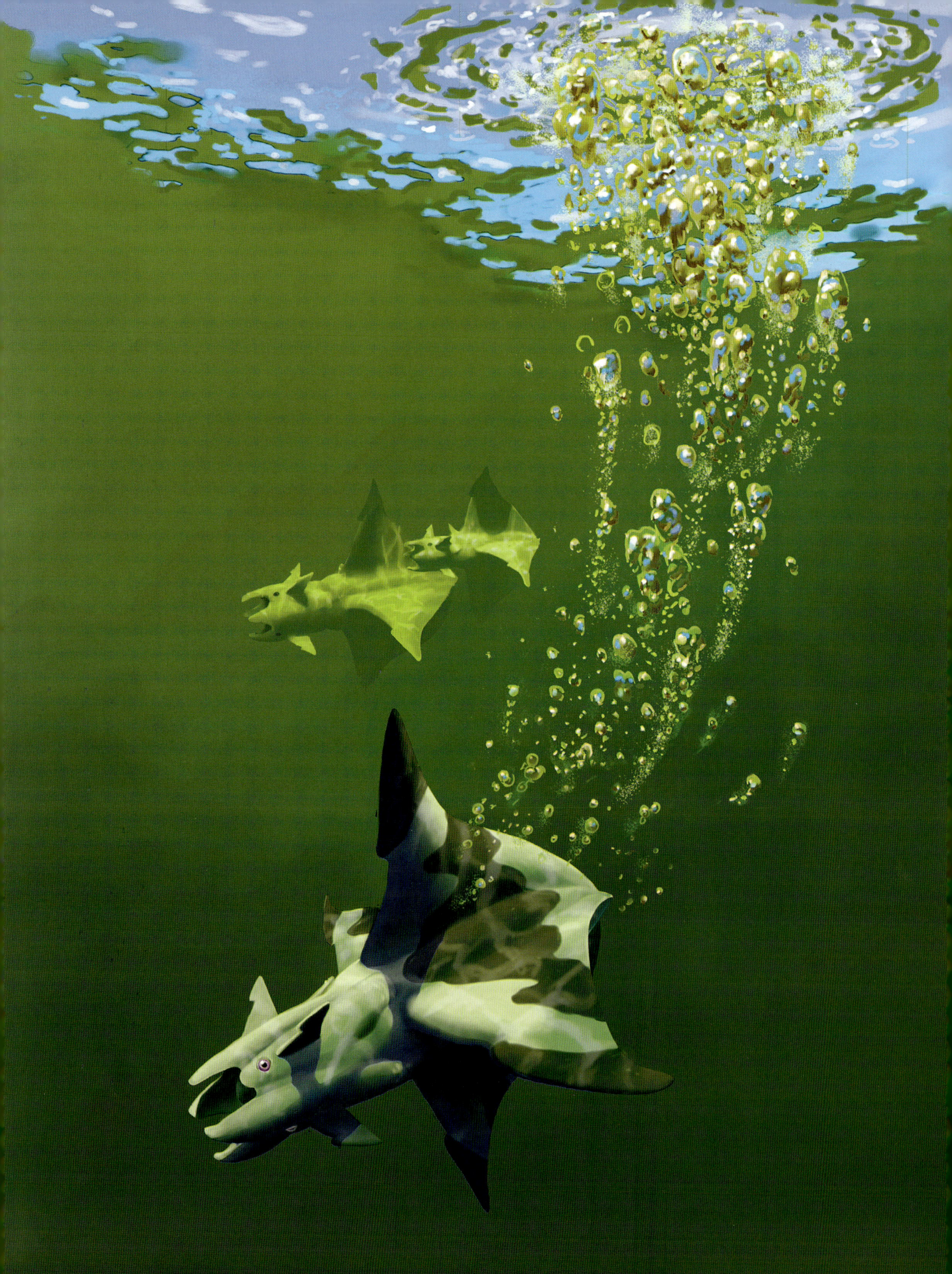

A Bevy of Capaespadas

Capaspatha augusta

The capaespada is the largest known cloakfish. Adult females reach a fin span of 7m. These massive animals move gracefully through the water, their four fins slowly undulating. Their body shape is typical of the short-sleeved group of cloakfish, in which the dagger and its attached fin membranes are relatively short. The fins make up for this in width. A comparison with long-sleeved cloakfish, having cloaks longer than they are wide, reveals that the difference in build influences the movement of the cloaks; whereas long cloaks may exhibit several waves at a time, this is not the case in short-sleeved cloakfish. Usually not even one complete wave is present, resulting in a movement of the cloak as a whole that closely resembles that of Earth's manta rays. In fact, the capaespada is sometimes called the 'mantacloak'.

Family Bevies

Capaespadas are usually found in extended family groups called 'bevies'. A bevy typically contains three to seven large females accompanied by a clutter of young and subadult animals, but no adult males – the latter are typically solitary. The finding that cloakfish can live in family groups came as a surprise to the scientific community, as cloakfish were regarded as not intelligent enough to exhibit such fairly complex behaviour. At present people wonder whether cloakfish in general have been underestimated, or whether some species are intellectually more developed than others. As a result, marine biologists try to investigate other short-sleeved cloakfish for similar traits, but so far without much success. As for capaespadas, the answer has not been settled conclusively what they gain from this living arrangement. There must be safety in numbers, but the details are not clear. There are obviously many predators in Furahan seas, but no-one has yet witnessed how capaespadas defend themselves against them. Perhaps the adults beat predators with their fins, but that is mere speculation.

◀ Juvenile cavorting capaespada.

Play

The image shows a young capaespada plunging back into the water, having jumped out of it for no apparent reason. Apart from jumping clear out of the water, young capaespadas may suddenly change pace and race between their elders' fins or chase one another. Why do they do so? Adults do not display such exuberant and energy-guzzling behaviour without any apparent direct benefit. For non-scientists, the answer is clear – they are playing!

Scientists have seriously and gravely studied why animals and people play. The layman's answer that playing is fun just rephrases the question, meaning only that evolution saw fit to equip young animals with a positive emotional reward to undertake play. We can only assume that it pays to hone your motor skills in a safe environment, rather than find them wanting in a dangerous situation later in life.

Regardless of such considerations, not even the most technically minded scientist fails to be cheered by the antics of young capaespadas, flitting through the water and jumping into the sunlight.

Disruptive Colouration

The image shows two adult specimens in the background, coloured an overall featureless grey. The young capaespada in the foreground displays a much more extravagant coat, with an overall white body and large black blotches on its fins. These make it more difficult to see the shape of the animal and break up its outline, something called 'disruptive colouration'. When the animal moves quickly the effect is enhanced. On the one hand this makes it difficult for predators to recognise the blotch for what it is, but on the other hand the stark contrast makes the animal stand out against the background, which could attract a predator. Given that young capaespadas are in fact blotchy, the benefits must outweigh the disadvantages.

▶ **Capaespada**
Capaspatha augusta
Name derivation: The name capaespada is derived from the Spanish *Capa y Espada*, literally meaning 'cloak and dagger' (the Spanish expression may have inspired the English one)
Capa (L.) (meaning 'cloak'); *spatha* (L.) (meaning 'sword'); *augusta* (L.) (meaning 'majestic')
Habitat: Open water
Distribution: All subtropical and shallow tropical seas
Mass: About 1600kg
Length: Fin width up to 7m

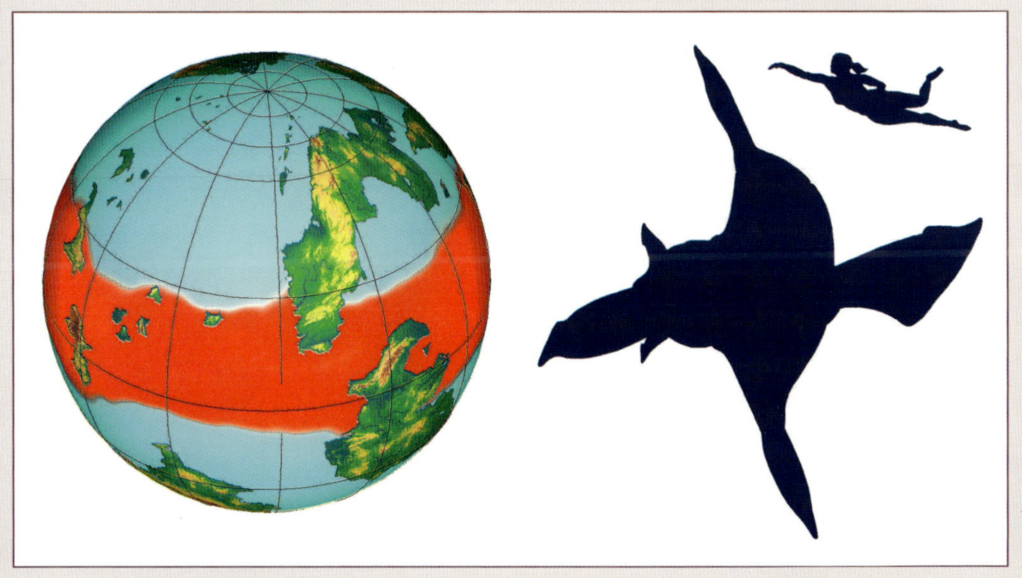

Cloakfish Anatomy

There are probably about 10,000 species of cloakfish, or about as many as there are species of birds on Earth. They are, of course, not true fish, neither in the Earth nor in the Furahan sense. Cloakfish are a Furahan clade, characterised as all clades are by a set of traits, or 'synapomorphies'. They are aquatic filter feeders with four-sided radial symmetry and camera eyes. There are additional traits but mentioning them makes little sense in a non-technical book. Each characteristic membrane – a cloak – undulates with waves travelling backwards, propelling the animal forwards. Having four membranes offers some interesting opportunities. For instance, adjacent cloaks can move toward one another and then almost envelop a volume of water. By ensuring that the propelled water can only flow backwards, the animal increases the efficiency of propulsion.

The prototypical cloakfish anatomy is best appreciated by studying the 'common cloakfish' shown here. The cloaks are relatively long and attached to a central spine, known as the dagger. Because the cloaks are long, more than one wave can travel along them at the same time. This is what makes the common cloakfish a 'long-sleeved' cloakfish. This pattern is geared towards steady straight swimming, a movement pattern well-suited to filter feeders. Do not underestimate what undulation can do. By altering the amplitude, speed, and even direction of the undulation, cloakfish can rotate along any axis and even swim backwards. The four front fins help not only to stabilise the animal but can in various species provide a sudden additional impetus.

Radial Symmetry

There are, sadly, no large complex animals with radial symmetry on Earth, but on Furaha such animals swim swiftly through the oceans, scurry quickly in all directions on land and shoot at will

▶ Detailed anatomy.

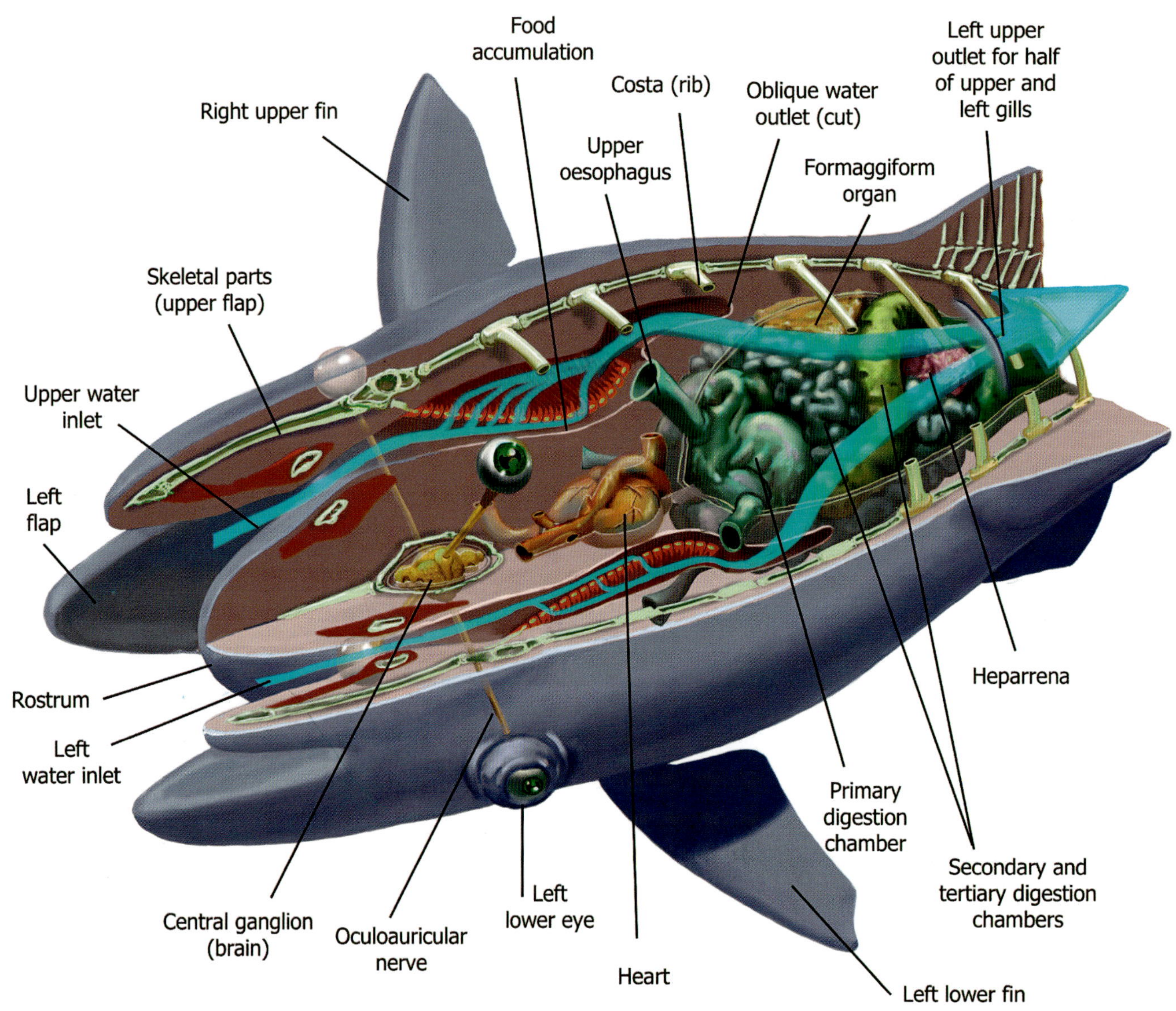

through the air. While cloakfish have four-sided radial symmetry, the symmetry is not complete. The internal organs are in part asymmetrical. Most cloakfish swim with the same side up and have light undersides and dark backs. Some derived cloakfish depart far from radial symmetry. Some are flattened horizontally, some vertically. In effect, by developing clear top and bottom sides, these animals are evolving towards bilateral symmetry.

Gills and Filters

All cloakfish are filter feeders. The front of their body consists of a central core, which is in fact the dagger extended forwards. A cylinder, the mantle, envelops this core. The dagger and mantle are connected by firm struts housing bones, blood vessels and nerves. The struts are positioned underneath the front fins. Water flows in between the struts, and is directed to the right, left, top and bottom gills. Oddly, the outlets are rotated so each of the four outlets receives water from two gills. The gills are parallel to the flow of water, so food particles do not get stuck on the gills but are collected further along, near an oesophagus. This is called 'cross flow filtering'.

Cloakfish are basically jawless, although some maintain that the four flaps at the front might as well be called jaws. Their function is to regulate water flow. In some species the flaps help maintain breathing when the animal is not swimming actively. Most cloakfish swim permanently though.

Camera Eyes

Cloakfish have camera eyes that offer better visual resolution than compound eyes. Many students of Furahan biology are surprised that many large complex animals on Furaha have compound eyes, while the relatively slow cloakfish have a better eye design. Well, evolution has no direction in mind when an animal clade comes into being. If the basic design is a compound eye, evolution can work wonders with it, but it cannot just scrap a design and start all over.

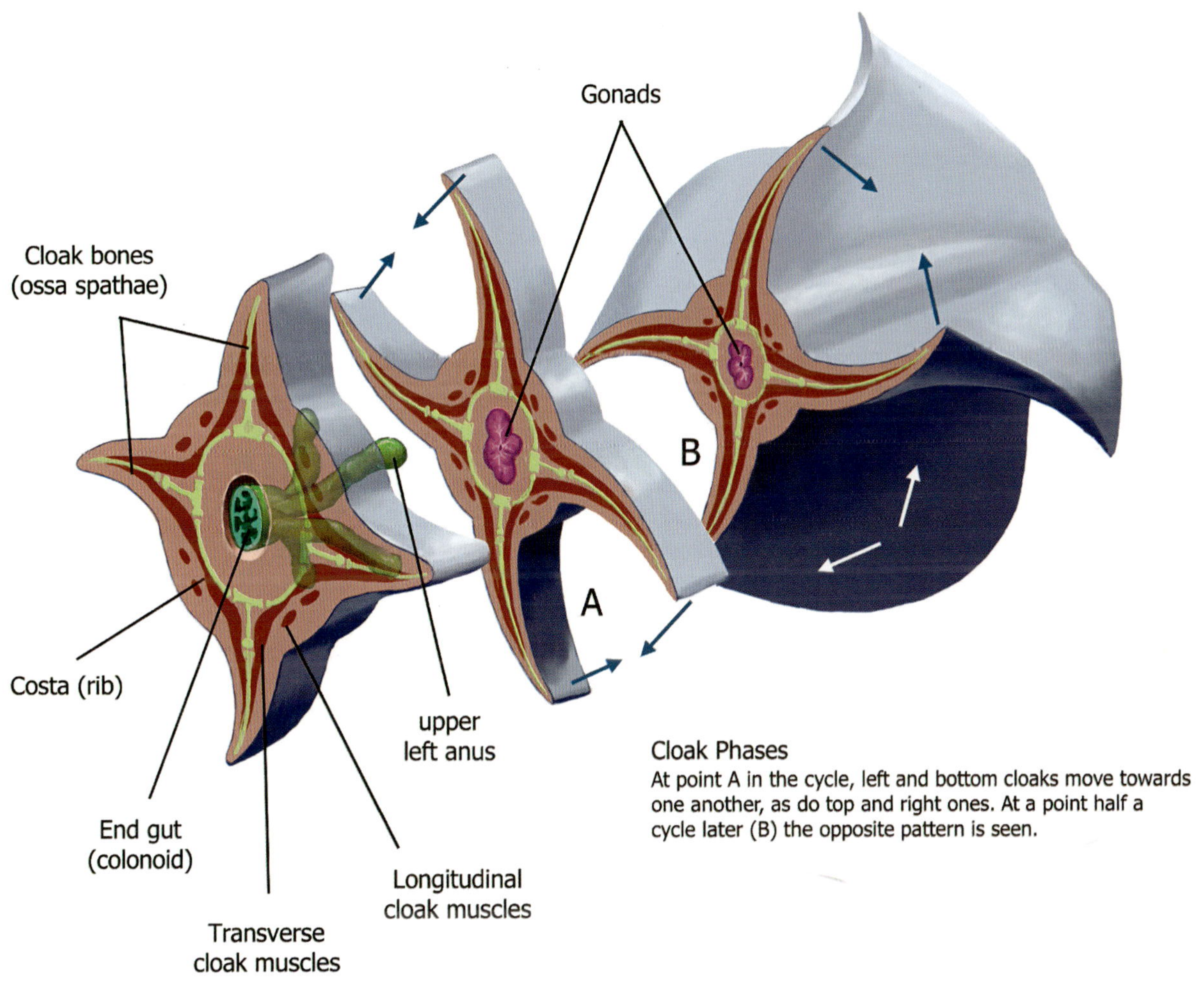

Cloak Phases
At point A in the cycle, left and bottom cloaks move towards one another, as do top and right ones. At a point half a cycle later (B) the opposite pattern is seen.

How Cloakfish Swim

Each cloak forms a thick ribbon that is attached to the dagger along one side. Thin flexible bones originate in the dagger and are embedded in the cloak: the 'cloak rays' or *ossa spathae*. These bones can swing to and fro, taking the elastic cloak membrane with them. The bones move with a small offset compared to their neighbours. If one bone is at its maximal position, the one before it will be just past that point, while the one after it is not yet there. These phase differences form waves travelling backwards along the cloak. The wave pushes backwards against the water, which pushes the cloakfish forwards.

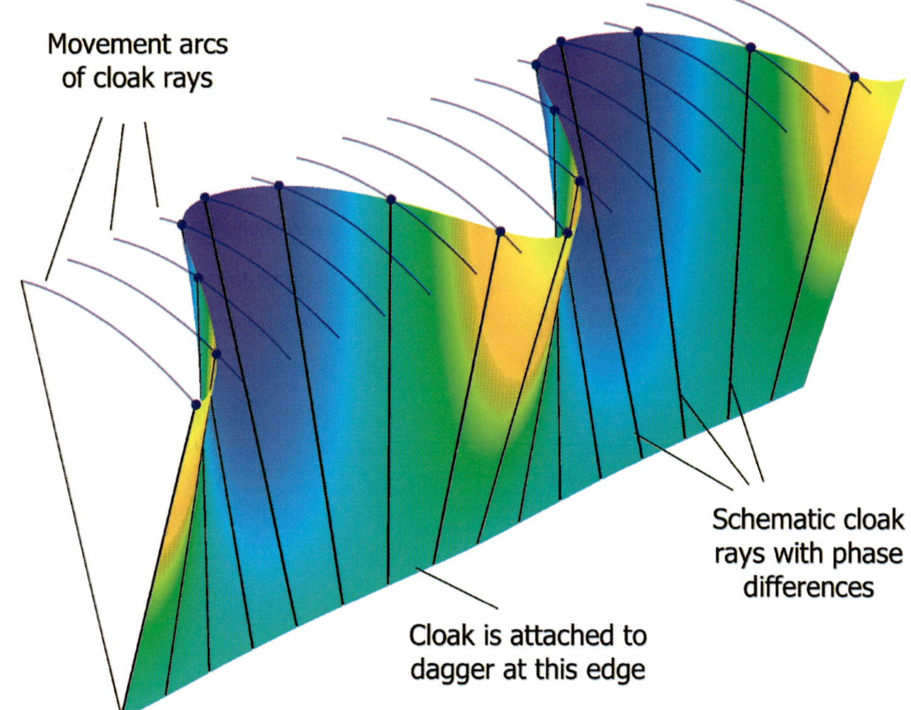

▶ Membrane movement.

▼ Waves along membranes.

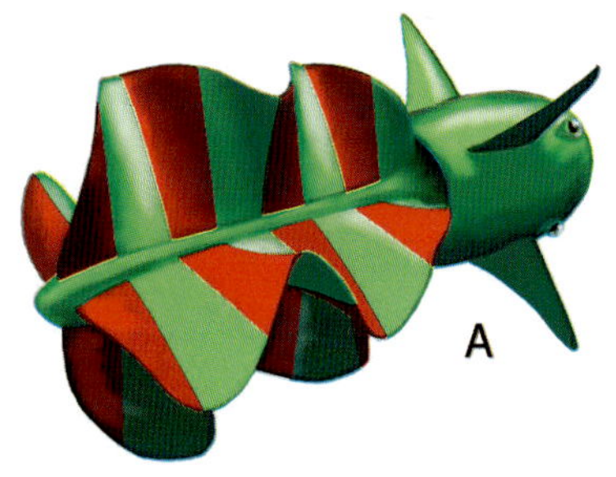

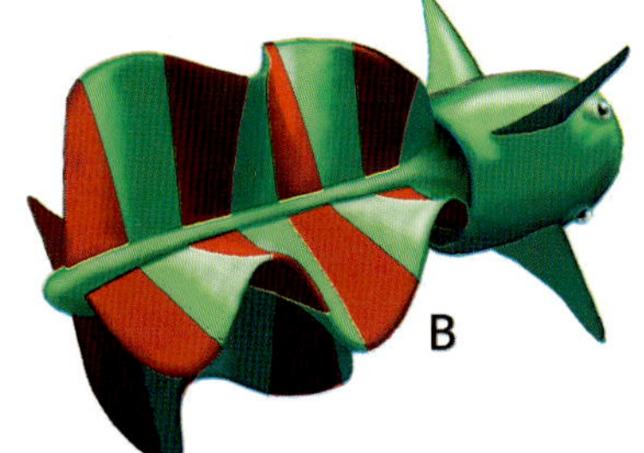

Wave Propagation

Image A shows the position of the waves at a moment just before that in image B. The red and green bars of the cloaks indicate the same anatomical regions in both images and help explain that all any bar does is to swing back and forth.

Here, the four cloaks differ in phase by multiples of 90 degrees. This helps to provide a continuous propulsive force.

Wavelength, Amplitude and Frequency

The top panel on the right shows the 'sour apple' species, with 1.25 waves on each cloak, while the bottom panel shows 2.5 waves per cloak. The fewer waves there are, the shallower the angles become between the membrane and the direction of movement. A shallow angle results in weak propulsion. That angle can be increased by using a larger wave amplitude, but there are limits to the elasticity of the membrane. Finally, the cloakfish can increase the frequency of wave propagation to speed up.

Of course, all this costs energy. As usual in biology, the optimal solution is almost always a compromise. Cloakfish solve an equation containing amplitude, length and frequency of cloak waves to swim at a given speed with the least energy.

▼ Wavelength.

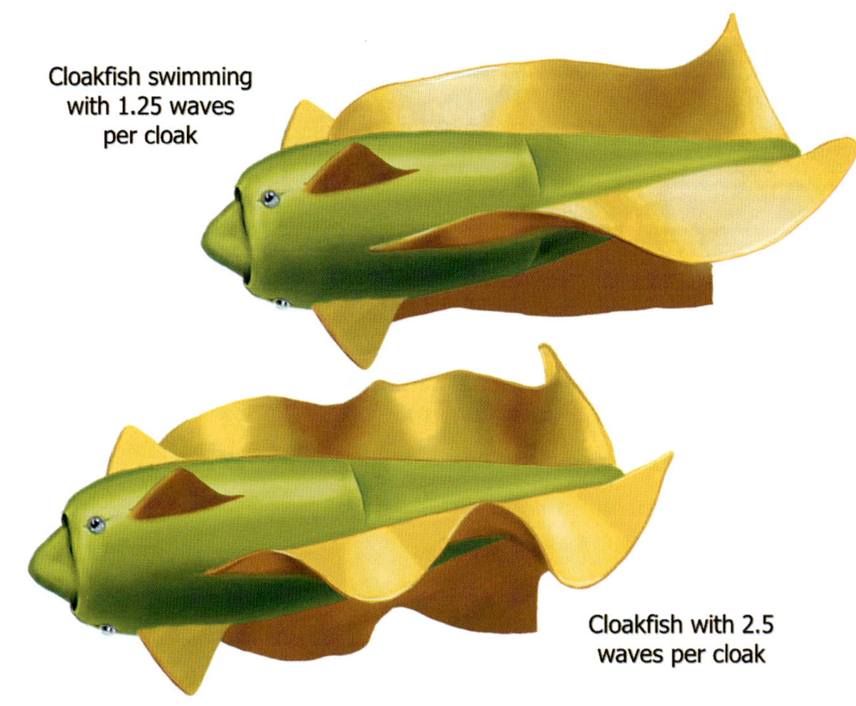

Swaying Curves

Cloak bones do not only turn back and forth at their base but are usually bent in those directions as well. The 'barred plumfish' on the right, a typical short-sleeved cloakfish, uses bone bending to increase propulsive force.

The plumfish shows another trick: wave amplitude is small at the cloak's front end, where the waves begin, but it increases as the wave moves backwards along the cloaks. This ensures the propelled water gets an additional push where it leaves the cloaks. The cloaks often approach one another closely at these ends, which accelerates the water caught between them, providing a final kick to help propulsion.

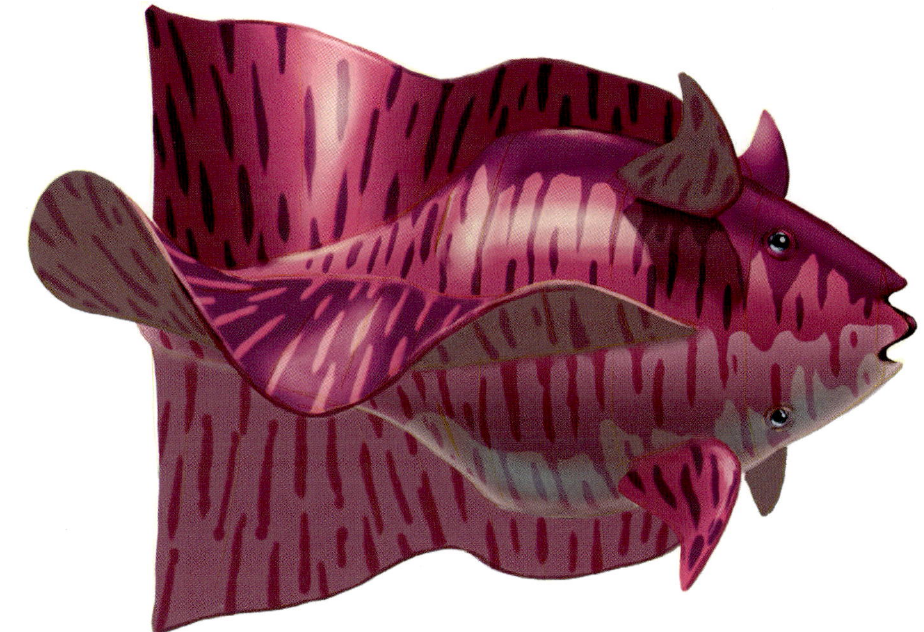

▲ Plumfish.

Ultrashort-sleeved Cloakfish

The large 'lone star' is an oceanic cloakfish with a very short body. Its cloaks differ quite a bit from the long undulating cloaks of its ancestors. This group is regarded by some as the most advanced group of cloakfish, even though 'advanced' is a difficult concept in evolution. Regardless, these cloakfish swim with remarkable efficiency and energy efficiency is a reasonable indicator of advancement.

The cloaks should now probably be called 'fins', because they resemble the slender and narrow fins of tuna or the wings of Earth penguins. They move like penguin wings too, not like typical cloaks. The latter undulate and each point of the cloak makes the same sinusoidal movement with a phase difference. But these fins move as a whole, with the entire fin turning as a whole. In fact, the fins do not simply push against water but they generate lift, meaning these cloakfish actually fly through Furahan oceans.

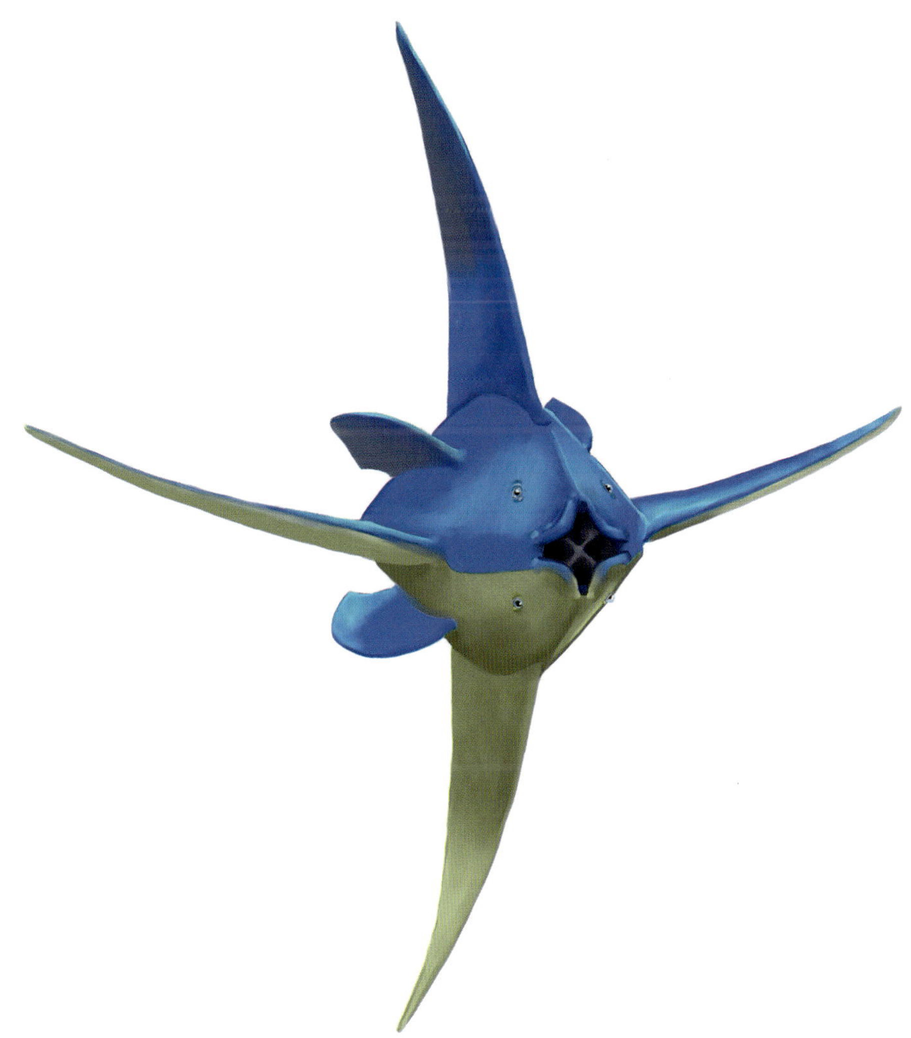

▲ Lone star.

POLYPODS

A LORICA OR TRILOBITE ON A DATE

LORICA SEGMENTATA

The lorica is a slow and, objectively speaking, silly creature. It lives in hot and humid jungles, scurrying over the ground. Its behaviour and environment have been well described by the inimitable Sigismunda Felsacker (*see* the 'Culture' section below).

There are good reasons why loricas can afford to behave in this carefree manner – they smell and taste extremely bad, and are well protected by their armour.

Loricas do not usually climb into trees because they have difficulty getting down again. When one is found in a tree, such as in this picture, it may be looking for a mate. If so, it will probably be flashing its white 'nose shields' in the mating pattern. Dr Felsacker was the first to find out that the intricate flashing shield movements are not random but depend on the circumstances. Apparently, these movements mean something to other loricas, but for humans the patterns are not easy to learn. An easier way to tell whether a lorica is looking for a mate is by listening – its mating gurgle is loud and unmistakable.

Trilobites?

Obviously, there are no true trilobites on Furaha. There used to be many on Earth, but they all died out 250 million years ago at the end of the Permian. Perhaps it was inescapable that lay people started calling these Furahan animals trilobites. The armour is divided into three longitudinal zones and consists of overlapping plates, indeed like true trilobites. Zoologists on Furaha and palaeontologists on Earth vehemently object against calling loricas trilobites. These professionals keep reminding the rather uninterested public that names should not be mixed between planets, that loricas are not exoskeletal but endoskeletal and so on, but to no avail.

What's in a Name?

Interestingly, the lorica has many different names in the languages used on Furaha. This may be because loricas are welcome guests around people's abodes at the South Palaeogaea station: they remove the locally very large slugs efficiently. Some call them 'droodles'. Other names include 'casque de boche' (meaning 'Hun's helmet'), Glattigel (meaning 'smooth hedgehog') and one peculiar name of unknown etymology 'darfader'.

From *Palaeo Days* by Sigismunda Felsacker

Droodles are bottom crawlies living in wet jungles. They feel right at home among the rotting leaves, mildew, spidrids, moulds and fungi, or whatever else is found squelching in the mud and sludge.

This habitat is so wet that humans quickly give up. The soaked detritus causes 'flying toe-rot' if you do not keep your feet dry and keeping anything dry is impossible where the driest place is best described as 'dripping' or 'hopelessly soggy'.

Still, droodles feel right at home. You can hear the little rotters humming contentedly to themselves, in the few seconds when the air is not filled with the maddening drone of a hundred species of spidrids and the hum of thousands of microtrapters trying to make their way into your ears and mouth.

But droodles like all this. Insignificant little creatures that they are, they behave in their diminutive world as the Lord of Creation, watching the carnage and decay around them with indifferent detachment.

At any rate, that was how I felt after two weeks in this saturated environment. Irritated and exhausted sums it up nicely. I know that you should not ever give in to anthropomorphic yearnings but, at the end of those wet weeks, I swear I felt that the droodles trudging through camp were making fun of me.

Yes, I knew that their peri-oesophageal ganglia were not at all sophisticated, but at the time I did think that they waved their silly shields around to get other droodles to laugh at me as well.

► Lorica waiting for the rain to stop.

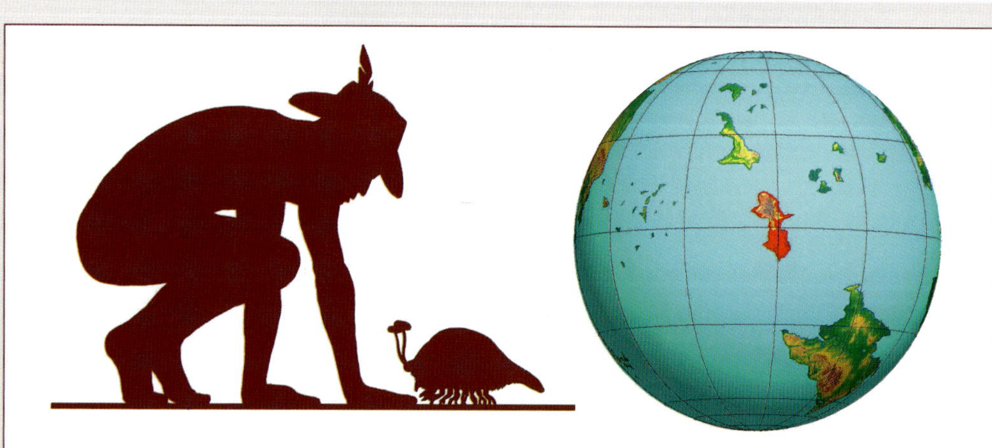

◄ **Lorica**
Lorica segmentata
Name derivation: *Lorica segmentata* (L.) ('Roman body armour with overlapping iron plates')
Habitat: ground level of tropical jungle
Distribution: Southern half of South Palaeogaea
Mass: 1kg
Length: 30cm
Vocalisation: Makes odd humming or grunting sound
Other remarks: Easily observed as it does not hide. Can curl up and roll into a ball if worried

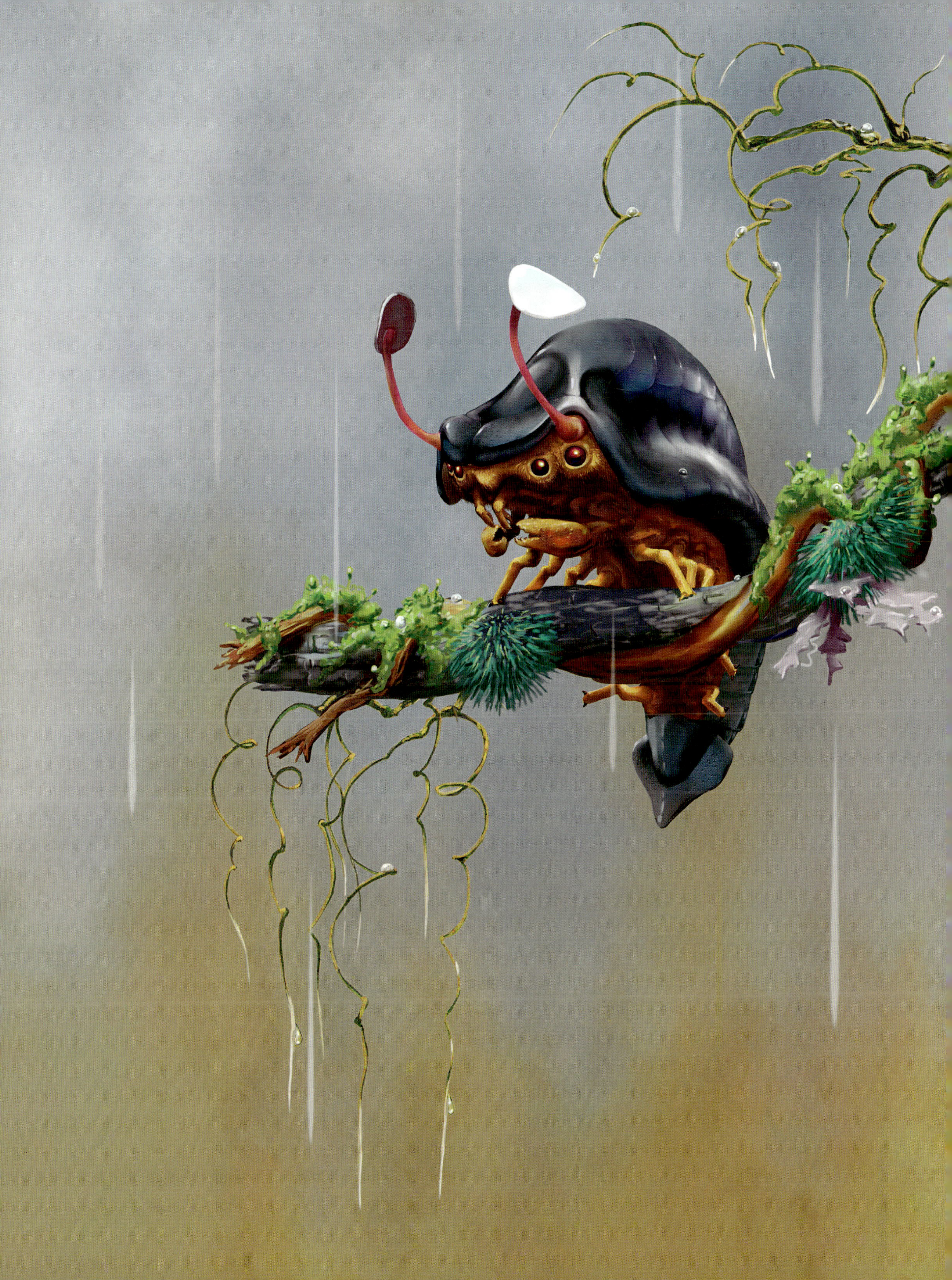

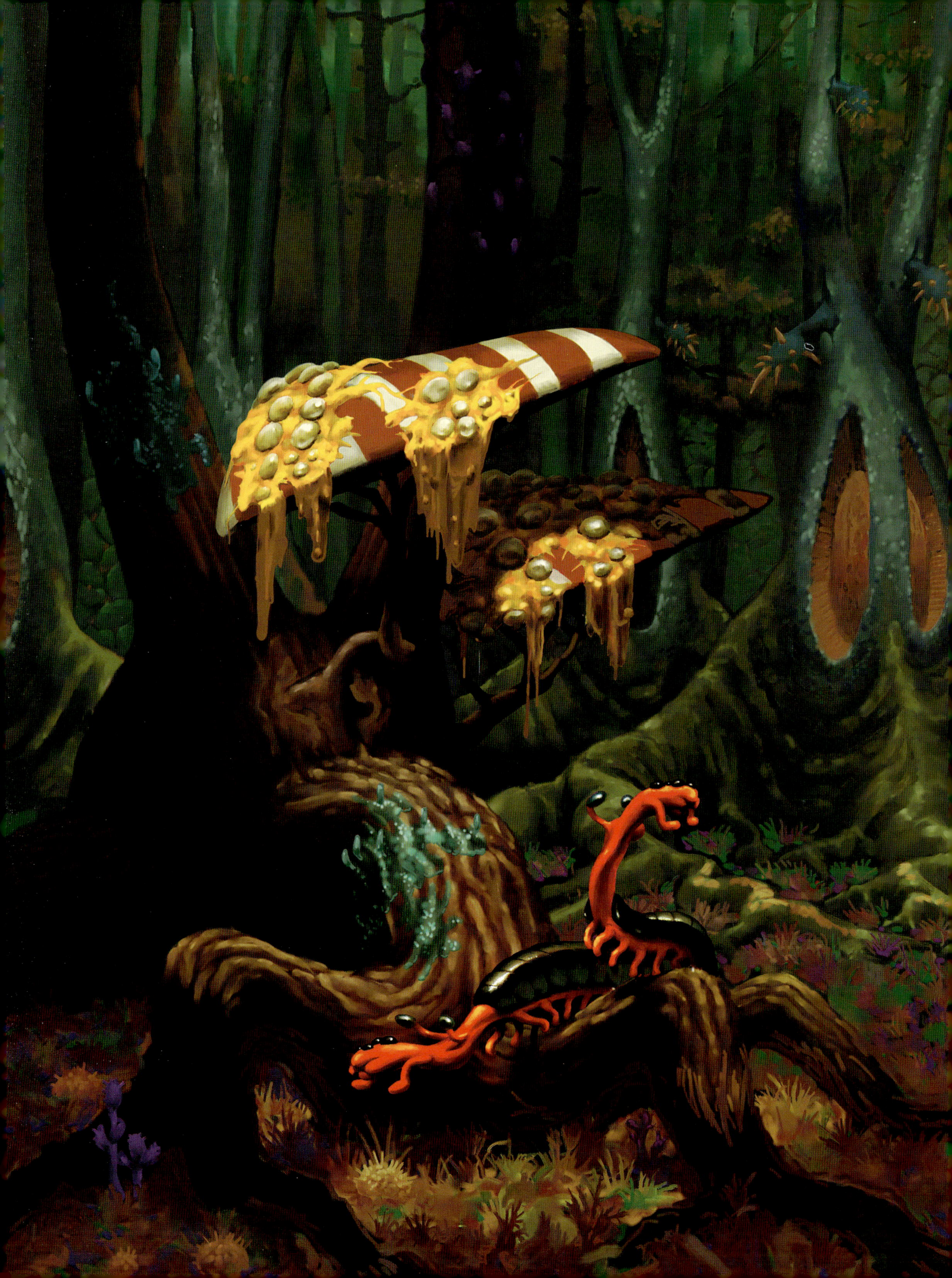

A Pair of Sunbathing Grouillards

Eructator olidus

Little light reaches the nether regions of the dense jungles of South Palaeogaea. The wind, never strong to begin with here, is stopped by the dense crowns of the trees. Down on the jungle floor, the humid, motionless air weighs heavily on the inhabitants of this oppressive biotope.

Forest life, evolving for aeons in these stable forests, reached ever higher and higher in its quest for light. The trees' upper crowns intercept most of the light, while those at the middle level seep up most of the remainder. The lack of light handicaps saplings on the forest floor so severely that an established tree must die and leave a gap before a sapling can start growing enough to climb out of its dark prison. In attempts to get ahead, saplings may start as parasites on adult trees, preferably those of other species.

Bomba Tree

The Bomba tree (*Pogocycus glaucus*) in the background uses a 'scorched-earth strategy' to fend off competition – it poisons the ground, making life difficult for anything but the hardiest plants. Such actions certainly harm the 'Barber tree' in the foreground. Barber trees (*Mansus pomovirgatus*), as mixotrophs, usually do better than plants on the forest floor because they do not depend exclusively on light for a source of energy. Still, this particular specimen has its share of problems. It is in danger of being overrun by commensals, parasites and saprophytes, boring into it, overgrowing its leaves, or stealthily tapping the sap from its veins. Slime carpets are spreading on its surface, eggscum (*Nemamucidus oioxanthides*) literally drips from it and a puckernet zwam has started working its way around one of its trunks.

Grouillard

The image shows a pair of grouillards taking advantage of an unusual spot of sunlight. These inhabitants of forest floors scurry through leaves and detritus, looking for edible titbits. Their red colour serves as warning to others.

In this gloom, humans do not see red very well, but that disadvantage does not apply to many Furahan predators. Grouillards presumably taste horrible and defend themselves quite well by spitting a disgusting secretion at their opponents.

◀ A pair of grouillards enjoying a sunbath.

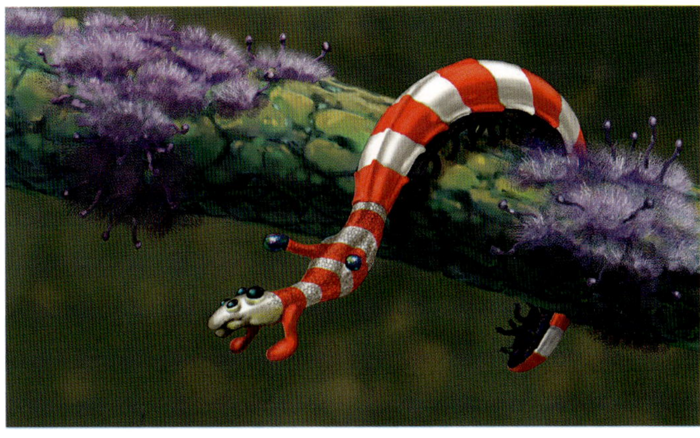

▲ This is a related species, *Eructator admonitionis*, or in common language the 'greater banded grouillard'.

Let Grouillards Be!

Sigismunda Felsacker provided some lively comments regarding grouillards (section taken from her classic book *Palaeo Days*).

> Beware of grouillard spit! One of the camp staff, Leonella Haddock, found that a grouillard had entered her cabin while she was away. She was inexperienced and foolish enough to try to frighten it out by shouting and waving at it, instead of waiting for it to come out in its own good time. The cornered grouillard was confused and agitated. After scampering about in the cabin, it suddenly froze, straightened its neck, pursed its lips - well, not really lips, but there's no better word - and spat. Three or four gobs of greenish spittle hit Leonella's head and chest. For a second both she and the grouillard stood there, frozen in place. But then the stench got to her, and she start vomiting immediately. The grouillard took the opportunity to rush out of the tent, cackling to itself. The spit proved to be extremely nasty stuff, for poor Leonella reeked for an entire week, in spite of splashing herself all over with perfumes and deodorants.

▶ **Grouillard**
Eructator olidus
Name derivation: *Eructator* (L.) (meaning 'spitter'); *olidus* (L.) (meaning 'stinking'); *Pogocycus* (meaning 'bearded palm'); *glaucus* (Gr.) (meaning 'blue-green'); *nemamucidus* (meaning 'threaded mildew'); *oioxanthoides* (meaning 'with yellow egg')
Habitat: Tropical forest
Distribution: South Palaeogaea
Mass: 2kg
Length: 50cm

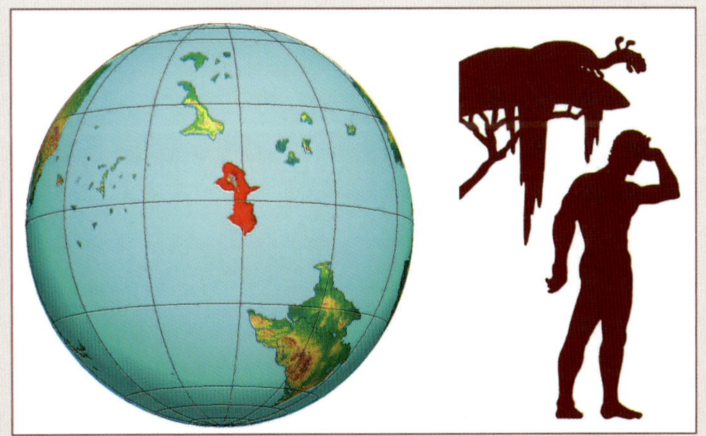

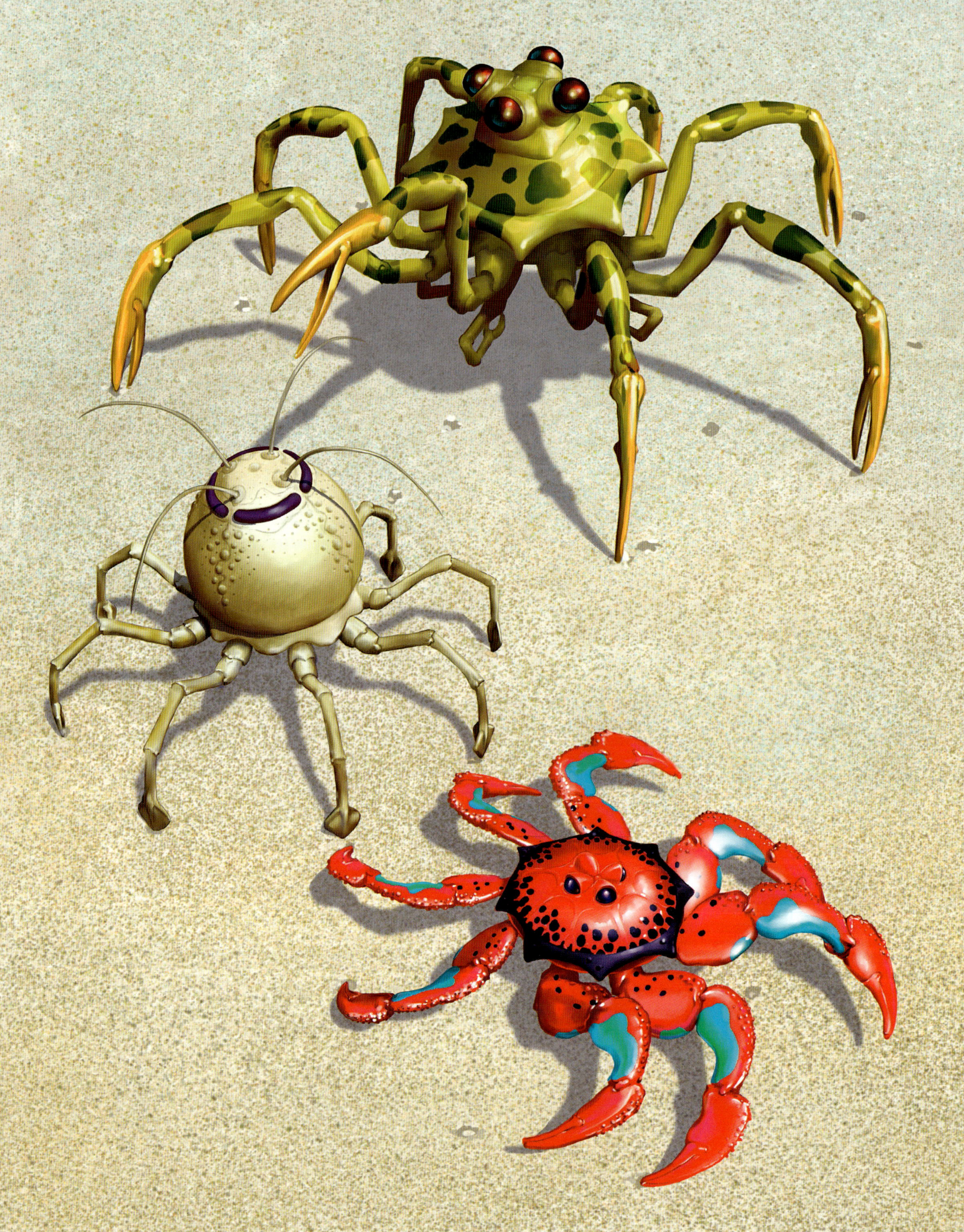

Spidrids

Forwards in Any Direction

Visitors just arriving from Earth may mistake spidrids for spiders or crabs, probably because spiders, crabs and spidrids all have exoskeletons and segmented legs. Spidrids also have pincers and faceted compound eyes. The majority of Earth animals are bilaterally symmetrical, so they have right and left sides, a front and a back, and upper and lower parts. Not so spidrids, as their symmetry is a radial one. This leaves them with a top and a bottom, but they have neither left nor right, nor do they have front or back. This frustrated anatomists who wanted to number the legs but could not decide where to begin counting. As the inset image shows, the eight-sided symmetry is not complete, as various parts of spidrids come in fours rather than eights. For instance, there are only four eyes at the top of the animal. The bottom part of the spidrid also carries only four eyes and is equipped with four clawed palps and one mouth tentacle.

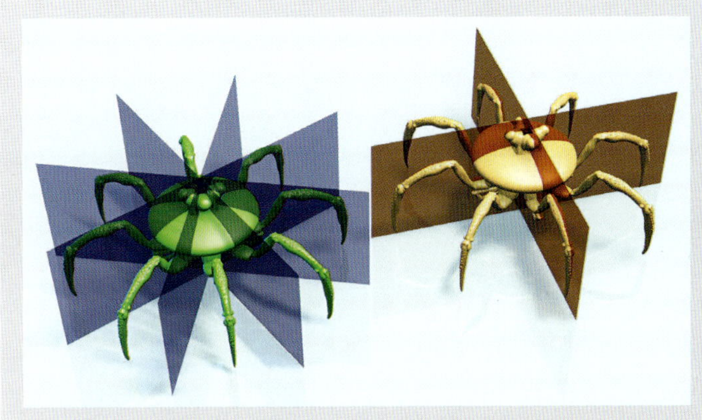

▲ **Lost in Symmetry**
The radial symmetry of spidrids is not straightforward. On the left, four transparent planes pass through a schematic spidrid, cutting it into eight pieces. Each piece holds one leg. But because spidrids have four eyes, each piece contains either one eye or no eye at all. Spidrids also exhibit four-sided radial symmetry as shown on the right. Each of the four pieces now holds one eye – well, two half eyes; and two legs – well, one whole leg and two half legs.

Spidrid Biology

Spidrids are terrestrial and as a group are quite successful. Their size varies from less than one millimetre to some 30–40cm. They are omnivorous. These three species, taken from different biotopes in a small region, illustrate their variability.

Spotty Woodrustler *Marmoratus nemoris*

Name derivation: *Marmoratus* (L.) meaning 'marbled'; *nemoris*, from *nemus* (L.) meaning 'forest'.

This is a woodland species, camouflaged to resemble spots of light among the foliage. Its build is close to the spidrid nonspecialist prototype. Note that this specimen nicely shows that the upper body, the cephalothorax, can move on the lower body, the abdomen.

Strandsprab *Ovum littoris*

Name derivation: *Ovum* (L.) meaning 'egg'; *littoris*, from *littus* (L.) meaning 'beach'.

This species is specialised to dig for wurms and the like in the sand on beaches when the tide is out. While there is plenty of food, the open environment exposes sprabs. This is why the strandsprab has the same colour as the sand, and why its 360-degree vision is well-developed. It can dig very quickly, using as many as five of its shovel-shaped legs at a time to disappear from sight. However, when underground it will drown when the tide comes in and its predators know this. The animal is found in groups of many hundreds of individuals, which may be a way to spoil its predators for choice.

◄ From top to bottom: spotted woodrustler / strandsprab / mad sickle.

Mad Sickle *Sicilicula insana*

Name derivation: *Sicilicula* (L.) meaning 'little sickle'; *insana* (L.) meaning 'frenzied' or 'maddening'.

This species represents a major spidrid clade. While 'square spidrids' move their legs in a vertical plane, the 'slanted spidrids' do not as the basic leg joints have tilted. The most likely reason for this is that the flexion and extension muscles can now more easily help with propulsion. Most 'slanties' are very flat and live in crevices. There are clockwise, 'clockies', and anticlockwise, 'antics', slanties. The direction is inherited, so each species has its own exclusive direction. It seems that the two types of slanties arose completely independently, so clockies and antics are not closely related. The mad sickle is very agile.

Please do not try to catch one. You are not likely to succeed and if you do, it will pinch you.

Man-Handled Spidrids

Of the three species shown, only the strandsprab belongs on the open beach. The other two would never congregate on the beach as shown. It is therefore likely that the illustrator placed the animals here. Note that the animals have begun to respond to the unnatural situation.

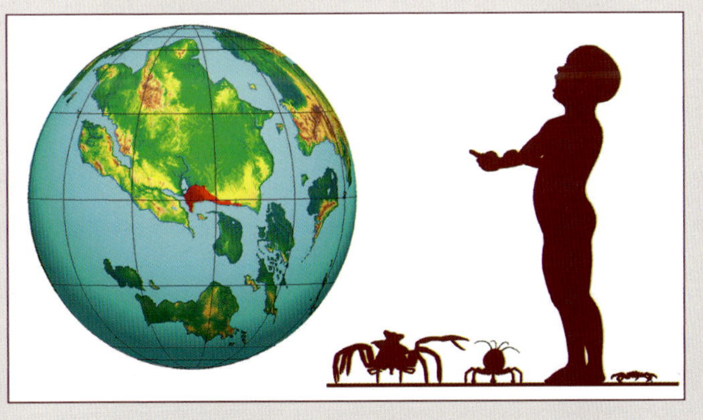

► Child learning about spidrids.

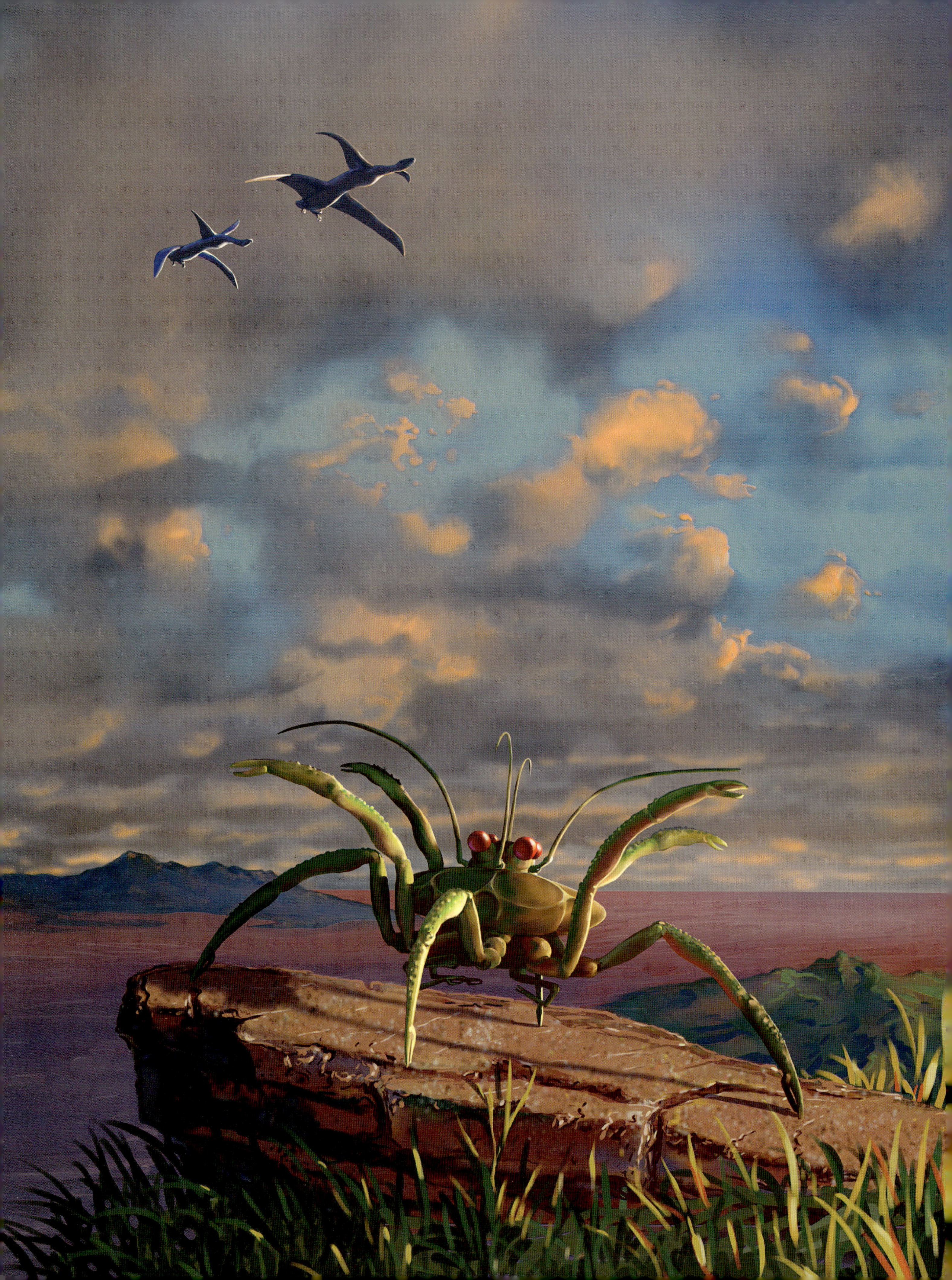

SUNDANCER AT DAWN

HELIOPHILUS RA

The sundancer owes its name to a peculiar habit. Before daybreak, it ambles towards an exposed spot, such as a rock or branch, and awaits dawn. When the sun Jua rises above the horizon, the sundancer lifts four of its eight legs. The raised legs slowly swing back and forth. That's it. As dances go, the performance is not exciting and the occasional antenna twitch does not make it more enticing. From time to time, it may alternate between standing and lifted legs. After 20 to 30 minutes of this, the sundancer abruptly walks away to attend to the day's business.

Why do Sundancers 'Dance'?

The reasons for this behaviour are as yet unknown. The sundancer's scientific name tells us, in Latinised Greek, that it loves the sun, with the added name *Ra* signifying the Egyptian god of the sun. Obviously, the people who named it did not believe that spidrids worship mythological beings. Spidrids are, by the way, about as smart as Earth reptiles, which on average are brighter than Earth arthropods.

Experimental Ethics

Their dancing behaviour must have another meaning. The most common speculation is that sundancers raise their legs to gather heat more efficiently. While this might make sense, other spidrid species do not exhibit similar behaviour. As yet there is no evidence that sundancers operate better with an elevated temperature anyway. It would not be difficult to design experiments to settle the issue – catch some sundancers, put them in a terrarium, insert temperature probes into their bodies, vary the ambient temperature and see how their behaviour changes. No doubt this is what some spidronomers (also known as spidridologists) would like to do, but the Institute's ethics rules forbid any experimental protocol containing the phrase 'stick a probe in it', at least when the animal falls into the 'might be able to suffer' category. Over Furahan history, the vague borders of this category at times forbade almost all animal experiments. At other times citizen-scientists inserted probes with gusto in any uncuddlable animal, rather like alien beings in cheap science fiction novels used to do to people. On Furaha, the irony was the probing aliens were humans.

Also Appearing

Two 'empress seasoars' fly across the view, which is not surprising, as they thrive in such coastal areas. One is large and glides serenely onwards, but the other flits around. This behaviour makes it likely that the small one is trying to woo the large one. As explained on the pages showing the 'porcelain seasoar', there may be more here than meets the eye. Seasoars are hermaphroditic and aim to get mutual benefit from pairings, impregnating their partner while being impregnated themselves. But a small seasoar may try to avoid being impregnated. Here, the large empress so far shows no interest whatsoever in the smaller one's antics.

Publication Ethics

Lacking direct temperature data, some limited remote sensing temperature measurements have been carried out. These showed that spidrids indeed warmed up appreciably while in the sun. These admittedly meagre results were reported by the hopeful young *Studiosa* Ulyana Fortescue in a journal of at best mediocre standing, appropriate for this modest pilot experiment.

Unfortunately, the simple experiment attracted the unwelcome attention of Preprofessor Emilio Porcone, infamous for his harsh treatment of students. He felt students were unworthy of his attention, but still spent much energy and time making their lives miserable. Some people should not be allowed to teach.

Porcone pointed out condescendingly that the rock the spidrid was sitting on probably warmed up too, but that this inevitable result of solar irradiation did not mean that the rock was better off.

Years later, Fortescue chaired a committee to decide whether Porcone should become *Professor*. The request was denied.

◀ Sundancer at dawn.

▶ **Sundancer**
Heliophilus ra
Name derivation: *Heliophilus* (Gr.) (meaning 'sun-lover'); *ra* (the name of the old Egyptian god of the sun)
Habitat: Variable, from rocky outcrops to woodland and forest
Distribution: South of Van Vogt Station on *Imparia Orientalis* (Mind you, the distribution areas as shown in this book very often rest on a large degree of conjecture: there is no way to describe the area of residence of so many animals to this degree of precision. The Institute simply does not have the economic means to send off thousands of bots for such a purpose. Then again, the Department of Cartography does not have that much to do and happily obliges.)
Mass: Up to 2kg
Length: Body diameter 20cm, leg span up to 45cm
Diet: Omnivorous

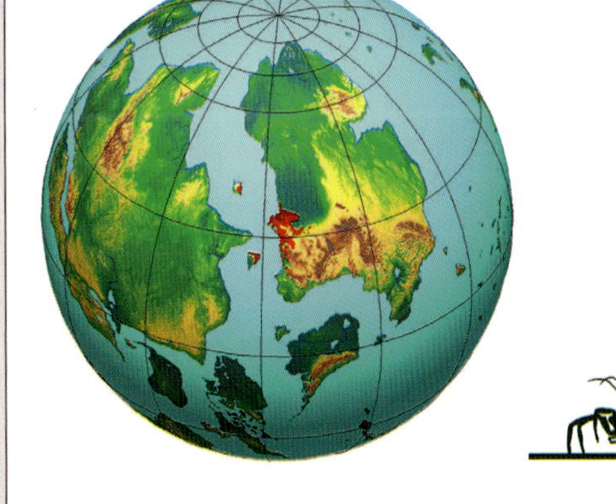

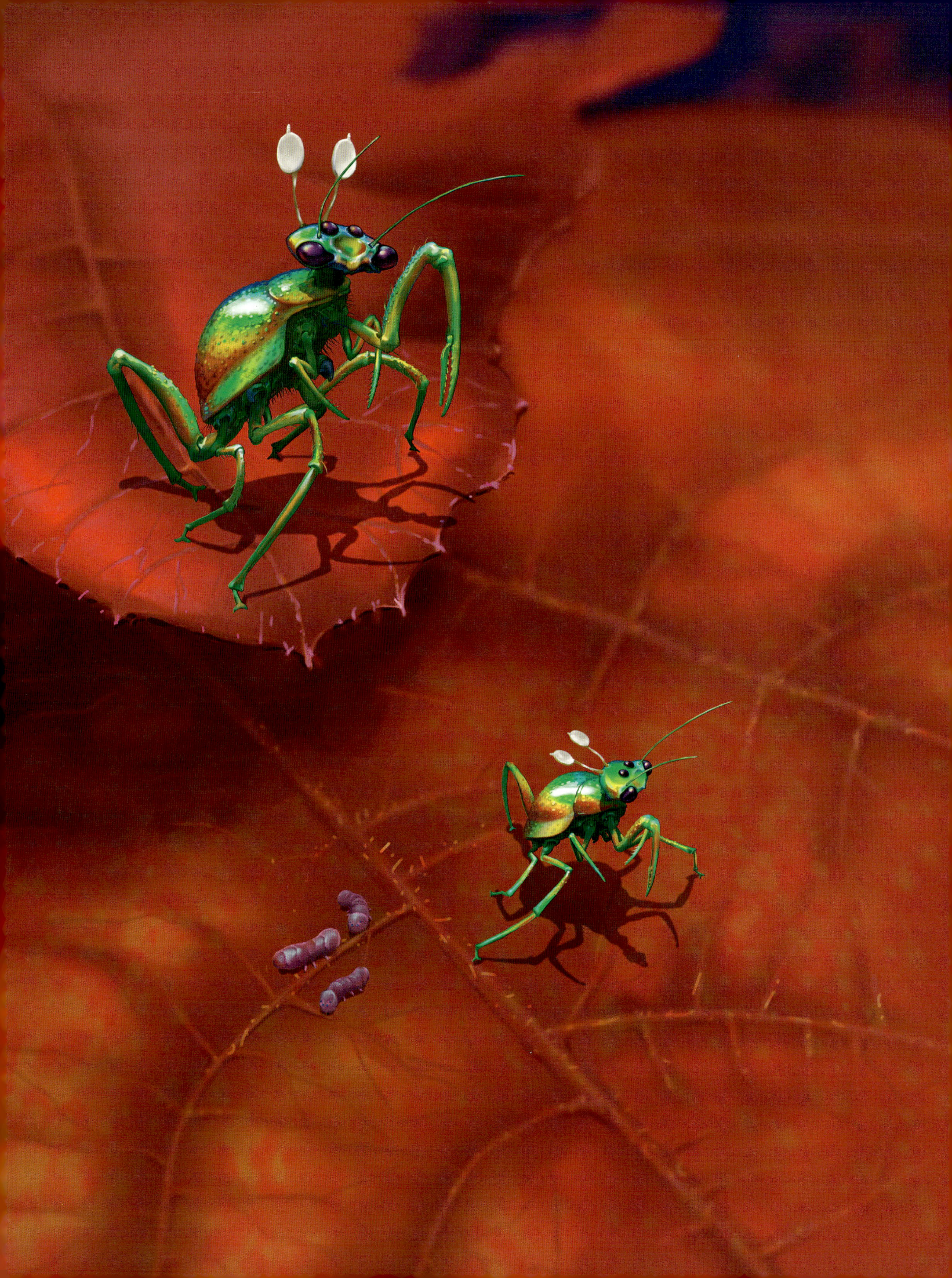

Prigoons on the Prowl

Perissus dougalii

These small animals, less than 2cm in length, are fierce predators, as frightening in their own world as Earth leopards or Furahan probers are in theirs. Whether prigoons should frighten anyone or anything depends on relative size – humans can relax.

Prigoons are ambush predators. They usually sit in hiding, immobile, until a suitable prey comes into jumping range. Catapulted by their powerful single hind leg, they jump as much as 15–20cm and usually land exactly on their prey, grasping it with their front claspers.

Secondary Bilateral Symmetry

At first glance these tiny exoskeletal predators seem just like other Furahan wadudu. But the prigoon is an anatomical oddity. Its eight legs, four eyes placed on a cupola, and mouth on the underside of the body, all show that it is related to spidrids. But it is not radially symmetrical at all; instead, it has bilateral symmetry. It has two pairs of walking legs and one pair of accessory grasping limbs, making up six limbs, instead of the typical eight for spidrids. The two missing legs are unpaired and found at the ends of the animal. The one hind leg is now the main jumping leg and the one front leg has evolved into a clasper. The clasper catches and holds the prey, so the beak-like mouth can disassemble the prey before eating it or, more often, while doing so.

◀ One prigoon sizing up another.

Cartouche

The prigoon features on the cartouche of the Procházka clan. This was chosen by Petr, who was the first of his clan to become *Profissimus*. The example shown here was sown into that first *Profissimus* Procházka's robes. At the time, only such high faculty wore a cartouche, but once introduced, cartouches tended to stay in use by clan members, even when the descendants of the original *meritorius* had limited personal merit.

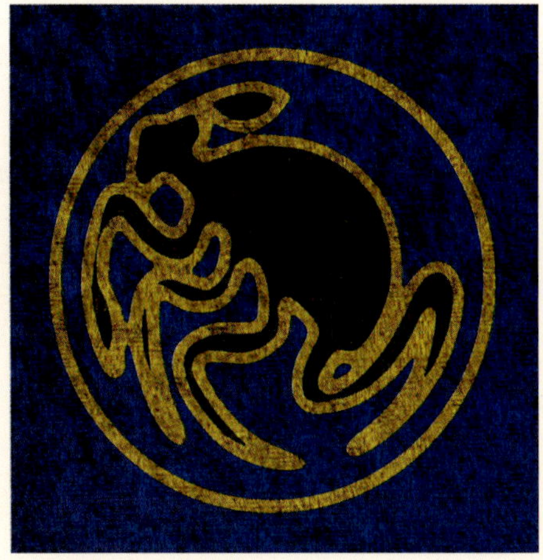

▲ Procházka cartouche.

Because of the bilateral symmetry this clade is often called the 'mirrored odd spidrid clade (MOSC)'. During their evolution, the plane of symmetry somehow ended up running through opposite legs. There is another group of secondary bilateral spidrids with the plane of symmetry between legs, called 'mirrored even spidrid clade (MESC)' spidrids.

Red Leaves in any Season

For animals that depend on ambushing prey, the prigoons in the main illustration do a poor job as they are in the open and their conspicuous colour clashes with the red leaves. The leaves are not red because it is autumn, but because they are part of erythrochrome plants, whose photosynthetic proteins do not absorb the red wavelengths of the spectrum. These red leaves differ from those of green chlorochrome and grey poikilochrome plants.

The large variety of Furahan leaf colours complicates life for animals with a need to camouflage themselves. If such animals regularly find themselves on plants of different colours, they profit from being able to change colour. Some can do so within minutes, but others, like the prigoon, take weeks. Interestingly, their offspring usually come into the world with the appropriate colour. This is not some Lamarckian trickery, but embryonic manipulation of the vast potential genetic colour gamut.

Mating, Cannibalism or Both?

Prigoons are only conspicuously iridescent, as shown here, in the mating season. As prigoons are cannibalistic when it suits them, the small prigoon in the image should watch out for the large one. That large one has its white head shields up and out, so there may be a happy ending yet. The signal these shields send is that the animal is, at present, more interested in mating than in snacking. But prigoon tempers vary and there is no guarantee that it will not like a snack after mating.

▶ **Prigoon**
Perissus dougalii
Name derivation: *Perissus* (Gr.) (meaning 'odd, in the sense of not even'); *dougalii* (L.) (after Dougal Dixon, Scottish writer); *prigon* (Russian) (meaning 'jumper')
Habitat: Forests
Distribution: Not well known, really. The map shown here is mostly guesswork. Prigoons do occur in the Archipelago, that at least is certain
Mass: Very low
Length (adult): 6–9mm
Diet: Mostly very small wadudu

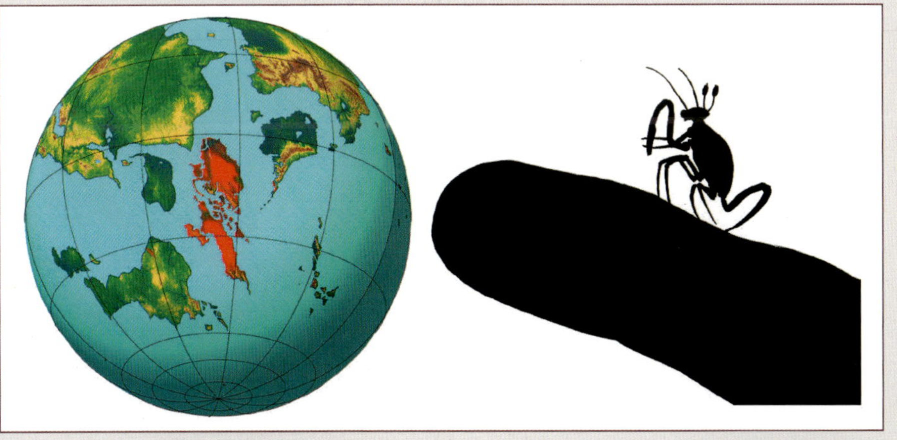

How Spidrids Walk

Spidrids, having radial symmetry, do not have hind, front, left or right legs as their eight legs are identical. When spidrids change direction, they do not turn, so a leg that was in front before may be on the left, right or on the back afterwards. Spidrid legs manage such changes with ease and yet each joint in their legs moves in just one direction. Knowing how they are able to manage this requires some understanding of their leg anatomy.

Turning this Way and That

The diagram at the right shows spidrid legs as mechanical devices. Each leg consists of five segments, although some clades have six or seven. Starting near the trunk, the segments are called coxa, femur, tibia, tarsus and pollux. (It is not known who named these segments, but the names suggest a haphazard combination of Earth insect, crab and human anatomy.) Of the five leg joints, the last three have the same task, so they can be dealt with together.

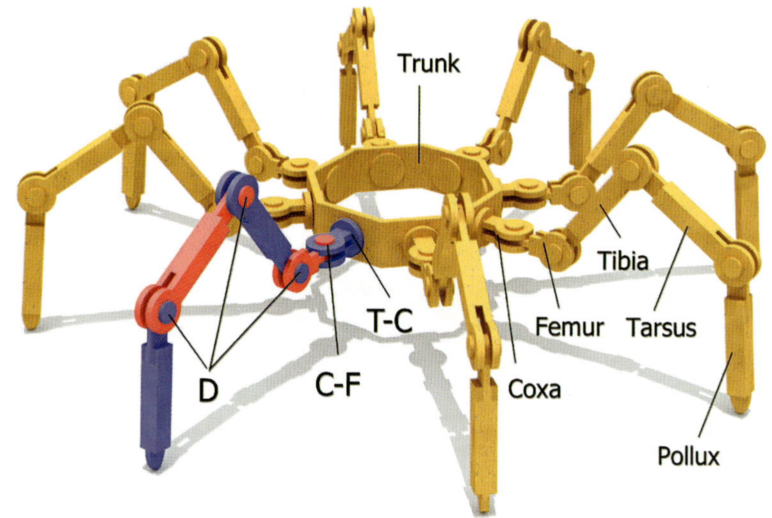

▲ Spidrid anatomy.

Joints

The trunk-to-coxa (T–C) joint connects the trunk to the leg. That joint's axis sticks out from the body in a horizontal plane, so the joint allows the animal to slant the entire leg. The panel labelled 'T–C' shows the endpoints of the resulting motion. Such movements are called pro- and retro-torsion.

The coxa-to-femur (C–F) joint has its axis vertically, allowing the leg to be swung clockwise and anticlockwise (helicte and antihelicte). These movements are important for walking.

The three distal (D) joints have horizontal axes – these are the femur-to-tibia, tibia-to-tarsus and tarsus-to-pollux joints. These joints allow the leg to be bent (flexion) or straightened (extension), which is as important for walking as the previous movements.

Spidrids have an exoskeleton, so their muscles are on the inside of the leg or inside the trunk, not shown here. There is one more leg joint that is not shown because it is not used for locomotion: the joint connecting the 'leg jaw' or 'chela' to the pollux.

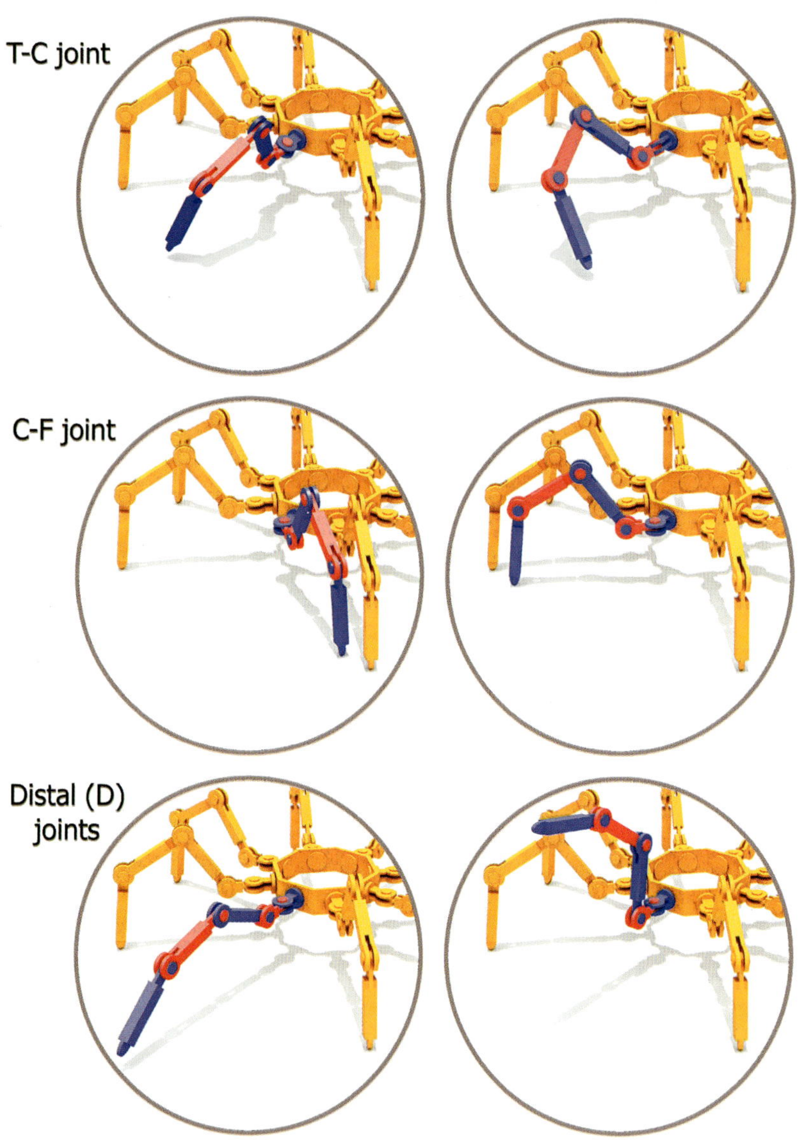

▲ Movement axes.

All Legs are Equal

This species, the 'cakewalk', is shown as if it walks on a treadmill, so the ground seems to be moving rather than the animal itself. It is walking towards the right (*see* the blue arrow on the ground).

The path of the leg's ends are shown as red lines. Note how all legs cycle through the same movement but offset from one another.

Which joints are used depends on the direction of movement. If the leg is parallel to the direction, the D joints are most useful. If the leg is perpendicular to the direction, the C–F joint does most of the work.

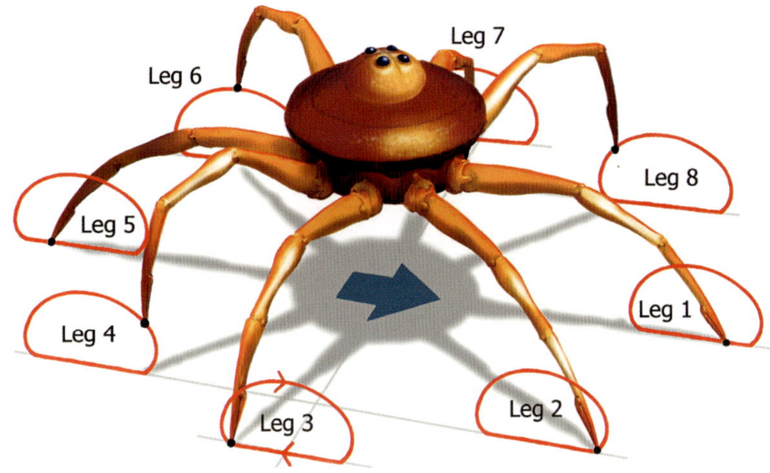

▲ Spidrid phase differences.

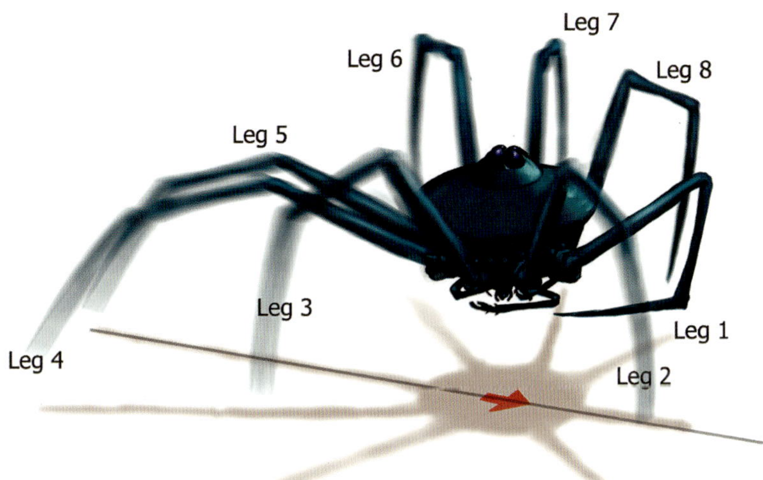

▲ Running spidrid.

Gait and Speed

All spidrids can run, but none as fast as the 'black streak'. At high speeds, cycle frequency and stride length increase, the time a leg is on the ground decreases, and the animals shift gears: they change gait.

While running, there are often moments when there are no legs on the ground at all, as shown here. The black streak goes further and simply lifts some legs altogether. The faster spidrids go, the fewer legs touch the ground.

The 'alternate ripple' gait is shown above, with always some legs on the ground. The most stable gait is the 'double table' one, in which there are either four or eight legs on the ground. This is a nice display gait: have a look at the 'sundancer' page for an example.

Wheelies

One of the more unusual spidrid gaits is the 'wheelie', or the 'tilted wheel of fortune'. This is only used by 'slanties' as a last resort when under attack. The animal then flips itself on a side and uses successive legs to push against the ground, effectively rolling. Slanties can achieve a considerable speed in this way and can even roll uphill. The downside is that their eyes also rotate continuously, making it very difficult to see where they are going. Apparently, if there is a serious risk of being eaten, fleeing without seeing where you are going may be wiser than just sitting it out.

Please do not scare slanties for fun. It may be fun to see them do a wheelie, but this is highly stressful for the animal.

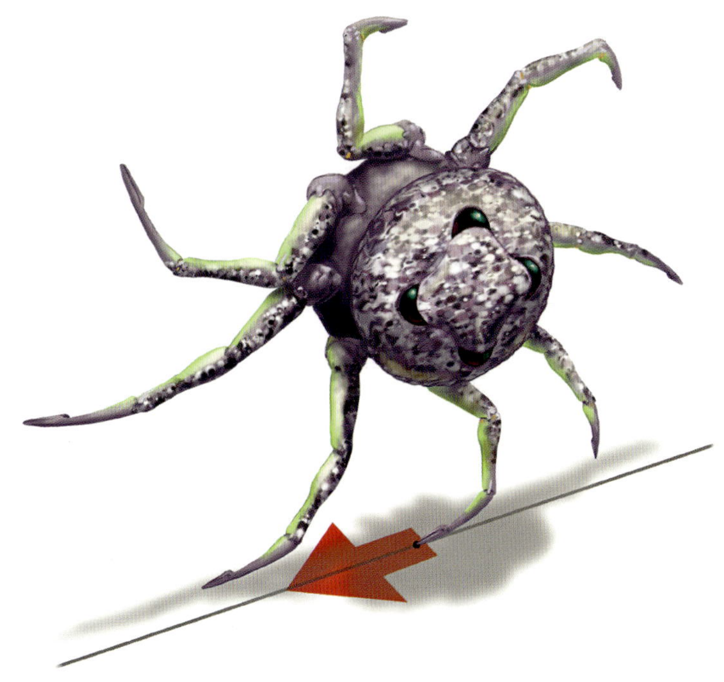

▲ Wheelie.

Tetrapters

The Red Baron

Dicella gampsonyx

A clade of tetrapters, the *Lestidae* (meaning 'plunderers') evolved to catch other tetrapters in mid-flight. Aerial hunting is a very exacting mode of life, so *Lestidae* must excel in many functions. In this respect they resemble Earth's dragonflies, which are also agile flyers with keen senses and quick reactions.

The species shown here, the Red Baron, is probably the largest of the *Lestidae*. The Red Baron has excellent eyesight. Its eyes are of the compound type, which means that their resolution is much poorer than is achievable with camera eyes. Still, like dragonflies on Earth, these animals somehow manage to distinguish small quick-manoeuvring prey from a visually rich and fast-moving background. They do not only have 360-degree vision in the horizontal plane, but their visual field also spans 180 degrees in the vertical direction, so they have no blind spots at all. Then again, neither do their prey.

The Red Baron is more manoeuvrable than other tetrapters, probably because their neural control is superior to that of less agile tetrapters. They swerve, turn and swoop at high speed and are a wonder to behold. They are also difficult to catch with something as clumsy as a butterfly net.

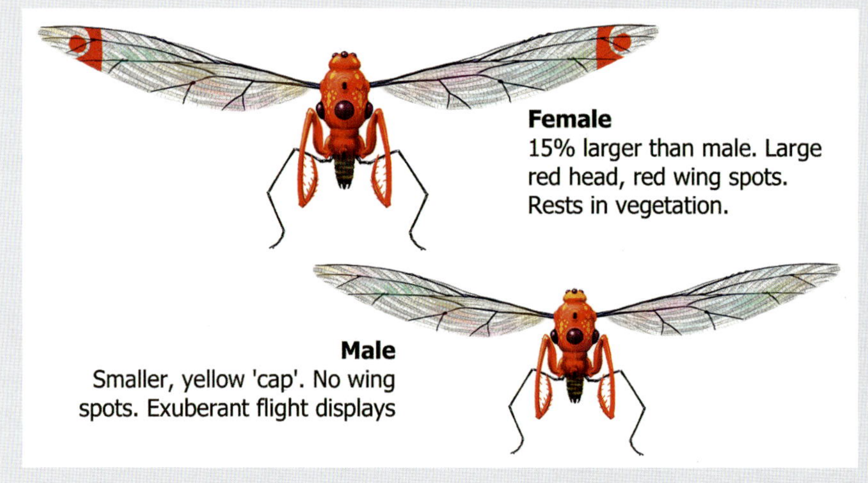

▲ **Field Guide Information**
This drawing is from the *Field Guide to Imparian Tetrapters* and shows the entry for the Western variant of the Red Baron. Note the sexual dimorphism. Legs and wings parallel to the viewing direction were omitted for clarity.

Catcher Legs

Finally, there are the legs. You might not expect legs to be a crucial factor in determining evolutionary success in an aerial predator, but they are. All tetrapters have four outer and four inner legs, but in most species, these do not differ much in shape. The *Lestidae* are predators, mostly because of their modified outer legs. These legs are no longer used for locomotion or as landing gear but are wholly devoted to prey capture. Each of these four 'catcher legs' looks like the front leg of an Earth mantis. In flight, the catcher legs as well as the landing legs are drawn up against the body. Just before engaging its prey, the Red Baron will extend and spread its catcher legs, forming a sort of net. The legs are closed at lightning speed, encircling the prey from four directions at once, leaving it little chance of escape.

Dinner is Served

Once the prey is caught, the Red Baron flies towards a nearby suitable spot and lands on its walking legs, looking a bit like a robot moon lander. It then bends its highly movable mouth tentacle to the prey and starts eating it. It does not usually bother killing the prey before devouring it. If the prey moves a lot, the Red Baron may tear off a limb here or there until it can eat at leisure.

► Red Baron preparing for next sortie.

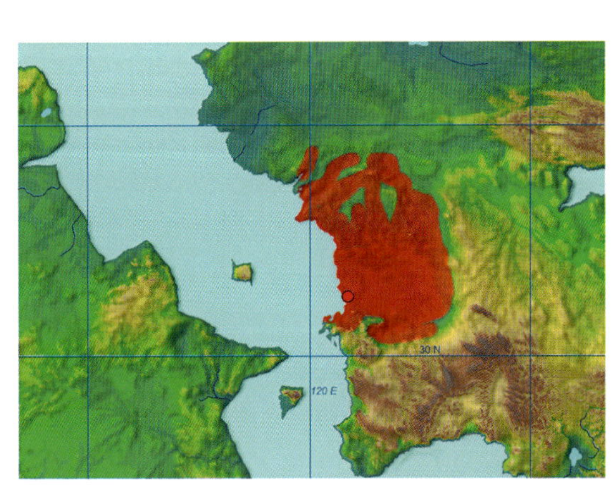

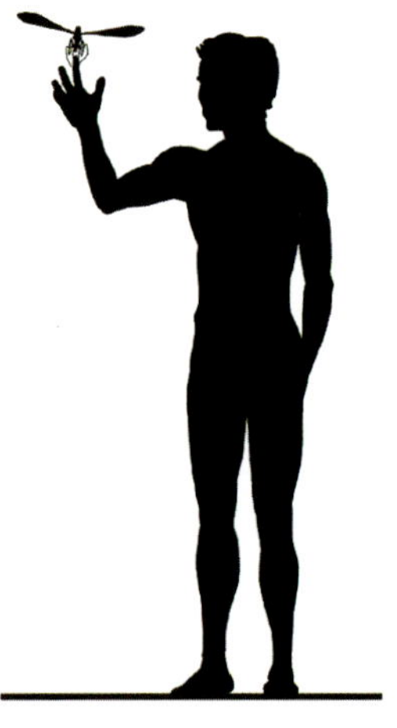

◄ **The Red Baron**
Dicella gampsonyx
Name derivation: *Dicella* (Gr.) (meaning 'pitchfork'); *gampsonyx* (Gr.) (meaning 'with crooked talons')
Habitat: Park-like open terrain
Distribution: *Imparia Septentrionalis*, well known around and in Van Vogt Station
Mass: A few g
Length: Wingspan up to 25cm
Diet: Mostly other tetrapters

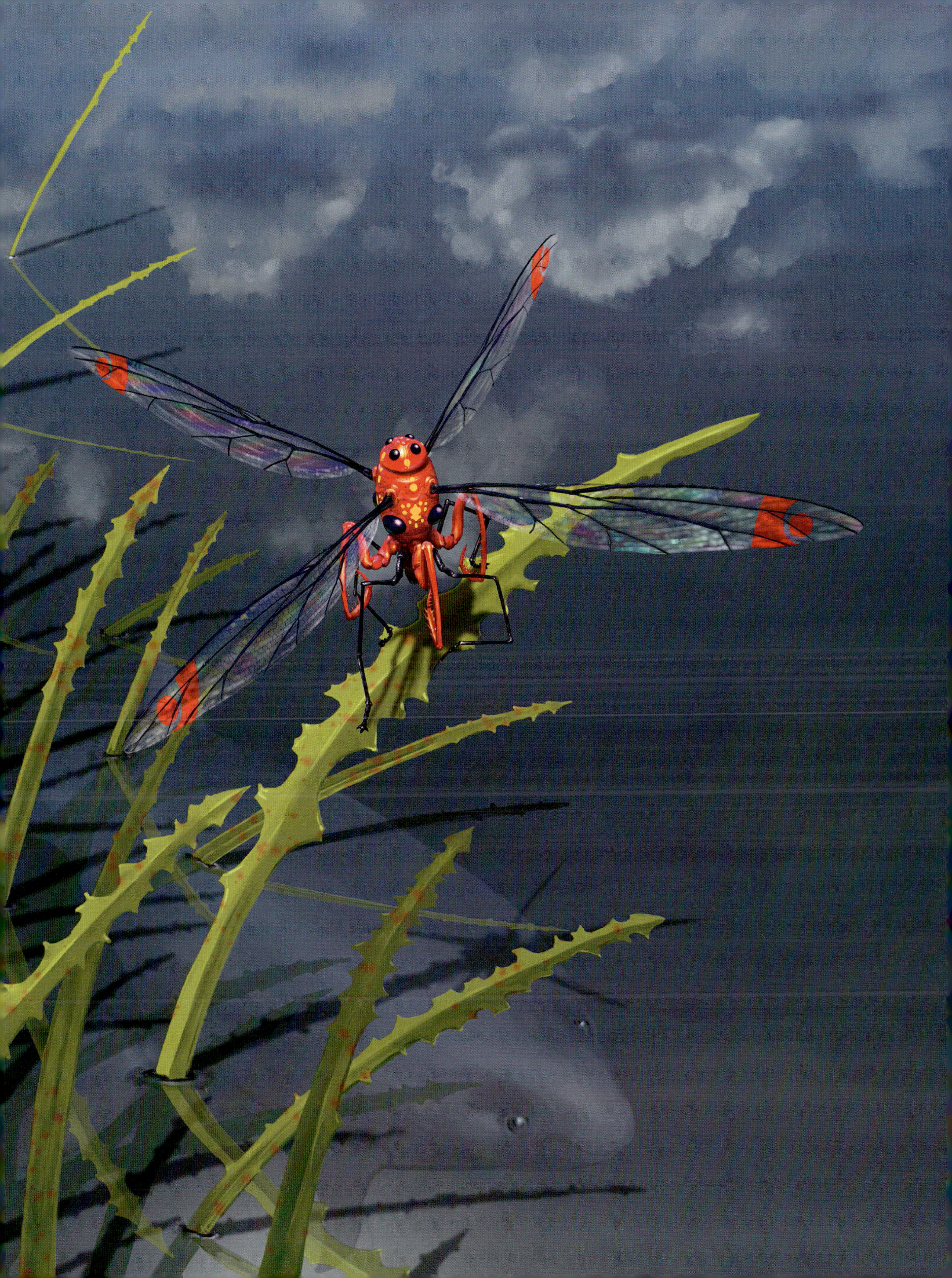

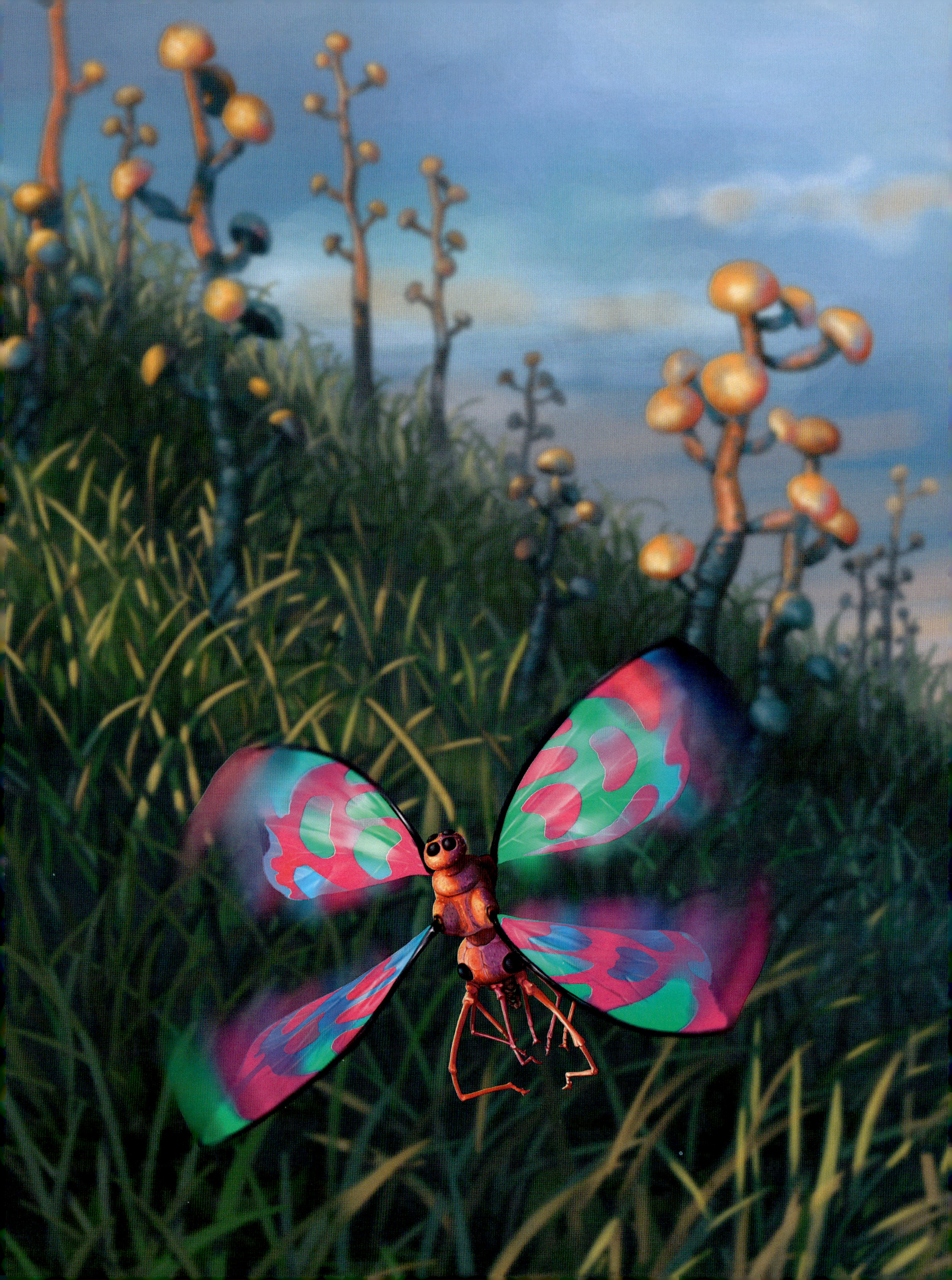

FARFALDEROL/ FAFADERAL/FAFERAEL

FARFALLA FULGOR

It is early in the morning, and it promises to be a warm day, even this early in summer. But it is not warm yet as the sun has only just emerged over the far hills. It is still too cold for the spore heads of the fungols to start releasing today's cloud of spores. The grass is still cool and moist, with only some blades lit by the first rays of the sun.

It is also still too cool for farfalderols to be fully active, but a few enterprising individuals, perhaps enduring the cool temperature a bit better than their fellows, have already taken to the air. These early risers flitter around looking for something to eat, or perhaps they hope to attract the attention of a potential mate.

Farfalloids

Farfalloids are large and conspicuous tetrapters. The name was loaned from an old Earth word for butterfly. That is not too surprising, as both Earth butterflies and Furahan tetrapters have large and brightly coloured wings. For some reason almost every language has its own separate name for butterflies, different from even closely related languages. Perhaps their bright fluttering brings out creativity in those who look at them. The early Horizonists on Furaha used words for butterfly from their original or adopted languages, so any farfalloid could be called 'kipepeo', 'kelebek', 'vlinder' or 'fluture' with equal ease. The scientific name 'farfalloid' was derived from one such word, 'farfalla'. In fact, the genus of the farfarderol is *Farfalla*. The species shown here has at least three different dialect names.

◀ Farfalderol flapping into the light.

Bright Colours

Farfalloids beat their very large wings relatively slowly. Smaller tetrapters, of fly or mosquito size, beat their wings as often as 200 times per second. Farfalloids may do so four to twelve times per second. A nice side effect of this low frequency is that their wing colours can be seen to good effect.

The colours of the farfalderol must tell us something about the visual system of whichever species they evolved to impress, which is probably farfalderols themselves. The pattern has a very strong colour contrast, with their bright cyan and pinkish hues. The light-dark contrast is however very weak – if the colours were depicted as shades of grey, they would hardly differ in tone. In man, visual acuity depends much more on contrast between light and dark than on contrast between colours, which is why pure colours with the same brightness appear to jump about. The high colour and low grayscale contrast suggests that the farfalderol visual system works in the same way. The spectacle is made even more striking because the colours in the pattern are reversed in two of the four wings. When these wings near one another in their flapping cycle, the flashing effect becomes even stronger.

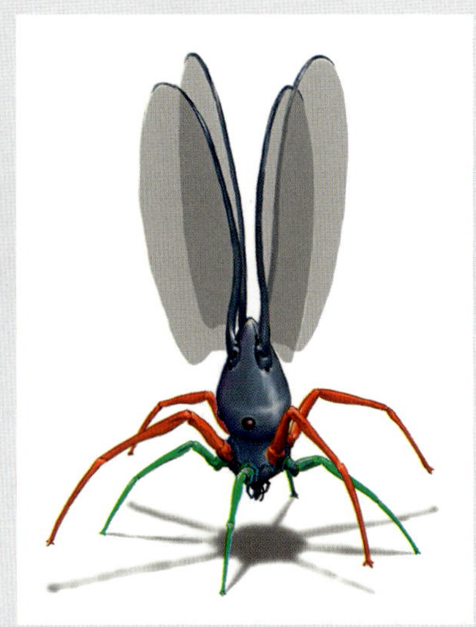

▲ **Tetrapter Walking Gait**
This sketch shows a walking tetrapter with several things worth pointing out. The first is the upright position of the wings when not in use. This upturned position is not adopted by all tetrapters and the choice may have to do with anatomical possibilities as well as available space.

For clarity, the large outer legs are coloured red and the smaller inner ones are shown in green. In this phase of the gait, the animal is standing on the inner legs only and the outer legs are in the air, swinging forwards (although 'forwards' does not really apply to radial animals; it here refers to the direction in which the animal is walking).

▶ **Farfalderol**
Farfalla fulgor
Name derivation: *Farfalla* (It.) (meaning 'butterfly'); *fulgor* (L.) (meaning 'brightness or flash')
Habitat: Open to half-open terrain
Distribution: *Imparia Septentrionalis*, northern part of the peninsula
Mass: No one ever bothered to weigh one
Length: Wingspan up to 15cm
Diet: Plant leaves

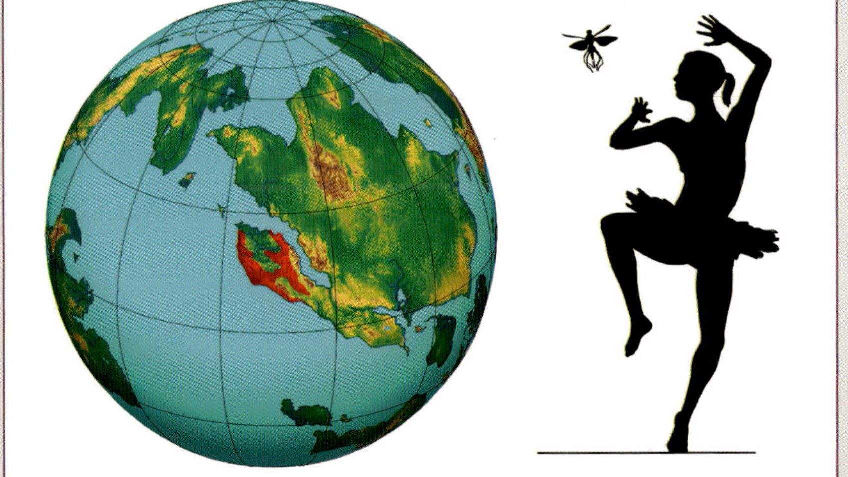

How Tetrapters Fly

Tetrapters are at present the only clade in the universe known to have developed radial flight. Their flight pattern is rather ingenious and rests on the so-called 'double clap and fling' mechanism. Each wing moves through a figure of eight pattern that involves movement around three axes at the same time. The wings are therefore as complex as that of many Earth insects such as dragonflies that share such triaxial movements. The first rotation, around a vertical axis, allows the wing to move clockwise and anticlockwise, while the second rotation lets the wing rotate up and down. The remaining axis of rotation runs the length of the wing and allows either side to be held up or down, much like humans can rotate their forearms to hold their palm up or down.

A Wing Cycle

The diagram below illustrates a wing cycle, starting from the top middle. The wings are held together at the end of the figure of eight. The leading wing edges are all held upwards. The next frame clockwise shows that the wings all move up, and then start to move away, which is clockwise for two wings and anticlockwise for the other two. At this turning point, the wings also start to move down and they all rotate around their longitudinal axis, so their plane is diagonal.

The rightmost panel shows the wings in mid-stroke, moving quickly in the clockwise and anticlockwise directions. They are held nearly horizontally and still move down. This is continued in the next panel.

The lowermost panel shows that the wings are now turning upwards again. They have rotated around their longitudinal axes, so they are again in a vertical position, just like in the first panel, but rotated 90 degrees. The remaining panels are the mirror image of the previous ones, but now the other side of the wing is held upwards.

Lift

Lift is obtained by the quick clockwise and anticlockwise strokes, aided by deformation of the wing plane. The secret of this movement lies in the moment when the wings almost 'clap' against each other (the top and bottom diagrams) and then 'fling' apart quickly. This provides strong aerodynamic effects, pulling and pushing the tetrapter upwards. Various Earth animals also use the 'clap and fling' strategy, but tetrapters are unique in having two such phases in each cycle, thanks to their four wings.

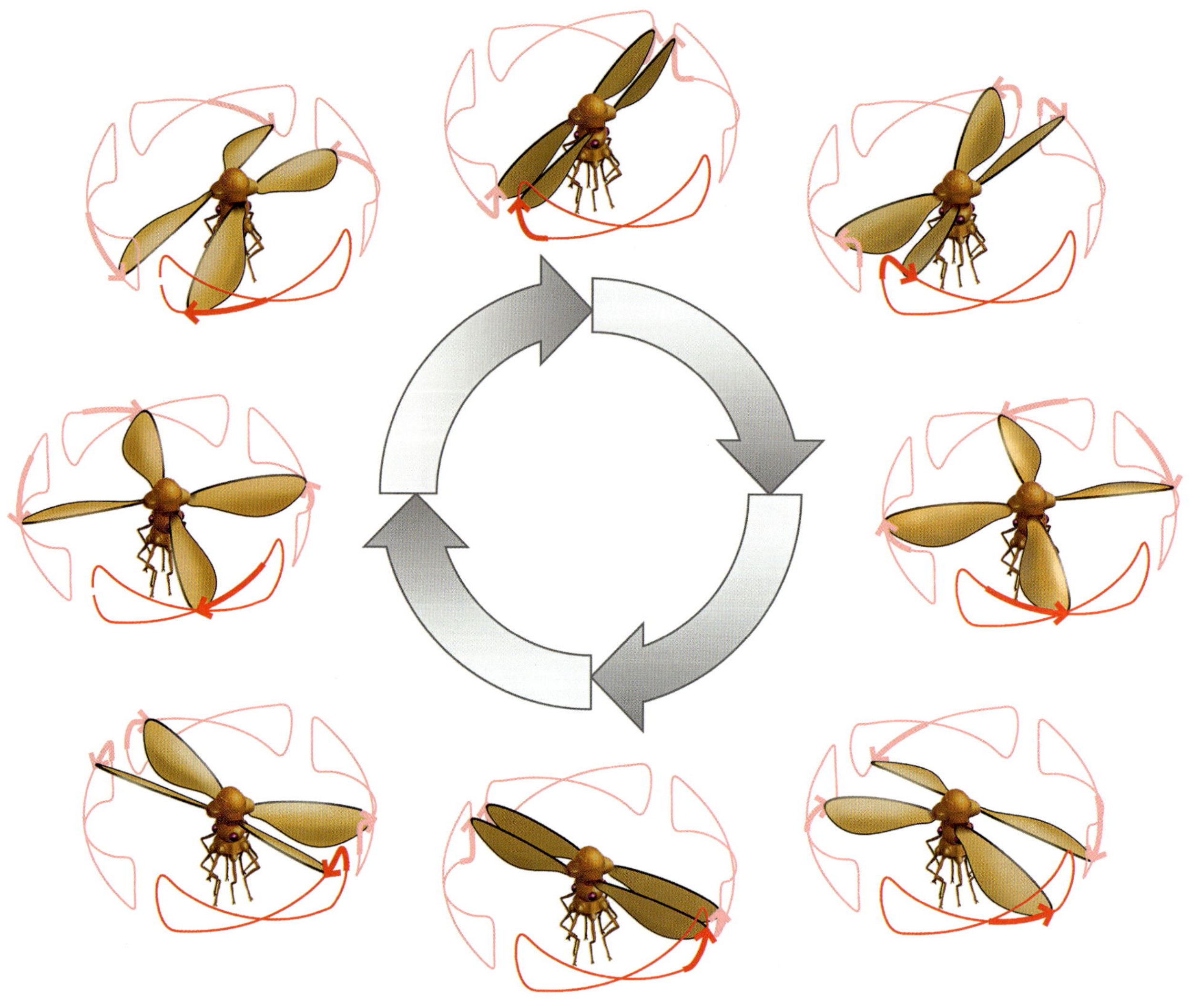

▲ Scheme of tetrapter flight.

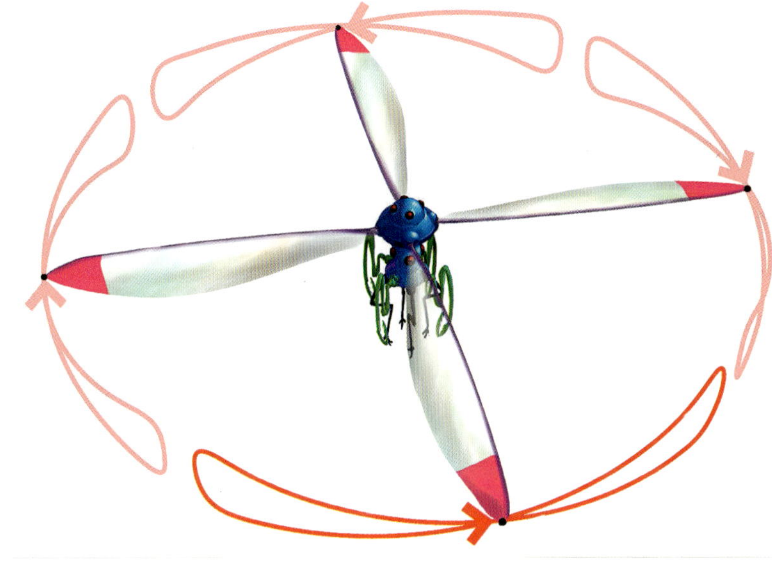

▶ 'Helicopter' Tetrapters

The movement cycles differ between tetrapter clades and species. The pattern shown here is adopted by many raptorial tetrapters (*see* the section on the Red Baron). The pattern is characterised by little downward movement. The wing's plane is held at a shallow angle during mid-stroke, so the wing is held nearly horizontally. The wings also very rarely touch one another in the 'clap' phase, which is thought to prevent wear and tear at the price of some lifting force. This pattern is thought to be optimal for larger tetrapters, although there are many exceptions to the rule.

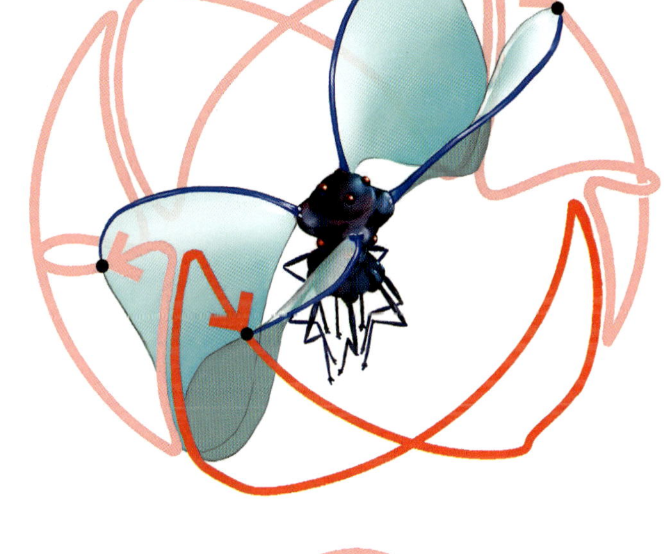

▶ Rowing Along

This pattern is nearly opposite to the one shown above. The wings move downwards and upwards through considerable distances, and the plane of the wings themselves is almost at a right angle against the direction of movement. Such a movement resembles an oar pushing forcefully against water. The clap phase often sees the wings actually touching, which increases overall lift. In the phase shown here, the leading edges of the wings are separating again, while the trailing edges are still very close to one another. The air that fills the widening gap can only come from above, which is thought to help create low pressure or a lifting vortex. Many fly-like tetrapters use this pattern, which creates an odd buzzing sound.

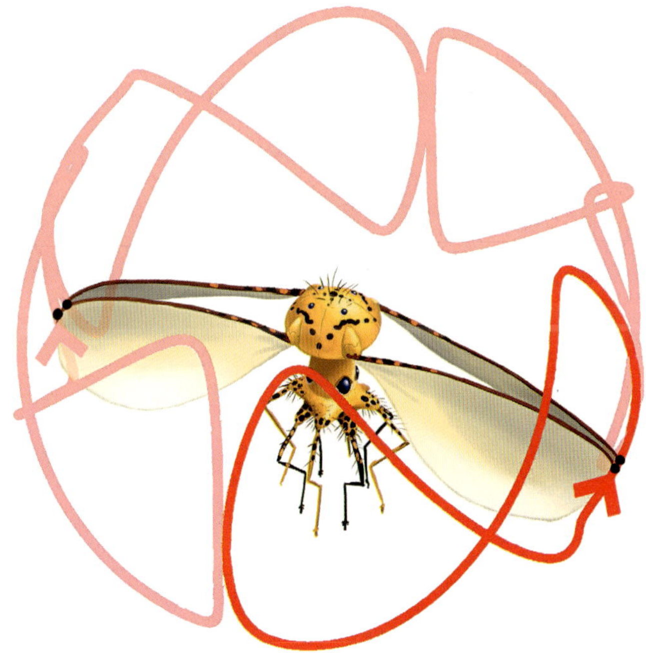

▶ Strong Supination

'Supination' means the turning of a wing around its longitudinal axis. 'Pronation' means the same, but in the reverse direction. In some species this turning is done with additional force at the beginning of the clap phase. This propels the air downwards; some air would probably move in the direction of the wing itself but meets the air coming in from the opposite wing head-on. Because the trailing edges meet earlier than the trailing ones, the two masses of air can only go in one direction, and that is down, providing lift. In some species, the wings are held below the attachment point to the body longer than they are above it. Strong supinators often have such droopy wings.

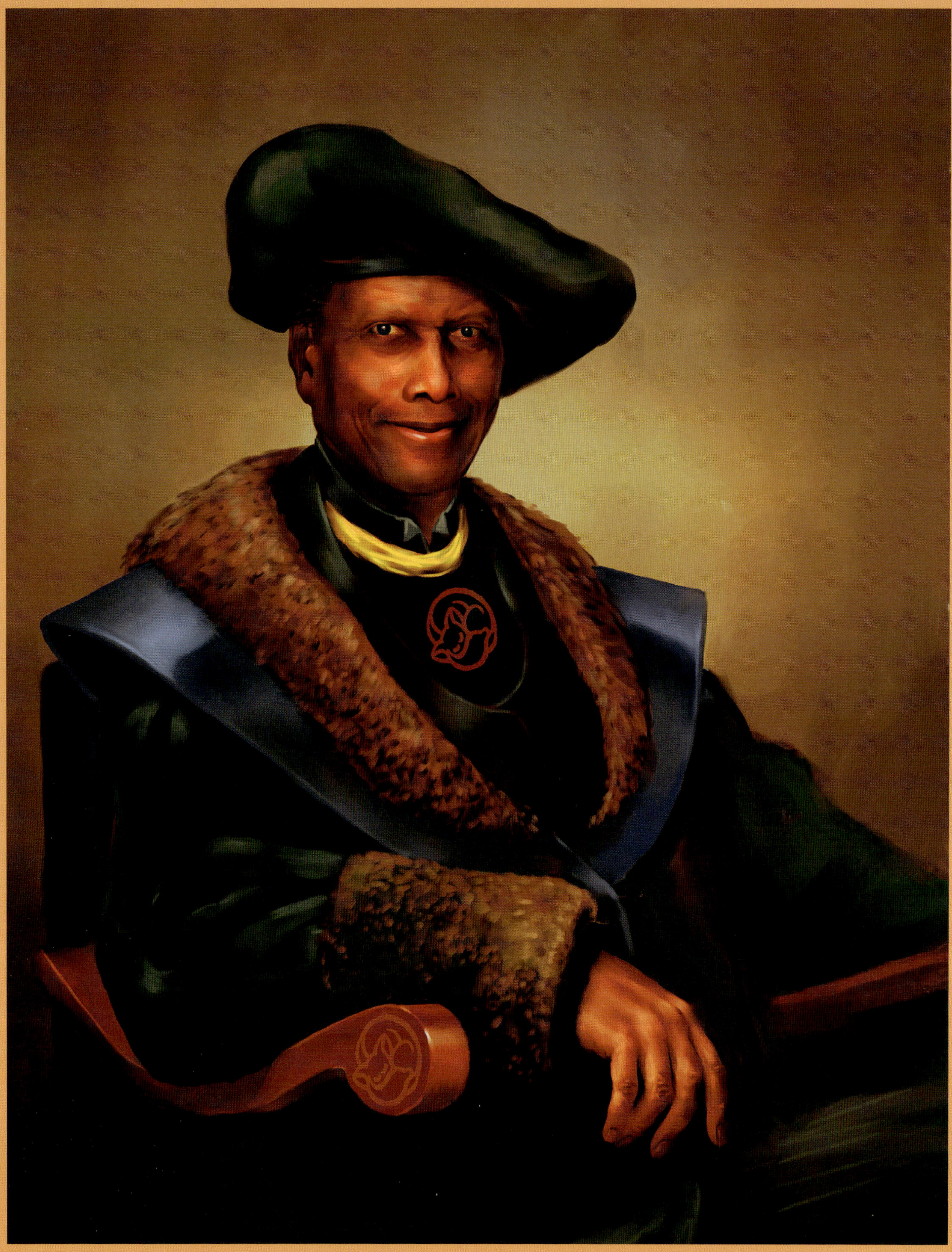

CHAPTER 6

SCALATES AND HEXAPODS

'Many people on Earth, sadly limited in experience, express surprise at finding that most large animals on Furaha have six or more legs. They assume, by force of familiarity, that four legs must be 'better', for which there is no evidence. It seems they do not realise that four-legged animals are in fact a small minority among Earth animals.

While more legs require more material and energy, they also aid stability and reduce the consequences of injury. Evolution always deals with a temporary fleeting equilibrium between changing demands, resulting in an eternal precarious balance.

This consideration may not convince those who think that four legs are best. Why then is there no general trend among millipedes to become centipedes? Likewise, centipedes do not tend to become hexapods, and tetrapods do not generally evolve into bipeds. The number of legs is apparently not critical for fitness. Furahan hexapods are hexapedal only because their Bauplan solidified in that shape.'

From *Worlds Apart: Natural Histories of Furaha and Earth* by Souren Nyoroge

◀ This painting of Professor Tigran Nyoroge was displayed in the Institute Hall. Tigran was a grandchild of Souren Nyoroge, the father of Furahan Biology. Some said that it would be impossible for someone with that surname to make a career, as the inevitable comparison with an illustrious forefather would always turn out to be negative. However, as a *Studiosus* young Tigran proved to be among the brightest of his year. He regarded status with mild amusement rather as something serious. He became a full professor, loved by his co-workers and students. At his farewell lecture, not all figureheads of the Institute rewarded him with their presence, but all his previous students showed up, a difference that says something about his priorities.

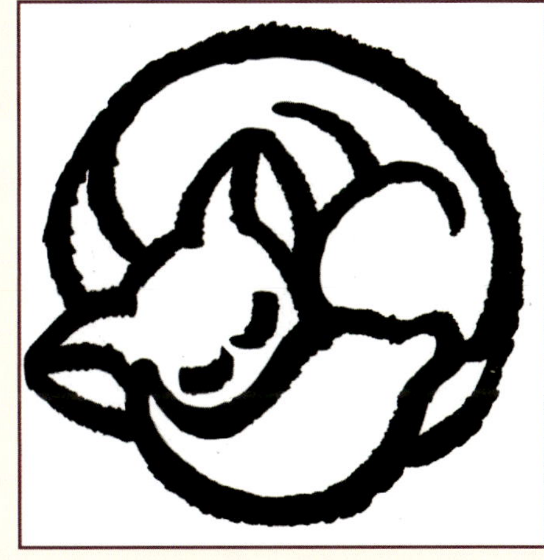

▲ The Nyoroge family cartouche is shown very modestly: Tigran wears a muted sartorial version on his shirt, but his chair also shows one, made of inlaid wood. Few families could boast personalised furniture in the Institute. The cartouche shows a sleeping Earth fox; it is rumoured that the animal only pretends to sleep.

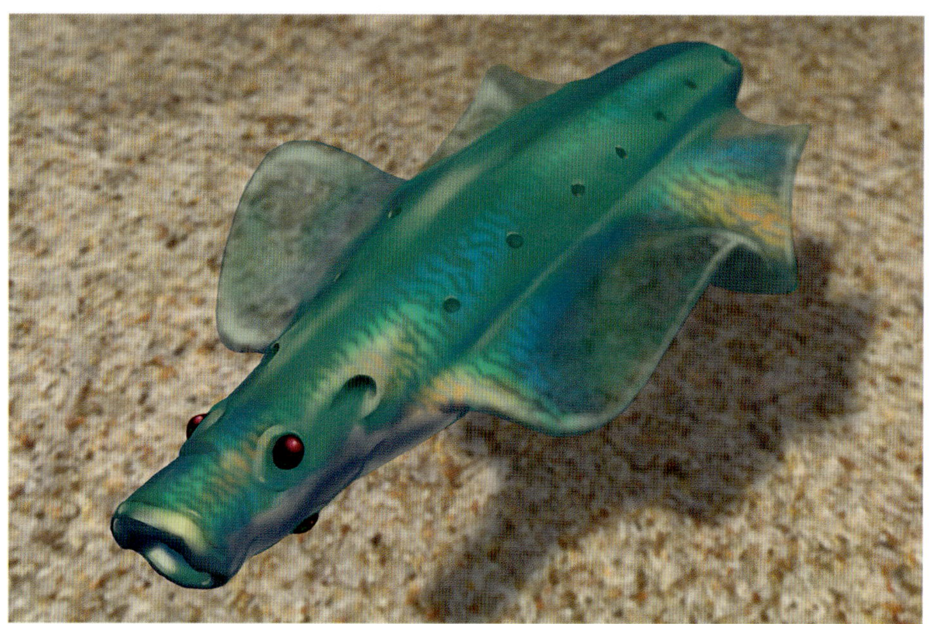

▶ A Living Fossil (Fishes I)
Fishes I are very close to the ancestral form of the scalate clade that includes hexapods. Their characteristics include the absence of jaws and four respiratory canals. The latter run from the front to the back of the animals, are lined with gills and have small lateral openings to the outside world. Fishes I swim with the aid of two undulating membranes.

There are only a few surviving species and these are all very small. This one, the 'flimb' (*Tapetulum volans*), exhibits some pretty iridescent skin patterning.

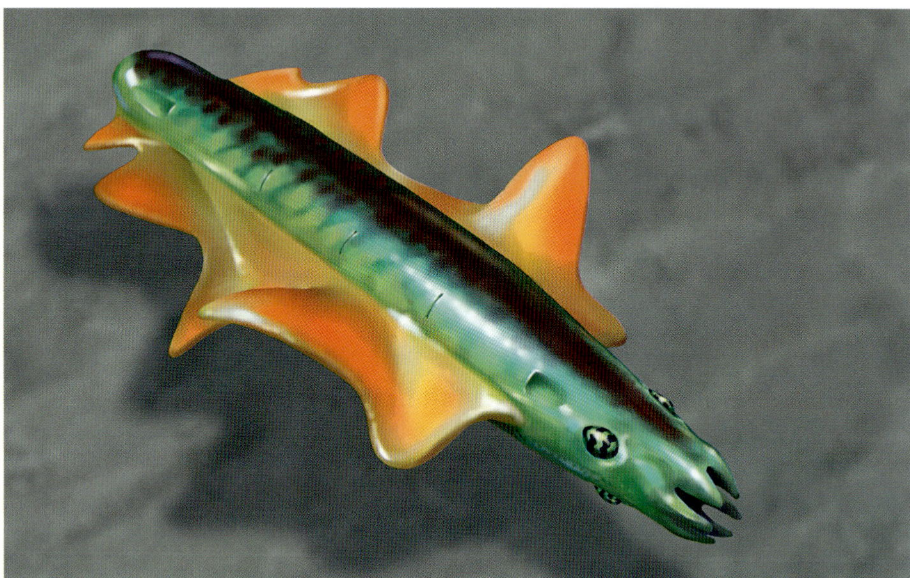

◀ Early Jaws (Fishes II)
Fishes II have primitive jaws, shaped as six finger-like organs protruding from the mouth rim. The species shown here, the 'mackleral' (*Vexilloscissus steyeri*), does not seem to have changed much in the aeons since its cousins went on to become Fishes III.

The lateral fins have become deeply indented, so some experts argue that these animals should be classified as Fishes III. Others sneered that any Fish with six jaws could only belong to Fishes II. The argument became quite heated.

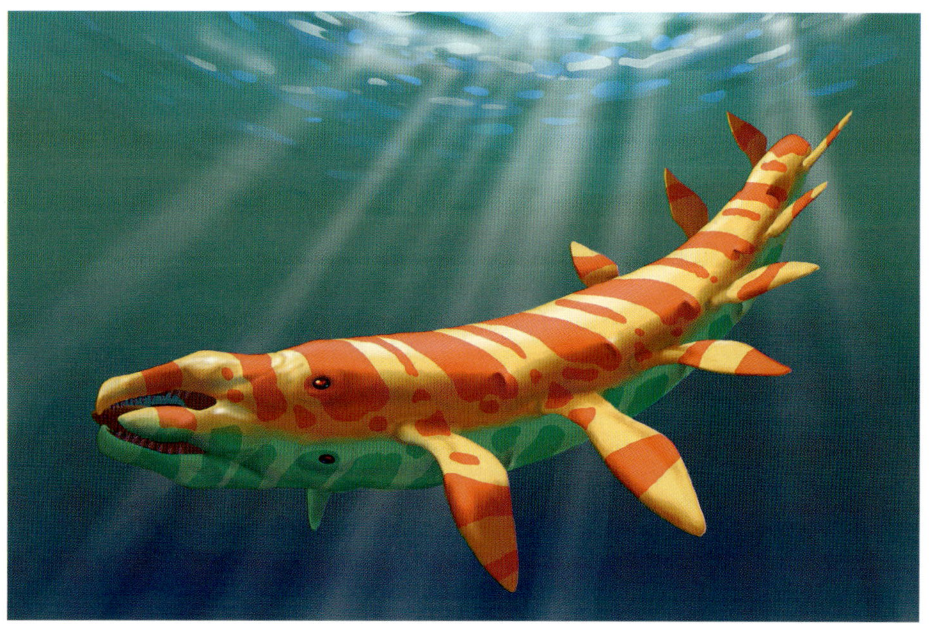

▶ Fins and Jaws (Fishes III)
Fusion of two upper and of two lower jaws left four jaws. That, as well as completely separated fins, define Fishes III. The species shown here, the Greater Snook (*Burseutomos boulayi*), is an ambush predator usually found near lake bottoms.

'Fishes'

Fishes I, II and III

The words 'fish' and 'fishes' are used as imprecisely on Furaha as they are on Earth. To begin with, the word 'fish' is singular as well as plural, leaving 'fishes' to indicate multiple species of fish. Early explorers made the problem worse by falling back on the age-old custom of labelling any animal living in water as a 'fish', so many Furahan 'fish' join Earth's cuttlefish, crayfish, starfish and shellfish in being called 'fish' while not being fish.

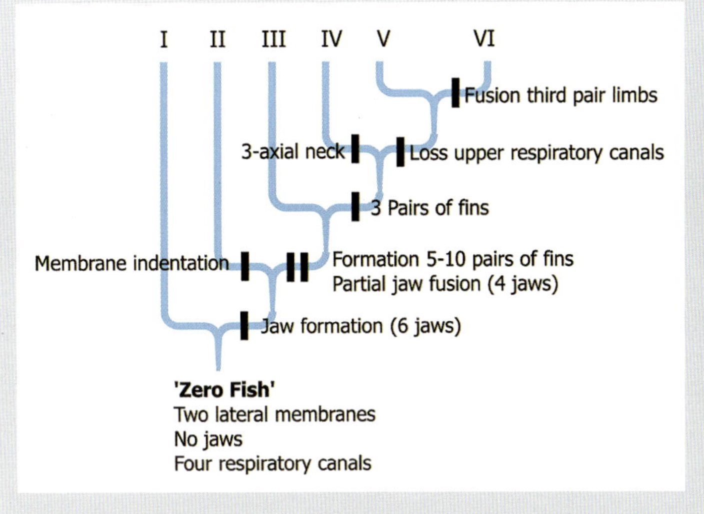

▲ **Fishes Cladogram**
This is a cladogram of the six groups of Fishes, showing some crucial characteristics. Many lay people are surprised to learn that terrestrial hexapods are not descended from the latest Fishes to evolve (V or VI), but probably from the much earlier group IV. But evolution does not stop when a group has split in two.

Fishes I

The fossil record of Nu Phoenicis IV is very incomplete, so most evolutionary relations are inferred from anatomical and molecular data. Luckily, a few species of Fishes I are still around. Some people proposed that the ancestors of Fishes I, the so-called 'Zero Fish', had four membranes instead of two and imply that these animals were not just 'Zero Fish', but also 'Zero Cloakfish'.

Fishes I have a tough rod ('chorda') at the base of both lateral membranes. These two rods are connected by other rods running left to right. Together, these rods form a ladder-like structure that gives the group *Scalata* its name, as *scala* means 'ladder'. The clade *Scalata* includes all Fishes as well as all hexapods.

Fishes II

The 'lateral fin theory' holds that indentations in the membranes of Fishes I allowed the parts in-between to move independently, which improved manoeuvrability and speed. This feature probably evolved together with jaws and predation. Without jaws to catch and hold prey, speed is pointless and without speed those jaws will have little to do.

Fishes III

In Fishes III, the two upper jaws fused, as did the two lower jaws, leaving four jaws. These are no longer finger-like, but are strong and well-muscled and are equipped with very functional teeth. The upper and lower jaws each bear a double row of teeth, whereas the lateral jaws have one row each. The fin membranes of Fishes III have split into five to ten pairs of fins.

► **Habitat**
The presumed biotope for Fishes I was the floor of shallow seas. The mackleral is rare and found only in brackish waters in estuaries at the Northern coast off Palaeogaea Superior. There are at least two types of Snook, although they are probably morphs, rather than separate species. Snooks are freshwater predators.

The scale drawing shows all three species along with a man in the universal position showing much excitement about the size of the fish just caught.

Names
Flimb *Tapetulum volans*: *Tapetulum* (L.) (meaning 'small carpet'); *volans* (meaning 'flying')
Mackleral *Vexilloscissus steyeri*: *Vexilloscissus* (L.) (concatenation of 'flag' and 'cut to pieces'); *steyeri* (after Jean-Sébastien Steyer)
Greater Snook *Burseutomos boulayi*: *Burseutomos* (Gr.) (concatenation of 'skin/hide' and 'well cut'); *boulayi* (after Marc Boulay)

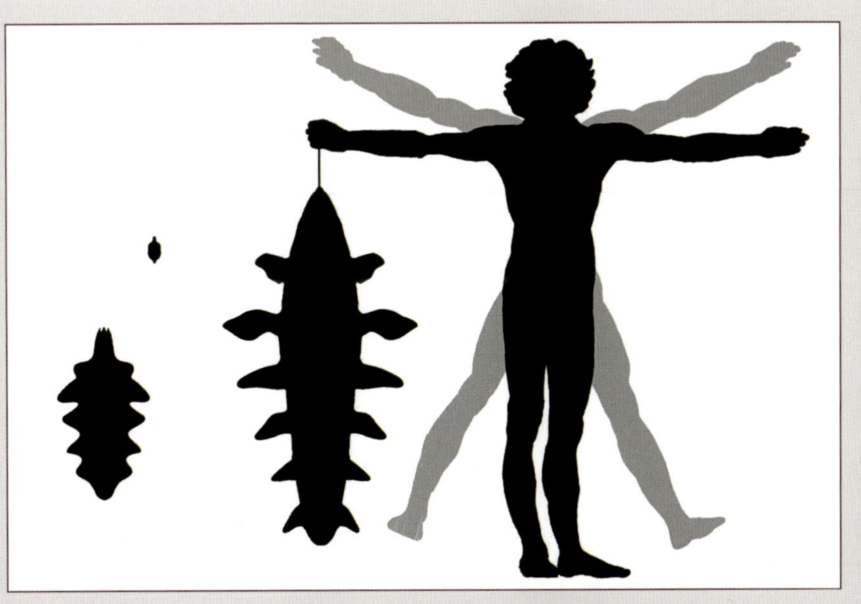

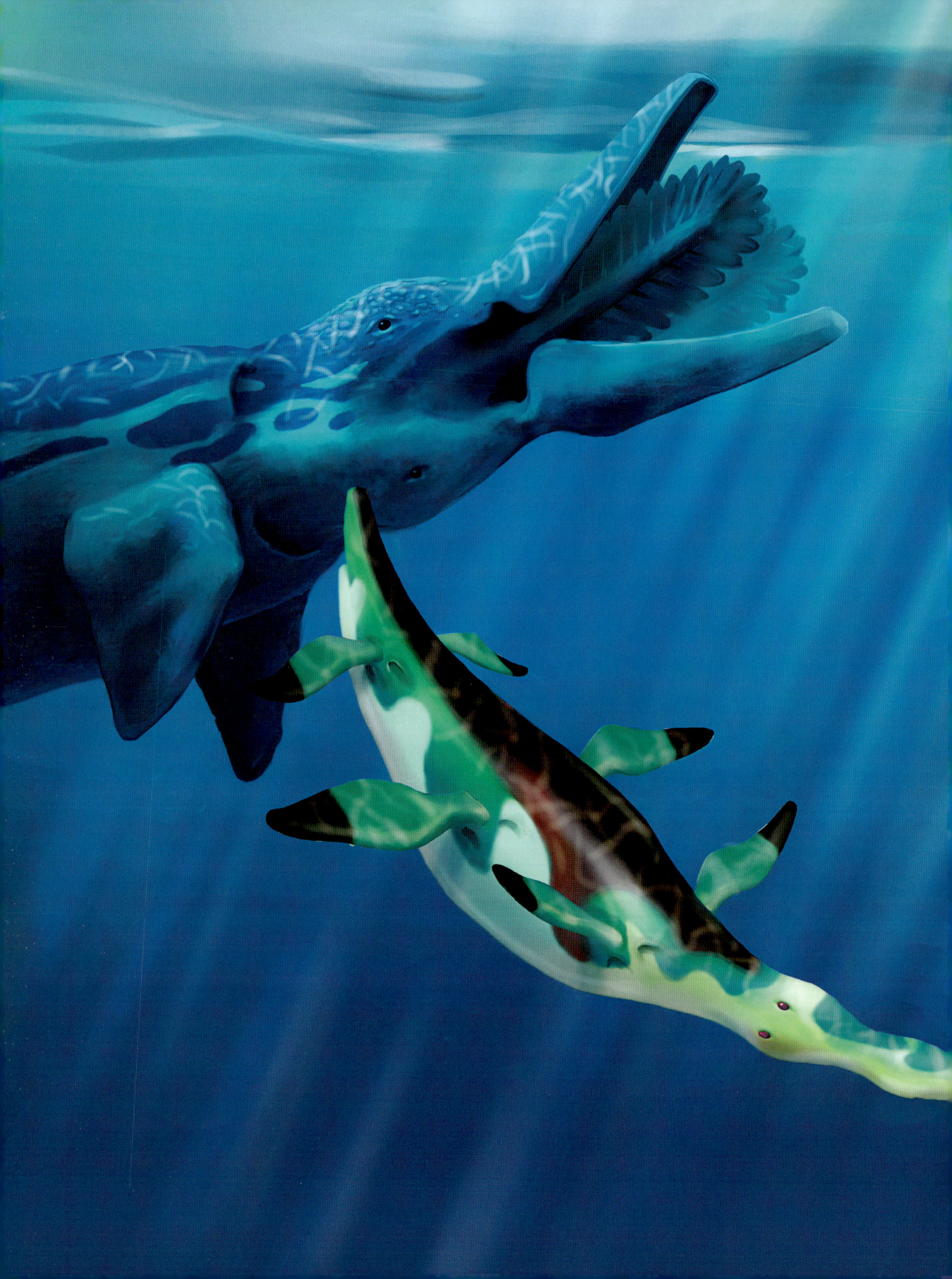

Fishes IV

Those that Stayed Under

Fishes IV receive much attention because some early group members climbed on land about 250 million years ago and became the ancestors of the important group of hexapods.

Many researchers thought they could tell why Fishes IV were predestined to step onto dry land, a process often labelled the 'Anabasis'. Some pointed to the characteristic necks of Fishes IV, thinking that the ability to stick their head out of the water would entice the animals to leave the water altogether. Critics were quick to point out that quite a few species of other Fishes look around above water with their upper eyes without ever showing any tendency to become land animals.

Other researchers said that the flippers of Fishes IV could easily evolve to withstand the pressures of weight. Their critics agreed but added that flippers of Fishes II, V and VI were just as sturdy, but that none of these left the water.

Yet others mentioned that Fishes IV had retained their ancestral four respiratory canals, while Fishes V and VI had lost the upper pair. They reasoned that this redundancy would allow early hexapods to adapt the upper canals to breathe air, keeping the lower canals to breathe underwater. Having left the water, the lower canals would atrophy, leaving hexapods with just the upper canals turned into lungs. While this ingenious explanation is correct in that hexapods indeed breathe with the upper canals, the theory still does not explain what enticed these animals to leave the water.

As there are no living 'in-between' animals, it remains unknown how Fishes left the water and became hexapods.

A Success as They Are

The defining characteristic of Fishes IV is not that they have a neck, but two! The first neck connects the body to the neurosensory cranium and the second connects the cranium to the craniognathes (jaw apparatus). The necks permit movement in all directions. Many Fishes IV are predatory, helped by necks that allow them to catch prey without having to move their entire body.

Shapes

The 'Blue Leviathan' (*Afflictor affluxus*) is cetiform, meaning whale-like. Its lateral jaws have a fringe of stiff feather-like protrusions, which are used to sieve small animals out of the water. The animal swims around with open mouth to trap small animals in its mouth. It then closes its mouth, forcing water through the sieves to exit the mouth through a 'fenestra' on each side. When not in use, the lateral jaws are tucked in and are protected by the closed upper and lower jaws.

The 'Green Snakefish' (*Cunagus dexius*) is a hunter, resembling a plesiosaur. Of course, plesiosaurs had two instead of four jaws, four instead of six flippers, one instead of two necks and two lungs instead of four respiratory canals. Still, the overall picture is that of a smooth-skinned body propelled by agile flippers, with formidable jaws at the end of a long neck.

Fishes IV and plesiosaurs are examples of convergent evolution, in which similar pressures modify the shape of animals until they look alike. Here, the pressure lies in a shared need to catch swift animals and the common starting points were a heavy and stiff body, with fins sticking out sideways. The animals' interiors were not submitted to the same evolutionary pressure at all and are thus strikingly different. The internal anatomy of Fishes IV represents a different kettle of fish altogether.

◀ Snakefish in front of leviathan.

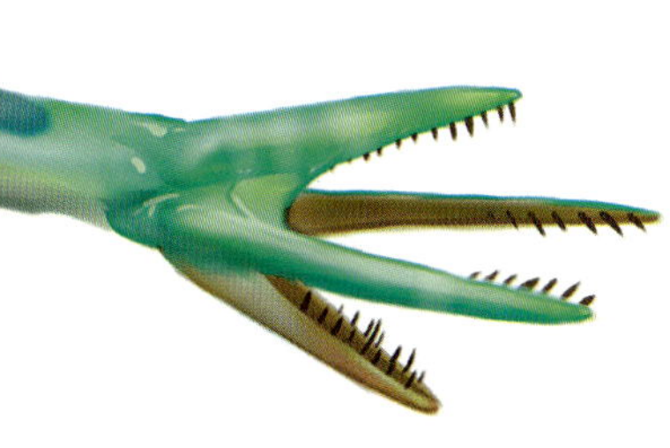

▼ **Green Snakefish**
Cunagus dexius
Name derivation: *Cunagus* (Gr.) (meaning 'hunter'); *dexius* (Gr.) (meaning 'adroit')
Habitat: Surface to middle depths, coastal waters
Distribution: Tropical (shown in blue)
Mass: Unknown
Length: 10m
Diet: Any animal that it can tackle

▼ **Blue Leviathan**
Afflictor affluxus
Name derivation: *Afflictor* (L.) (from *affligo* meaning 'to crush', hence the nickname 'crusher'); *affluxus* (L.) (from *affluo* meaning 'to glide or drift quietly')
Habitat: Surface ocean waters
Distribution: Southern temperate to warm seas (shown in red)
Mass: Unknown
Length: 30m
Diet: Small animals in shoals

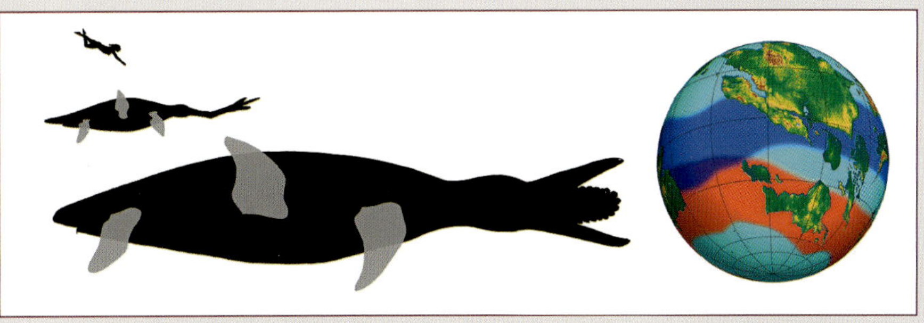

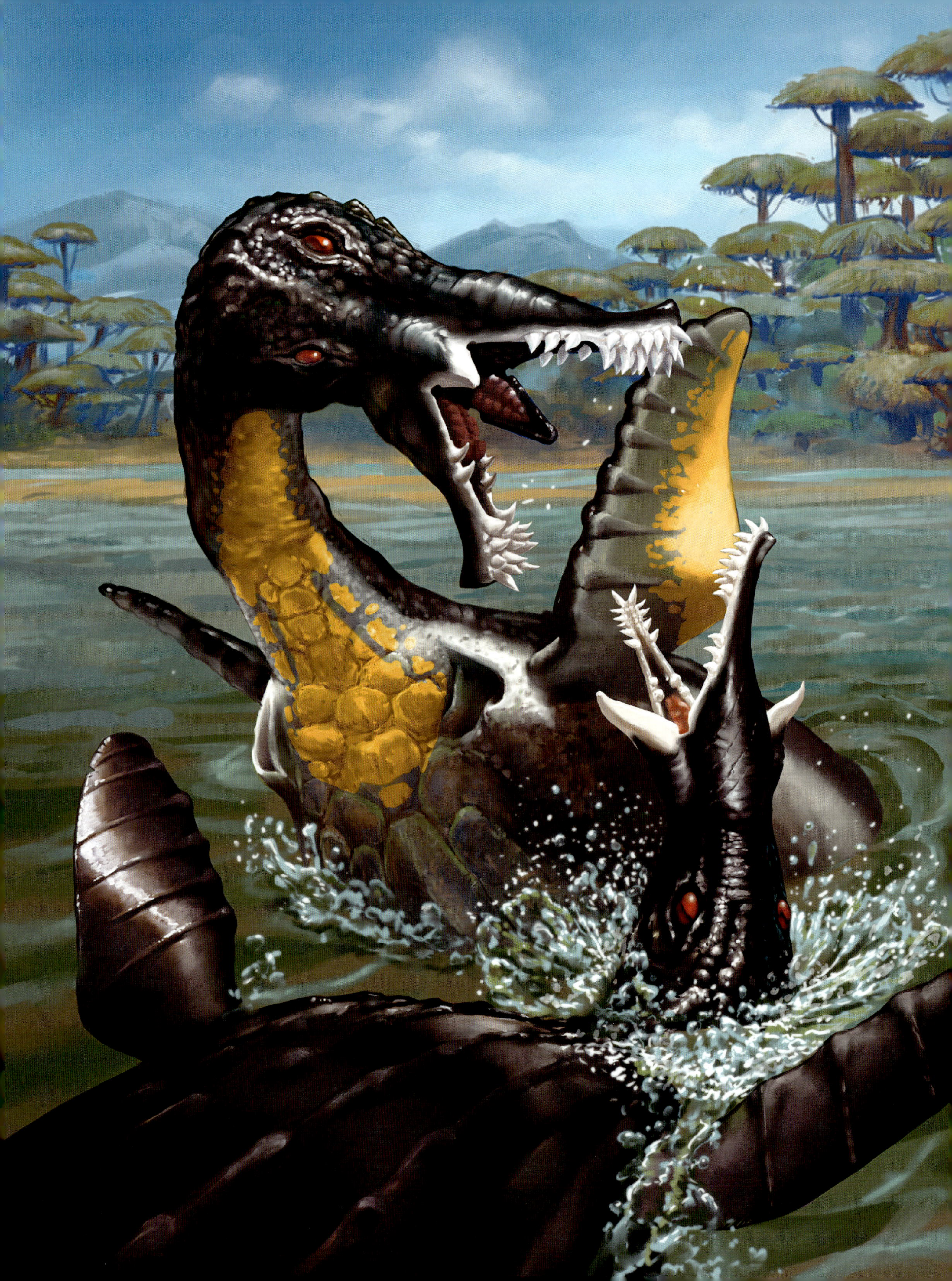

A Sawjaw Stand-off (Fishes IV)

Serrabucca piscivora

It may take a while to get used to the shape of Furahan creatures. A common reaction is to use words like 'monster' for Furahan animals with less than cuddly features. If you do so, you are immediately recognised as an uncouth off-worlder. Furahan citizen-scientists pride themselves on being scientific and open-minded. Still, they have their weaknesses too and tend towards scientific over-correctness to set themselves apart from less enlightened souls. Local researchers may wax lyrical over shapes that others may find disturbing or even repulsive. Off-worlders and citizen-scientists alike agree that the spectacle of a conjugation of two sawjaws in the mating period is an impressive one, but they may do so for quite different reasons.

Sawjaws belong to the group known as 'Fishes IV' and may represent an early offshoot: the truncocranial neck is well developed, but the craniognathal one is short and hardly movable. Some say this is a secondary trait, though.

The sawjaw's teeth are well-formed to get and keep a grip on slippery animals. The 'caterpillar ridges' inside the mouth push any morsel of food towards the gullet. Note the first lower respiratory inlets just before the first pair of flippers.

Is Swimming Allowed?

The pleasant surroundings where sawjaws live suggest swimming in the lovely warm water. However, sawjaws are stealthy and pose a very real threat. They cannot digest humans but find that out too late for the human involved. Some discussion arose as to how the wisdom of whether to swim in such areas should be communicated to the public. The legal minds of the Magisterial Board phrased their view as follows:

> A failure to indicate that swimming in sawjaw waters carries a risk of grievous bodily harm is likely to result in legal liability leading to considerable financial risk to whichever party neglected to communicate said danger to the offended party, provided the victim's heirs can foot the legal bill.

However, the more biologically minded Decanate overruled this reasoning, stating that:

> A failure to grasp that large powerful predators hiding underwater are dangerous constitutes a lack of primary survival skills of such magnitude that the consequences of this deficiency belong to the realm of evolutionary biology rather than of legal responsibility.

In laymen's terms, does the Decanate's ruling means that swimming in such waters is allowed? Yes, certainly! But do you think that it is wise?

Writing in Style

Sean Nastrarruzzo, later famous for his thornbush studies, dabbled for a while in nature writing. He was not successful. As a warning, we will provide an excerpt from his book *Face to Fang*.

> The rains swell the river, and all of gorged and intumescent nature becomes restless! The previously placid sawjaws now breach often and raise their gleaming flippers to the golden light of the dazzling sun. During the incessant Rain Time their beauty is austere and brooding. Now, when everything swells, shudders and steams vigorously, their raw muscular power is magnificent. Their teeth, capable of goring an infant marshwallow's soft pallid entrails, or of reducing a mudbutt with one snap to a splatter of tiny bits of crimson meat and a spray of scarlet blood, sparkle in the rays of light, as the animals bellow and shake their ponderous heads forcefully to and fro. The rancid smell emanating from their turgid gorges washes over the water, rousing others to join in the fracas. Their wide-open, cavernous, gaping jaws glitter with rivulets of water and phlegm. Booming and protracted burping sounds cascade over the banks. The sawjaws raise their ridged and scaly backs to dizzying heights into the humid air. Sometimes they spread no less than four of their six flippers in the air at the time, before they flounder back with a thunderous splash! The larger combatant rises into the fumes and dazzles his antagonists' compound eyes with flecks of light from white teeth, sparkling with the reflected sun. The lesser bucks spit frothingly and hiss in frustrated awe. The sights and smells are stirring and arousing but are as nothing compared to the actual Mating Meeting.

Enough said?

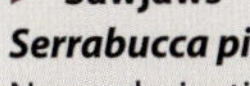
◄ Two sawjaws.

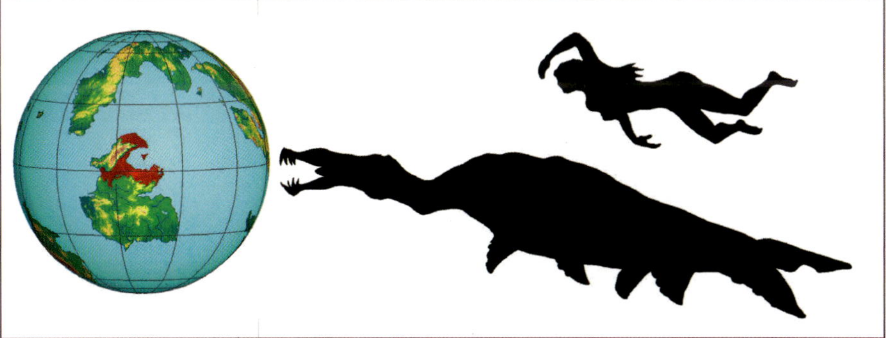

► **Sawjaws**
Serrabucca piscivora
Name derivation: *Serrabucca* (L.) (meaning 'saw-mouth'); *piscivora* (L.) (meaning 'fish-eater')

There are probably six species with about twenty sub-species.

Habitat: Warm water rivers and lakes. One species ventures into the sea from time to time
Distribution: Northern Meralgia
Mass: Up to 1,000kg
Length: 4m
Diet: Anything of a vaguely animal nature, including carcasses
Vocalisation: Incredibly loud booming sounds in the mating season

FISHES V

YET MORE FISHES

Fishes V, rather like Fishes IV, do not look a great deal like fish on Earth. Apart from linguistic problems, there are cladistic problems. Fishes on Earth include sharks as well as bony fishes, even though these are quite distantly related. Following cladistic rules all descendants of a given species fall under the same category as that ancestor, meaning that all Earth vertebrates are fish, in addition to being, for instance, mammals. Obviously, some cladistic advice is not heeded by the general public.

All Earth fishes have one thing in common, which is a vertical tail that swings from side to side. Not so on Furaha: Fishes V propel themselves with flippers and have no tail to speak of.

Fishes V have a scalate skeleton, which means that the body cannot bend sideways. This does not make them clumsy at all, because their six flippers allow incredible manoeuvrability. Note that their respiratory system followed limb formation, so there is one water inlet per flipper. In a development that was to reoccur in Fishes IV and VI, the water outlets are fused on each side.

'Fishiness'

Fishes V do share some features with Earth fish. For instance, many species have 'counter colouring', meaning their undersides are light and their upper sides are dark, which helps to hide their round shape. Others have a silvery reflective skin, helping them to become even less visible under water.

But they do not resemble Earth fish at all in how they move. Their bodies never bend sinuously from side to side. With their remarkably stiff bodies propelled by flippers, they move more like turtles, plesiosaurs or even swimming beetles.

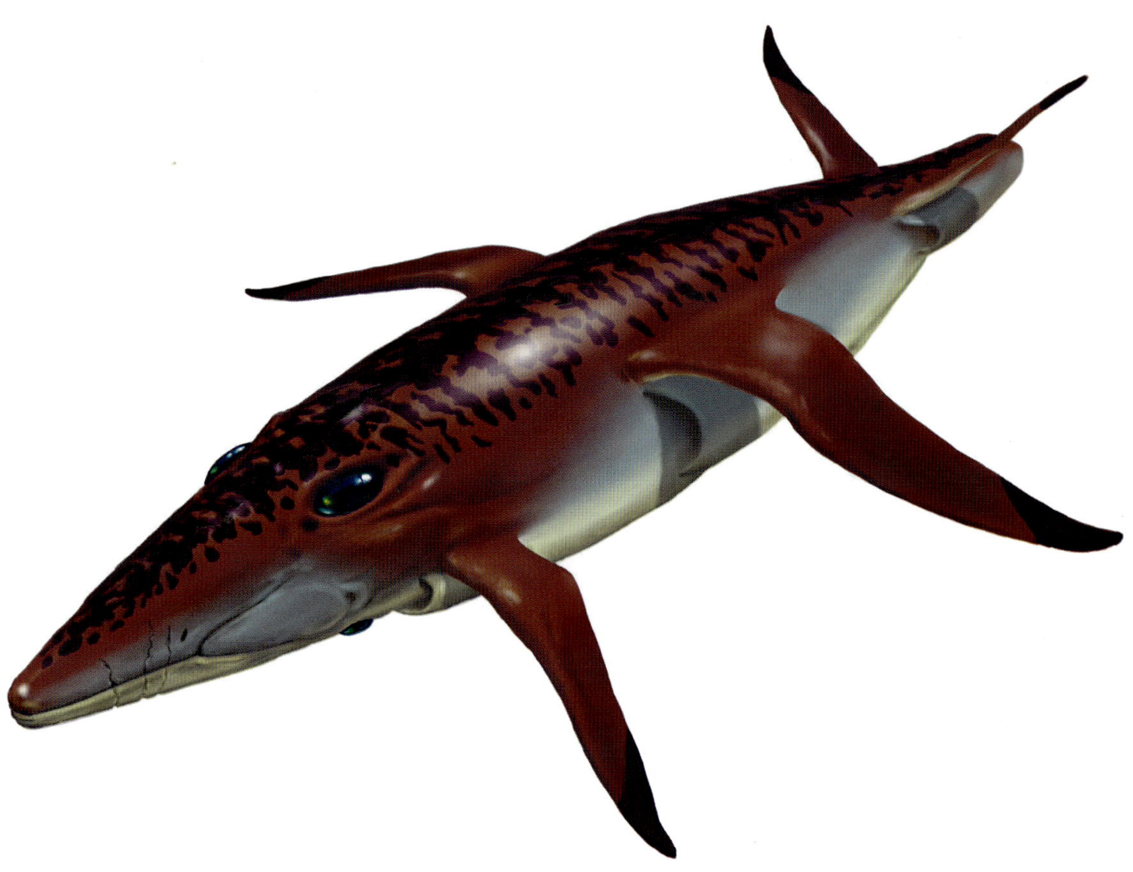

▲ **Brownfish**
Furs macer
Name derivation: *Furs* (L.) (meaning 'thief'); *macer* (L.) (meaning 'thin or lean')

Most brownfish species are not brown. The most common one, plentiful in coastal waters just off the coast of Nexus, gave the group their name. Most brownfish are generic predators living in open but shallow waters. They are agile swimmers and have large mouths, earning them their local nickname 'bassmouth'.

Brownfish serve well as prototypes of the Fishes V anatomical body scheme. The six flippers, six respiratory inlets and lack of a neck or tail tell them apart from Fishes V and VI at a glance. Note that the shape of the jaws means very little for determination, as jaw shape is notoriously variable in all Fishes.

Habitat: Surface waters to depths of about 50m
Distribution: Peri-archipelagic waters
Mass: Up to 20kg
Length: Up to 1.5m
Diet: Any animal of suitable size

► **Cowfish**
Praestolator paulisper
Name derivation: *Praestolator* (L.) (meaning 'he who stands ready'); *paulisper* (L.) (meaning 'a short while')

The cowfish probably owes its name to its docile nature. Cowfish are not easily perturbed. As is often the case with herbivores, they do two things at a time: looking out for predators and munching away steadily, working their way through swaths of floating water plants. They are slow and feature on the menu of various predators.

Habitat: Warm sheltered rivers and lakes
Distribution: Middle Palaeogaea
Mass: 60kg
Length: 1.7m

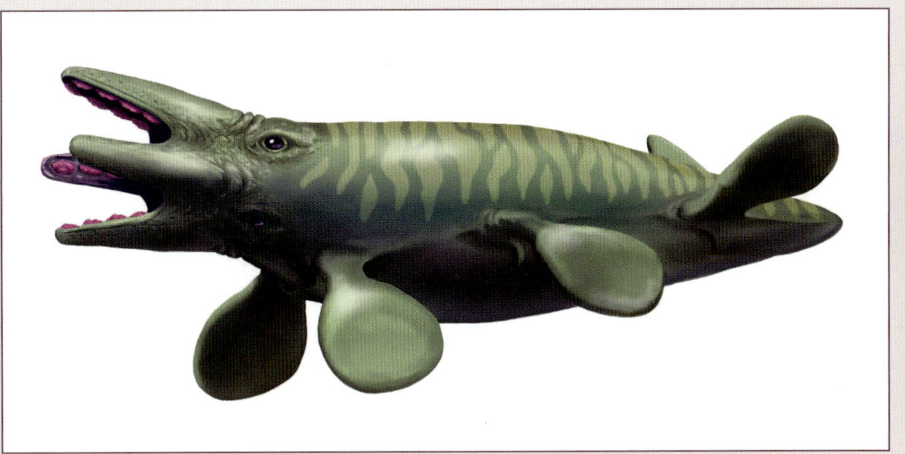

► **Witvis**
Ochrotes polychromatus
Name derivation: *Ochrotes* (Gr.) (meaning 'pallor'); *polychromatus* (Gr.) (meaning 'many-coloured')

This oddly named fish owes its name to its unusual white colouring, made lively by a touch of iridescence. Witvis exhibit some of the variability of Fishes V anatomy. The middle pair of flippers does almost all the propulsive work, while the other pairs serve for steering. The little dorsal fin is a simple skin outgrowth.

Witvis live in deep waters of frigid near-polar waters surrounding Bogoria. Not much is known about them, as their environment is not easy to investigate.

Habitat: Cold seas
Distribution: Bogorian sea
Mass: 18kg
Length: 80cm

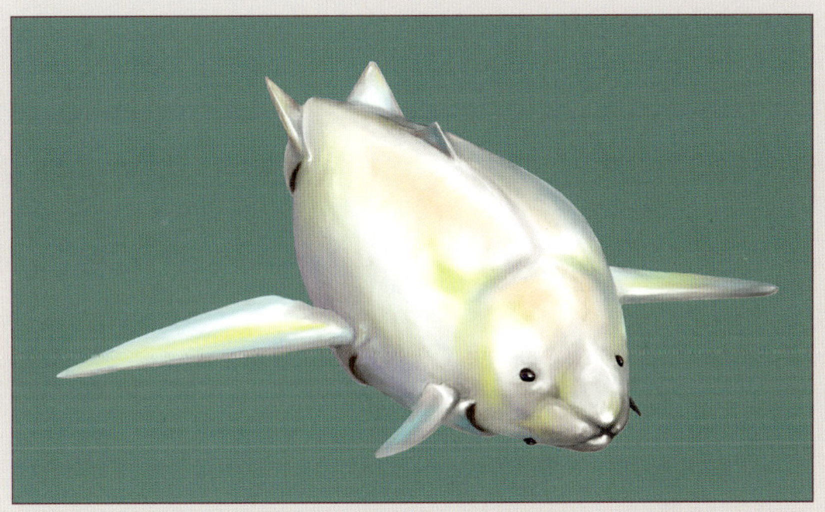

► **Beetle Fish**
Cantharichthus platyleuius
Name derivation: *Cantharichthus* (Gr.) (a combination of words meaning 'beetle' and 'fish'); *platyleuius* (Gr.) (combining words for 'flat' and 'smooth')

The flat beetle fish shown illustrates how Fishes V differ from Earth fish. Its body is stiff but streamlined in all directions. The animal can change direction nimbly with a deft stroke of a flipper. Its body looks like that of a water beetle and that likeness earned it its name.

Habitat: Cold seas
Distribution: Bogorian sea
Mass: 3.5kg
Length: 40cm

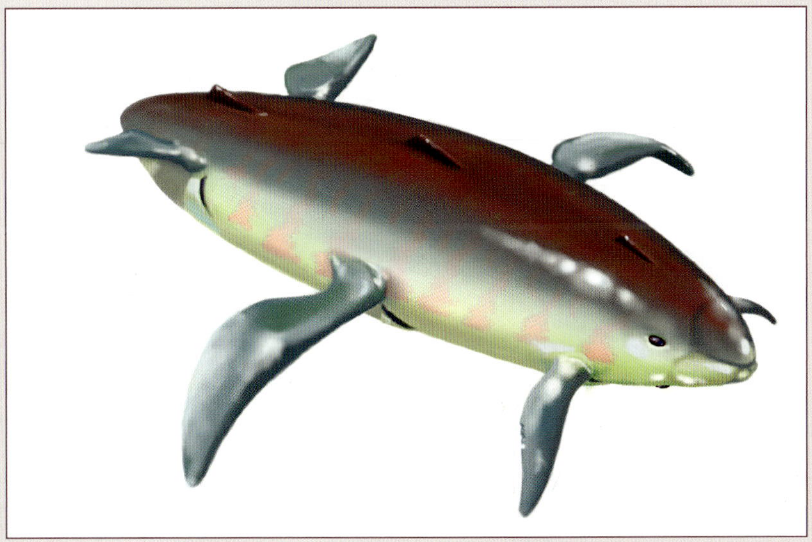

Fishes VI

A Whale of a Fish

Fishes VI look like Earth whales because of their horizontal fluked tails. These are derived from the third pair of ancestral flippers and only in Fishes VI do the three pairs of flippers have different functions. The first two pairs are stabilisers only and the third pair, the tail, is the only one used for propulsion.

Anatomists, by the way, protest against the use of 'tail' for these animals, saying that the structure in question is not an unpaired median structure but consists of fused limbs. Anatomists feel that everyone should know this; others could care less. Anyway, Fishes VI use both the upbeat as well as the downbeat of their tail for propulsion.

Moving quickly through water is a universal problem, for which the universal solution is streamlining, which is why whales and Fishes all look like, well, fish. Similar considerations apply to propulsion and stability in water, hence fins and flippers. Note that the fins of Fishes VI are secondary outgrowths of the skin that can occur almost anywhere on the body.

This page shows a few of the myriad species of Fishes VI that evolved to fill virtually every aquatic niche on the planet. They range from microscopic animals to the largest animals on the planet. The examples illustrate the middle size range of Fishes VI, with each example ten times the size of the previous one.

► Goppy
Goppies would make nice aquarium fishes, if keeping animals captive would not be frowned upon on Furaha. This diminutive freshwater species is on the menu of many other fish, but is still a predator itself. It is seen here eyeing some tiny but tasty water fleas. In the background a shoal of barches swims by. Barches prefer larger prey but will snap up a goppy if one is dumb enough to present itself.

► Trench Gobbler
The gobbler is a typical abyss species. In an abyss, the only light is produced by life forms. The barren ecosystem is based on a trickle of organic material from above, so animals need to conserve energy. There is no need to swim fast habitually but if an opportunity to eat presents itself, it must be jumped upon. These two drives resulted in very odd shapes. The Trench gobbler has very long lateral jaws to grab anything possibly edible. In this image, it is attacking something with tentacles that might be a larval Cthulhu. The larva has just emitted a cloud of bioluminescent ink to try to escape, a trick that seems to be working.

► Strider
Striders are fast! Half the mass of their long bodies is made up by strong tail muscles. Their lateral jaws have evolved into long saws. Normally the two lateral jaws close together in front of the upper and lower jaws, where they hardly disturb the flow of water. Once a strider approaches a shoal of prey, it opens these pointy jaws, equipped with small but sharp teeth. The strider slashes through the shoal at high speed, returning to pick up injured or dead fish. Its favoured prey are sardoons, themselves too quick for most other predators. Note that striders have sexual divergence: the male is larger and has longer fins.

Goppy (A)
Garus garus
Name derivation: *Garus* (L.) (meaning 'fish', source of the Roman sauce 'garum'); *Perca* (L.) (meaning 'perch'); *blanda* (L.) (meaning 'alluring')
Habitat: Small streams, ponds and lakes
Distribution: Rivers and tributaries
Mass: A few g
Length: 4.5cm

Trench Gobbler (B)
Forcilla depresensa
Name derivation: *Forcilla* (L.) (meaning 'little fork'); *deprensa* (L.) (from the verb *deprendare*, meaning 'to intercept or catch')
Habitat: Deep sea
Distribution: Trench (otherwise unknown)
Mass: about 2kg
Length: 45cm

Strider (C)
Sector sagax
Name derivation: *Sector* from *secare* (L.) (meaning 'to cut', hence 'he who cuts'); *sagax* (L.) (meaning 'sharp')
Habitat: Open ocean
Distribution: Anterior ocean
Mass: about 2kg
Length: 450cm

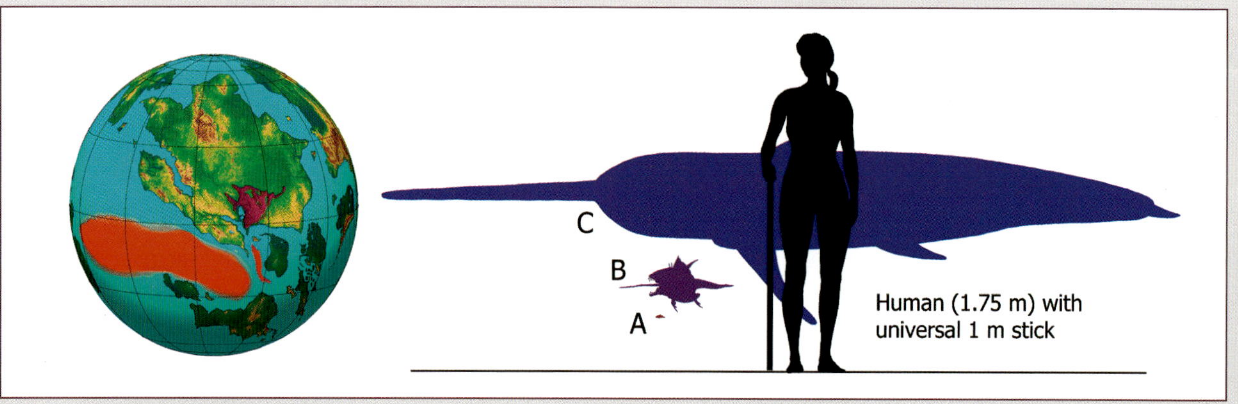

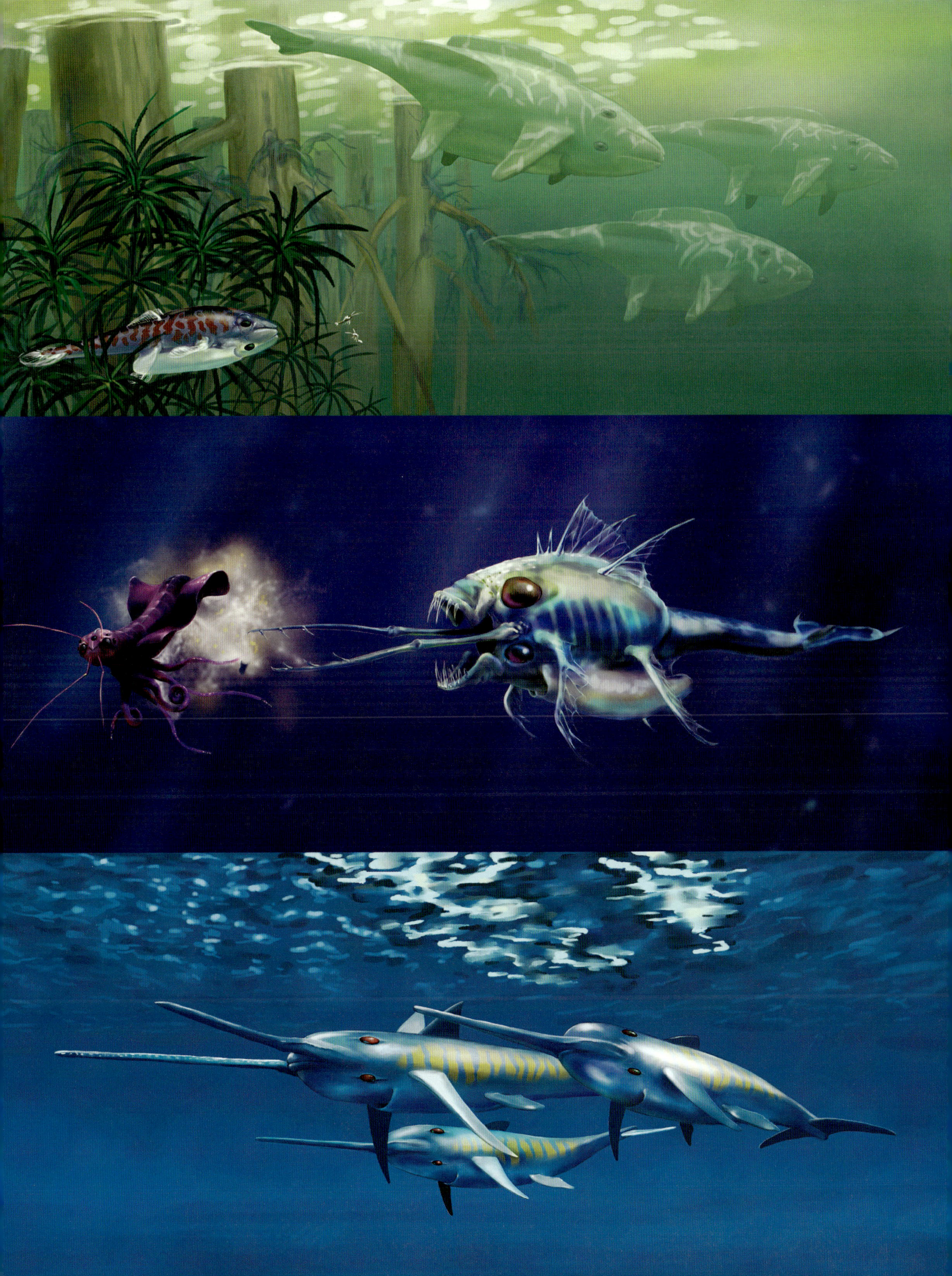

GRAPPLE (FISHES VI)

FORPEX ATTAT

At first glance the scene of the large image looks like it takes places on Earth – something like a fish or a small dolphin jumps out of the water, against a background of leaves and stems that look much like they do on Earth. The resemblance is not only there but is in fact to be expected. Given similar physical circumstances, similar evolutionary pressures are likely to result in similar solutions.

Grapple Anatomy

The animal jumping out of the water is a grapple, a species belonging to the very successful clade Fishes VI. The grapple has a torpedo-like smooth shape that is widest about one third of its length, with fins placed around that area and a fluked tail that provides propulsion on the upstroke, as well as on the downstroke. The shape and smoothness are no coincidence. On Earth, bony fish, sharks, ichthyosaurs, plesiosaurs and whales all converged to similar body shapes. Water opposes fast movement and streamlining minimises the energy costs, making streamlining a universal answer to a universal question.

Obviously, that is as far as the likeness goes. Fishes VI after all have four jaws, four eyes, a respiratory system that is wholly separate from the digestive tract and, of course, two side chains instead of a backbone. This is definitely not a Terran animal.

▶ A grapple showing off.

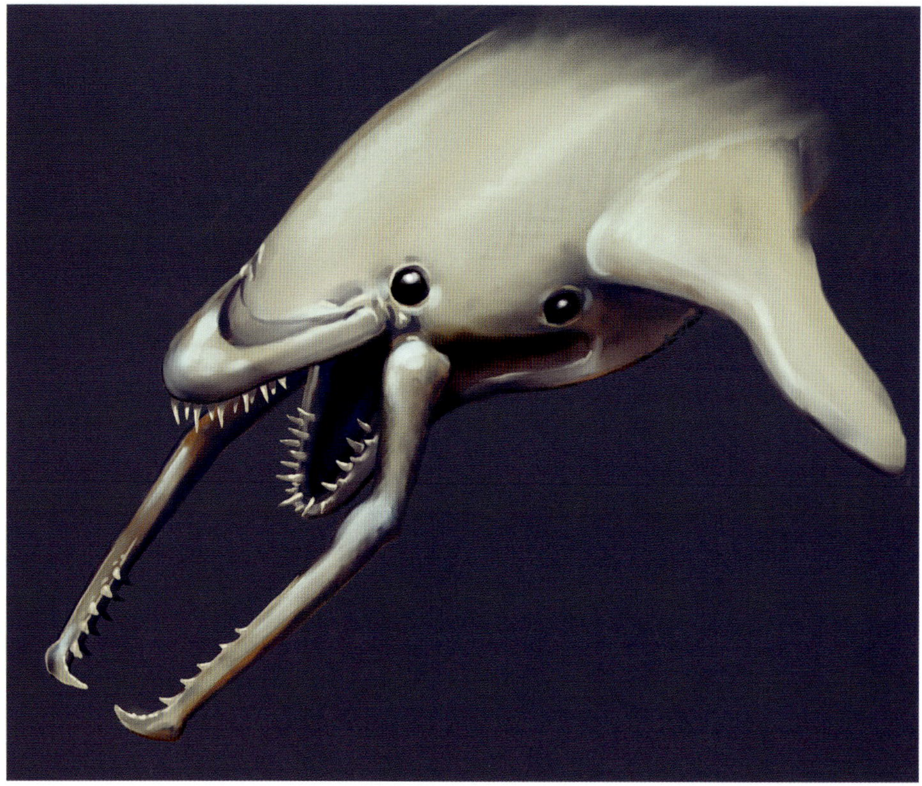

▲ The upper and lower jaws each have their own joints with the anterior cranial ring. The lateral jaws are normally stored in a groove in the side of the skull. To catch prey they are first abducted, anteflexed and extended, before being adducted forcefully. The 'teeth' are in fact skin structures. Their jagged appearance is typical.

Reaching Out

Predators need to catch their prey. Moving faster than the prey is one way in which this can be done, but the considerable resistance of water to acceleration makes this a costly solution. Although grapples are quite fast, their main trick is to accelerate only the part of their body that actually grabs the prey. Plesiosaurs on Earth and Fishes IV on Furaha evolved long necks and lightweight heads to do just this, but the grapple uses its lateral jaws. These are elongated and end in a series of inward-pointing sharp prongs. On making contact they hold the prey long enough to be pulled towards the upper and lower jaws that make short work of the struggling prey.

The grapple lies in wait or creeps up imperceptibly to its prey, only to lunge at it suddenly, making it a successful surprise predator.

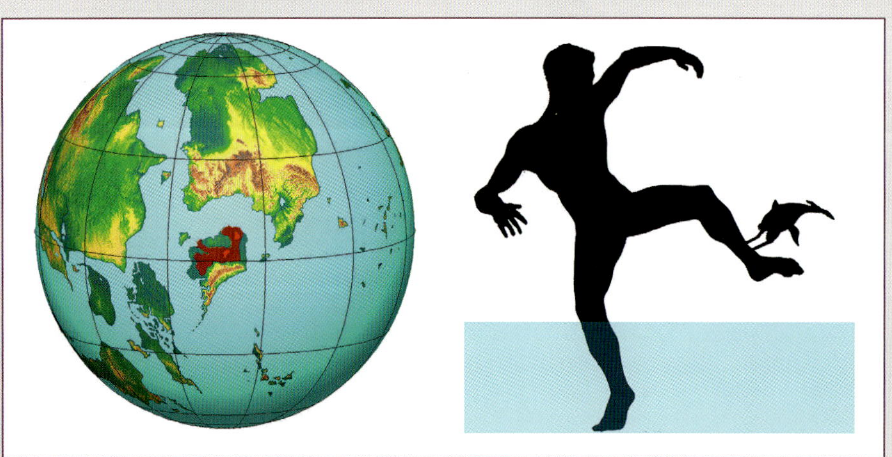

◀ **Grapple**
Forpex attat
Name derivation: *Forpex* (L.) (meaning 'shears or tongs or pincers'); *attat* (L.) (referring to Ah! Oh! Alas! – an expression of sudden surprise, fear or warning)
Habitat: Rivers, pools, lakes with abundant cover
Distribution: Northern regions of Tendaguru
Mass: 1–8kg
Length: 30–60cm

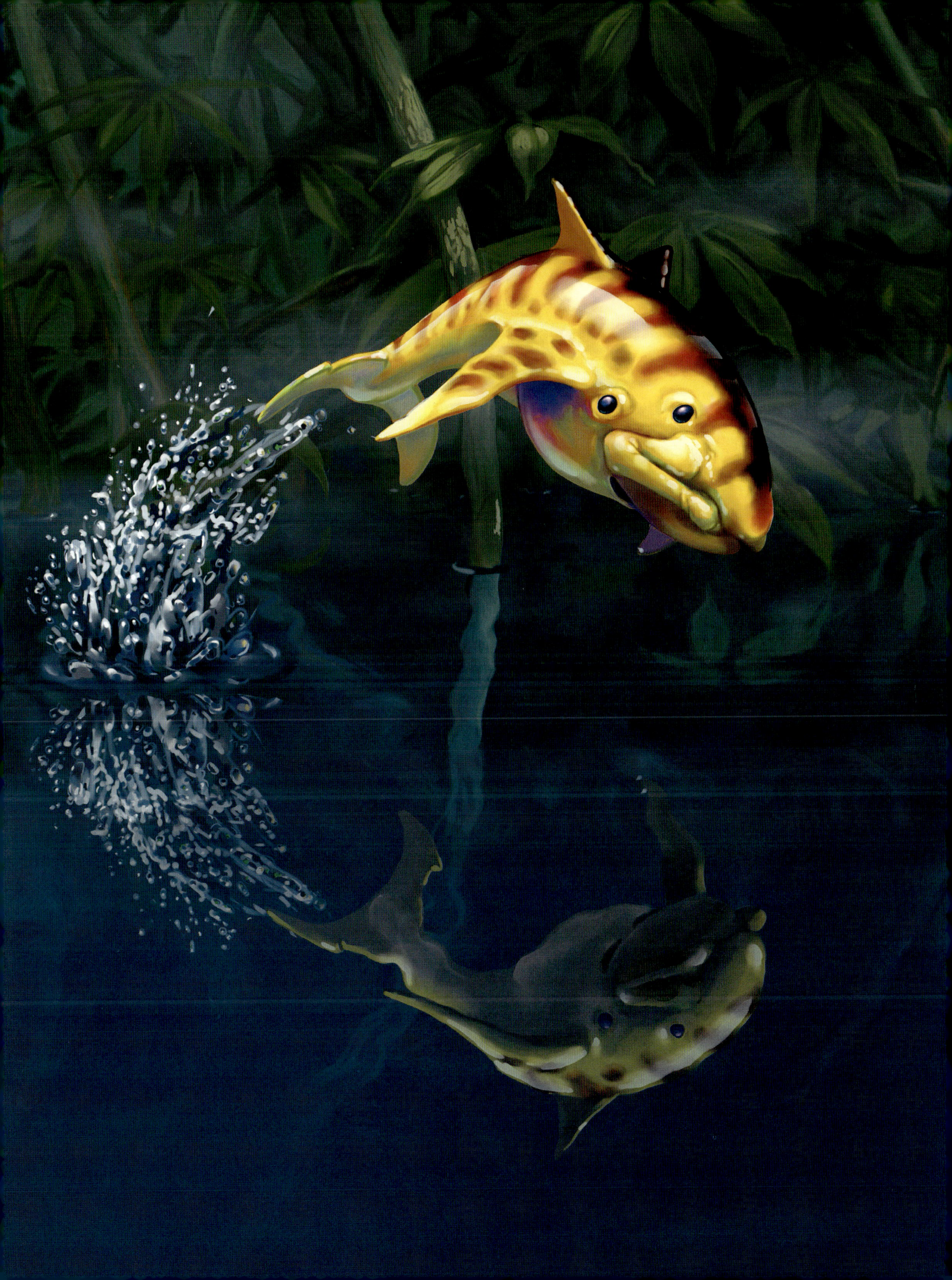

Hexapods

Hexapod Anatomy

Fishes I to VI are basic marine scalates and are already discussed. It is therefore time to focus on hexapods – terrestrial scalates. The scalate skeleton started with rods supporting the lateral membranes on the left and right sides of the body. These rods evolved into a series of 'rail bones', connected by transverse 'rung bones', forming a ladder. This is what, in Latin, gave the clade the name *scalata*.

No Sinusoid Slithering Here!

In hexapods, the scala still forms the main body scaffolding. The nodes where rail and rung bones meet act as joints, allowing the body to roll up and down or to be skewed diagonally. The one movement that a ladder skeleton does not allow is sideways bending. No scalate animal can perform the kind of sinusoidal movement that is typical of Earth's fish or snakes.

Necks as a Sneaky Solution

Hexapods evolved from Fishes IV, which already had elongated necks and heads. Gathering food with just a part of the body uses less energy than moving the entire body towards each mouthful. This energy-saving trick is more useful for large than for small animals. Good examples are Earth's sauropod dinosaurs that used small heads at the end of long necks to forage a large area while standing still, and elephants, using long trunks for the same purpose.

Fishes IV evolved long necks to this end, in the process acquiring separate 'heads' to house the brain plus main sense organs or the jaws. These necks offered Fishes IV another advantage – as their necks are a fusion of the right and left front ends of their ladder skeleton, the necks can move in all directions, including from side to side.

Fins to Legs

The lateral membranes were originally supported by elastic bones that divided one or two times. These bones only had to swing up and down. When the membranes evolved into fins, more was required of these internal fin skeletons. Fins must be able to move up and down, fore and aft, and must also be able to rotate along their longitudinal axis. These three axes of movement are best combined right at the start of the fin, where it joins the body. Fishes accordingly evolved three-axial ball and socket joints at their hips.

The rest of the fin at first only provided internal stiffness, but that changed when Fishes IV crawled ashore and became terrestrial. The fin skeletons then had to carry weight, had to end in a foot and the main direction of movement was fore to aft rather than up and down. All these new demands must have induced quick evolution.

Other Traits and Oddities

Although hexapod jaws show a wide variety of anatomical trickery to help gather food, the jaws never grind food. The reason for that is that the basic scalate Bauplan includes a separate 'gnathoskeleton' that includes grinding plates in the stomach.

Another oddity is that hexapods use the original upper pair of respiratory canals to breathe with and have lost the lower pair. (This also happened in Fishes V and VI, but that was probably a separate event.)

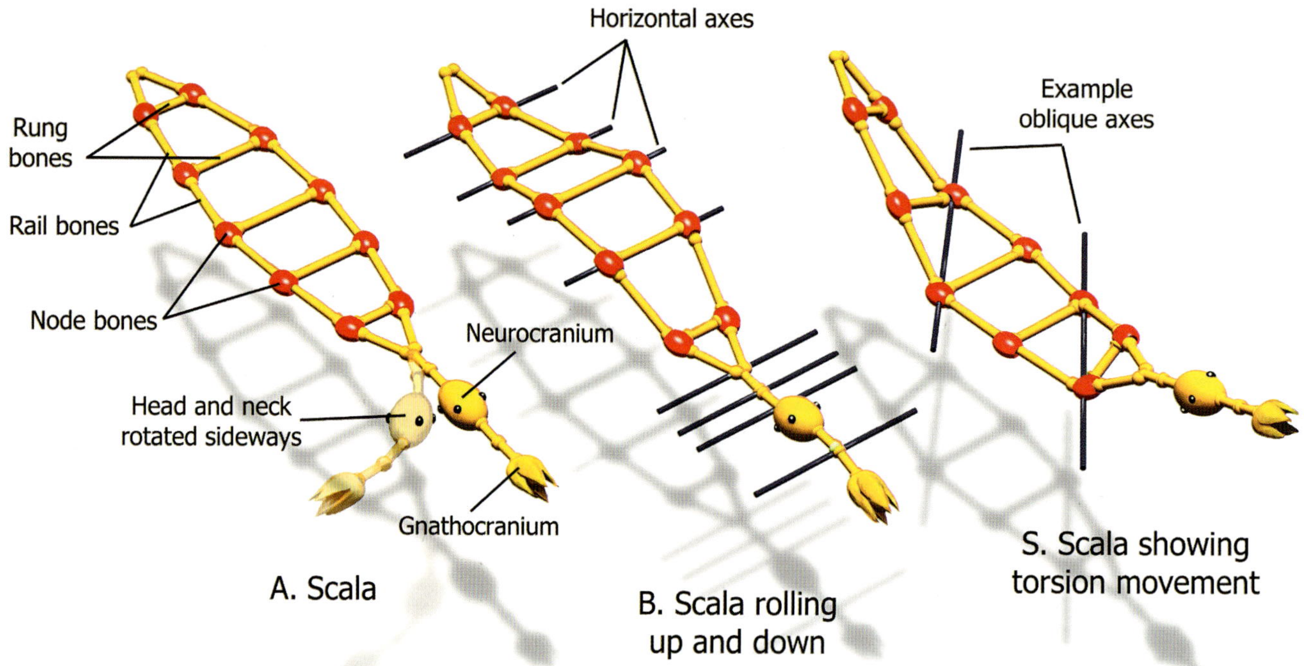

▲ The ladder skeleton does not allow the body to bend sideways, but the head and neck can do so (A). Movements within the body are restricted to combinations of rolling up or down (B) and torsion (C).

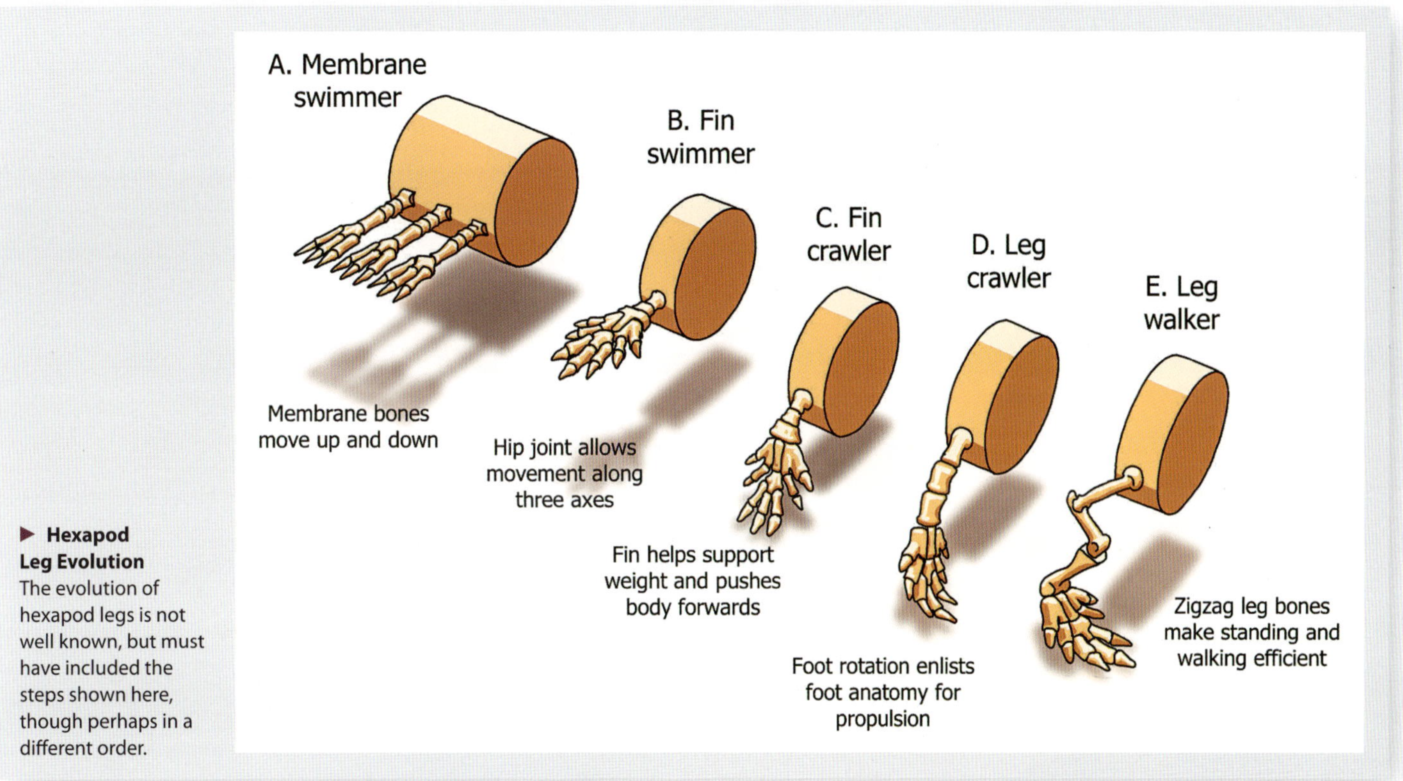

▶ **Hexapod Leg Evolution** The evolution of hexapod legs is not well known, but must have included the steps shown here, though perhaps in a different order.

▼ This schematic view gives an impression of the basic hexapod skeleton, consisting of the scala (yellow), gut jaws (blue), fin skeleton (red) and secondary trunk hoops (green).

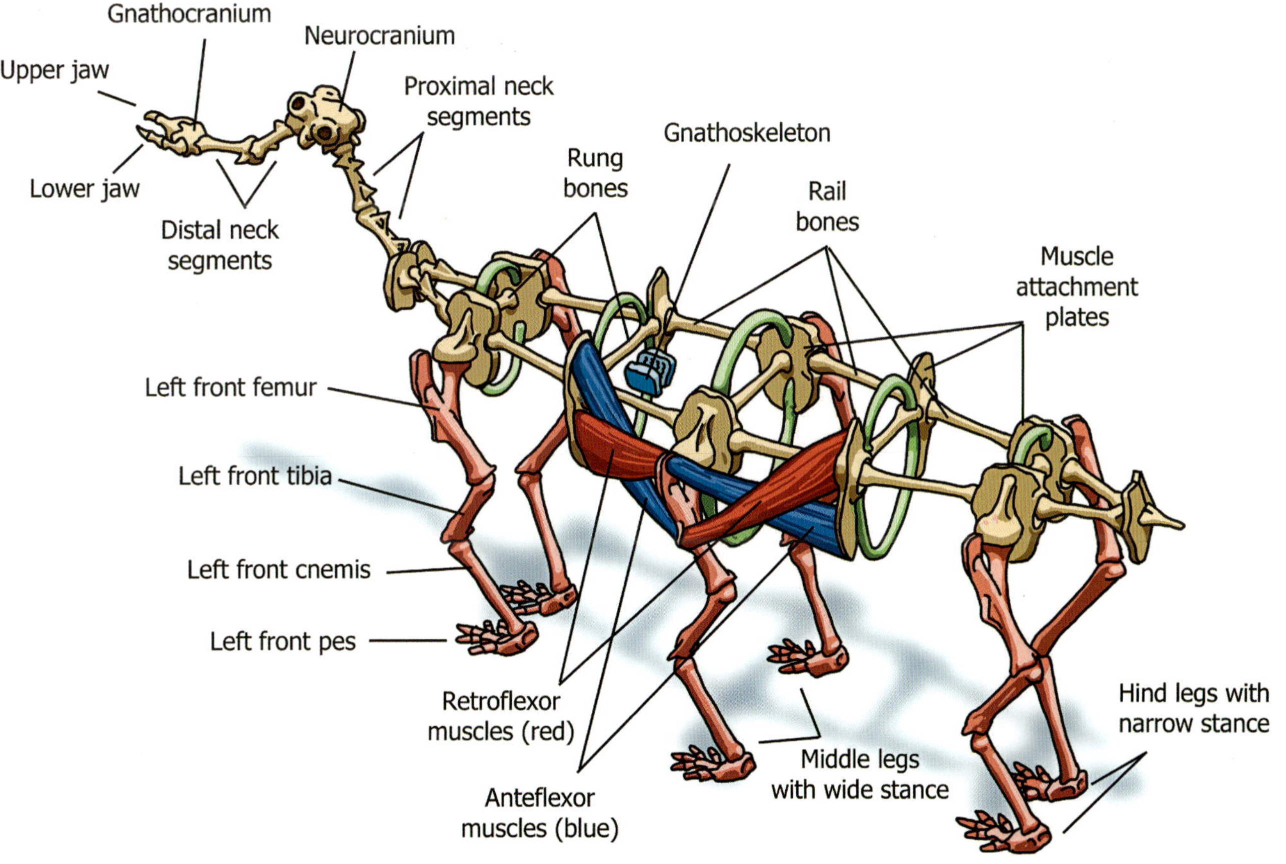

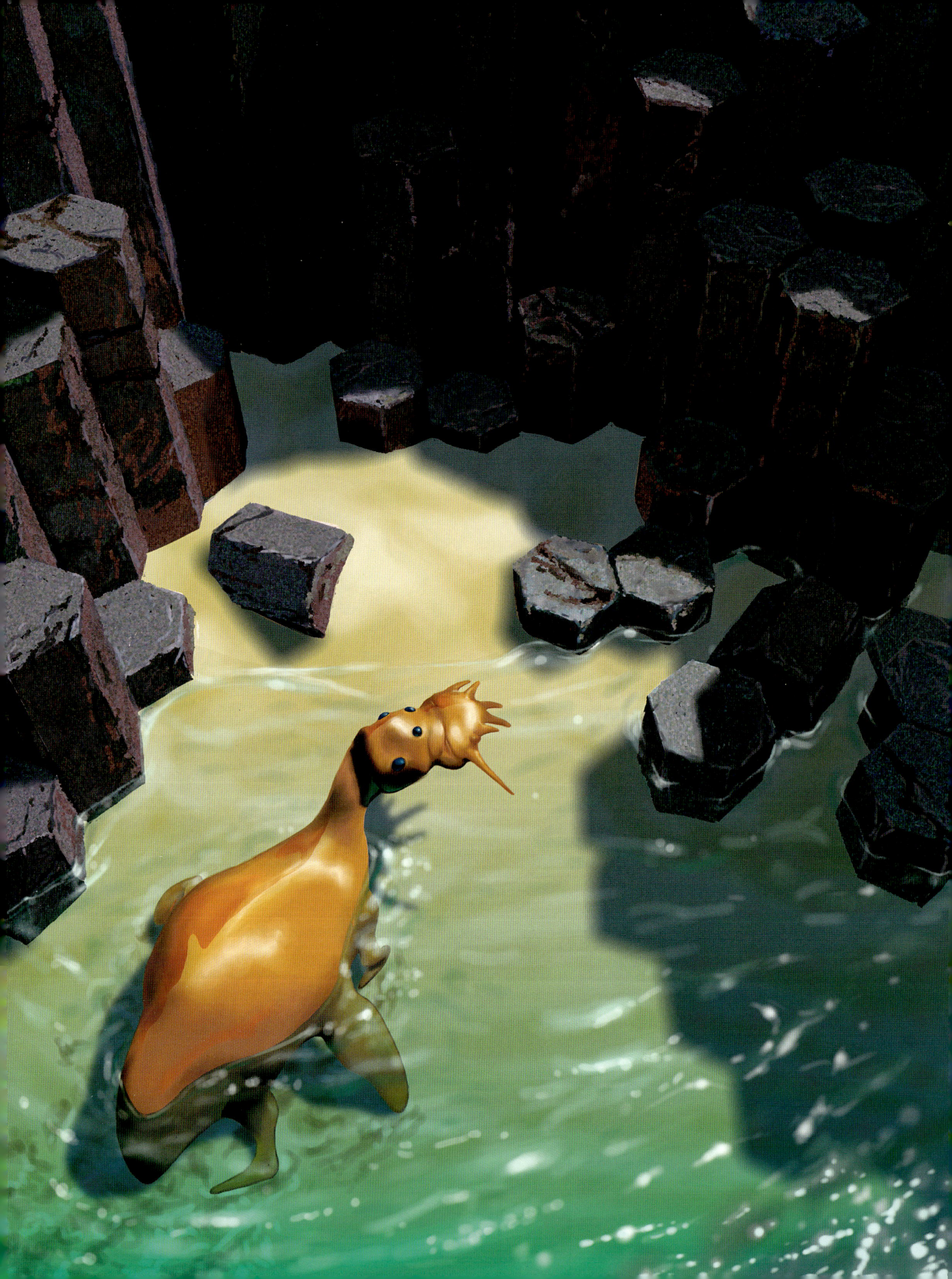

A Wetsloth Out of Place

Tardabarba succulenta

Like Earth seals, Furahan seals are descendants of terrestrial animals that returned to the sea. They responded to the demands of swimming in the usual and universal way as they became streamlined. Again, like Earth's seals, the animals are not fully aquatic, because they breathe air and procreate on land. Their legs must therefore function in water as well as on land. Furahan protoseals had it easier than their Earth counterparts, as they had legs to spare, in a way. The first and third pairs of legs can be swung directly underneath the body and have two functional toes. The remainder of the foot has become a flipper, bent backwards while walking. The second pair of limbs is hardly useful on land, although you can still see stubs of its toes. These legs are now agile flippers, placed almost near the perfect attachment point, one third along the spindle-shaped body.

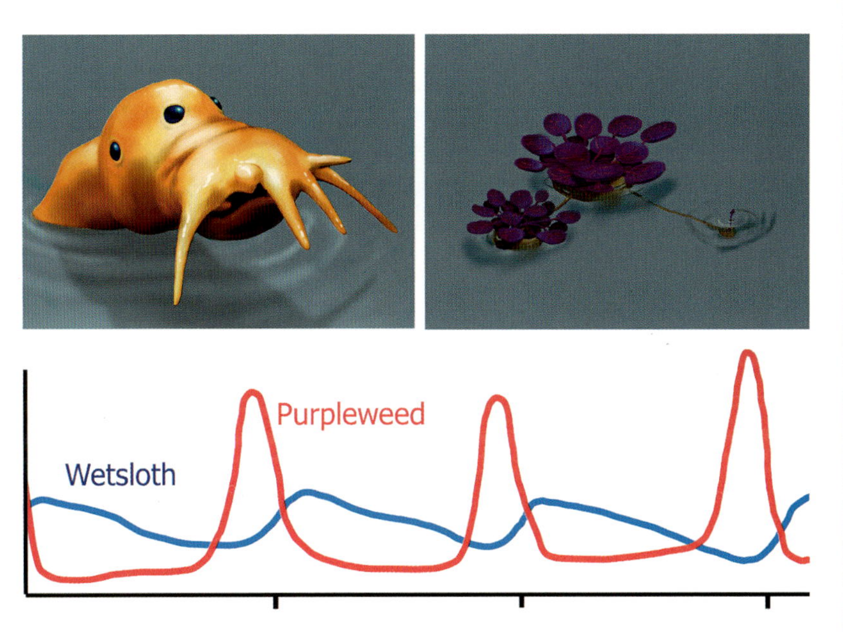

▲ **Population Dynamics**
The lower graph shows the estimated biomass of wetsloths and purpleweed over a period of some 10 years in the lower basin of the Palumbo River. Both populations undergo swings spanning several years. The relation is complex, as wetsloths eat other plants besides purpleweed and, in turn, purpleweed is also eaten by other herbivores.

Oscillating Numbers

Wetsloths are herbivores, usually found in swamp areas of tropical river estuaries, where they lead a peaceful existence processing the never-ending supply of purpleweed, a drifting water plant. If purpleweed growth is unchecked, these plants can clog waterways and cut off the light from deeper levels of the rivers. Wetsloths and purpleweed usually balance each other, so the waterways stay open and wetsloths do not starve.

◀ A wetsloth looking in the wrong place.

Occasionally the numbers oscillate out of control. Oscillation is quite natural in any feedback system with a built-in delay. In this case the delay is likely caused by the different duration of reproductive cycles of sloths and lettuce. The oscillations wax or wane in amplitude until the swings culminate in either a Weed Glut or a Sloth Plague.

On the Lookout

In case of a Sloth Plague, overpopulation causes sloths to venture out into the sea. They are not really suitable for a marine life and will quickly turn toward the shore, looking for an agreeable swamp or clean river not yet teeming with other munching sloths.

The particular wetsloth in the image is looking for a new home in the wrong place. The basalt columns are devoid of life, showing that this tidal zone is subject to heavy pounding by the waves. This environment is much too demanding for the soft-bodied wetsloth.

▶ **Wetsloth**
Tardabarba succulenta
Name derivation: Tardabarba: from *tardus* (L.), slow, and *barba* (L.), beard; *succulenta* (L.) full of juice
Habitat: Warm swamps, slow-flowing rivers
Distribution: Most numerous around Palumbo and Marsu rivers, elsewhere in smaller colonies at river estuaries in Auralgia
Mass: 2–40kg
Length: 60–90cm
Vocalisation: Above water hissing whistle, below water elaborate range of sounds, mostly splotting or whorbling in nature
Peculiarities: Slow walker on land, but reaches surprisingly far into interior

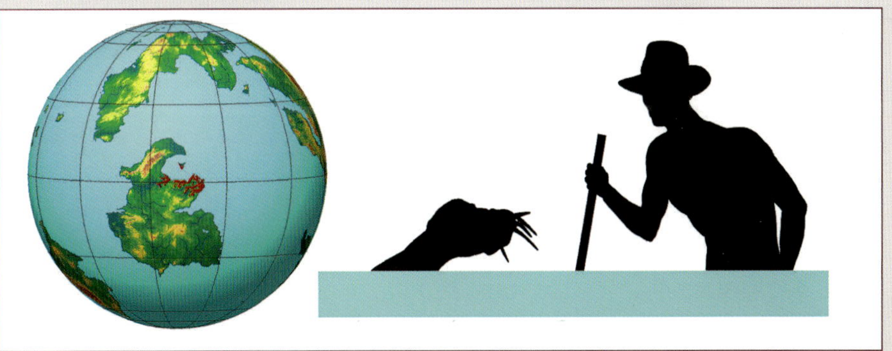

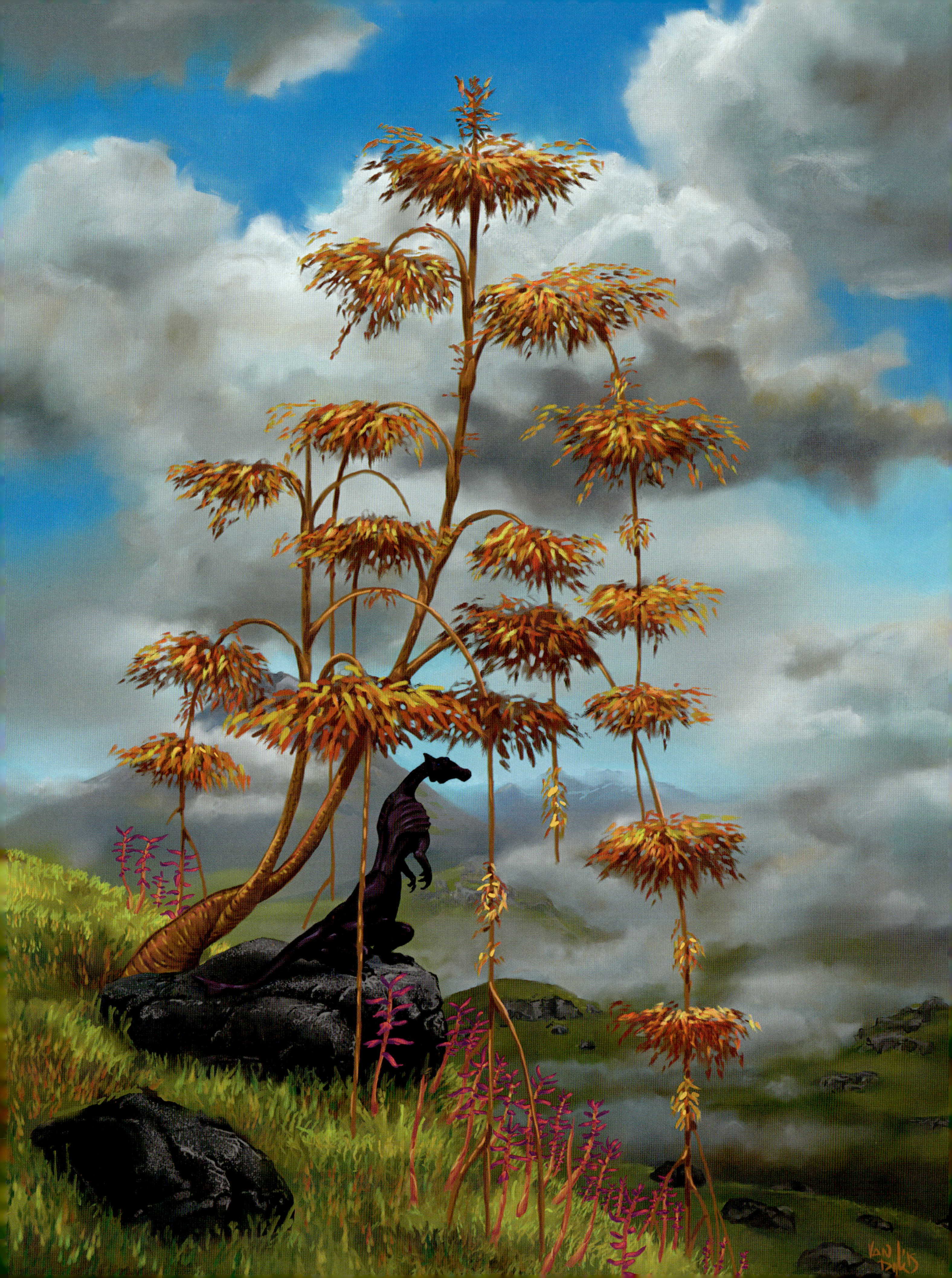

Snafe

Factotum sequax

The animal standing under the umbrella tree is a snafe, a remarkable animal for several reasons. Snaves (the plural is not 'snafes') have various anatomical peculiarities resulting from a very long time of isolated evolution. As such, it is a typical example of a Meralgian hexapod. As was explained in the planetology section, Meralgia has not been in close contact with other large land masses for some 180 million years. This period saw a fairly rapid spurt in the evolution of terrestrial hexapods. An Earth analogue might be the transition of reptiles into mammals. On Furaha, one anatomical scheme came to dominate all of Furaha.

All of Furaha? Only one small continent holds out, where hexapods with different anatomies evolved. The departures from the main pattern are not necessarily better or worse, just different.

Equal but Dissimilar

The snafe is a good example of the differences between Meralgian and main groups of hexapods. For one thing, snaves have tails, whereas mainstream hexapods do not. Snafe legs move in a different manner, although that is not apparent in this view, where the animal crouches on the ground.

There are similarities too. Snaves use only four of their six limbs for locomotion, freeing the front two legs for other uses, here manipulation. The principle of freeing front legs from locomotor use is called 'centaurism'. On Earth centaurism is common, with kangaroos and humans as mammal examples, but many arthropods are centauric too, having put front legs to other uses, usually involving eating. But centaurism is not a characteristic that sets Meralgian hexapods apart. Most Meralgian hexapods are not centauric, while almost all mainstream predatory hexapods are. A typical Meralgian trait is that the freed front legs often show polymery, meaning that there are more segments than in locomotor leg, so the legs become nearly tentacular.

Snafe Intelligence

Snaves are agile and swift. They can dig as well as they swim, dive, run, climb and jump. Snaves digest plants, tubers, mixotrophs, small animals and carrion. They can grasp objects with their flexible forelimbs. In short, they are agile non-specialised omnivores with free grasping forelimbs. According to one theory concerning the evolution of intelligence of mankind, this combination should ensure the emergence of intelligence, but snaves do not support this idea. In fact, intelligence in the snafe's case is more conspicuous by its absence than by its presence.

Umbrella Tree

The umbrella tree shows a peculiar Furahan tree trait – its branches grow down to the ground where they anchor themselves by forming roots. These anchored branches send up new stems on their own, which in time will also send out new parabolic stems and so on. The umbrella tree also exhibits a 'clumping' leaf pattern, in which leaves grow in clumps at regular intervals along the stems. This growth pattern is not peculiar to Meralgia, suggesting that it arose long before Meralgia became isolated.

◀ Snafe pondering life.

Getting Across

People tried to prove snafe intelligence several times. The zoologist Akkaatta Szilagyi spent about ten years studying the subject before she disgustedly moved on to megarusps. She considered her snafe years 'the biggest mistake of my life,' writing the following:

> It is not uncommon to find a snafe standing still on a hillside. In the beginning of my studies, I wondered whether it might be contemplating the nature of its past or future actions. Later it dawned on me that a snafe staring into the distance is probably not lost in thought; it usually is just lost in the general sense.

For acute observations on the subject of snafe intelligence, Szilagyi's concise remarks are unparalleled, tainted though they may be by personal discomfort. As her pupil Oleksandr Kokkinopoulos later recounted in *Speaker for Akkaatta* (p.19):

> 'Sasha,' she said, 'when you are trying to convince yourself that drooling can be construed as an attempt to communicate, you know it is time to reconsider alien as well as human intelligence.'

▶ **Snafe**
Factotum sequax
Name derivation: *Factotum* (L.) (literally 'fac totum' or 'do all'); *sequax* (L.) (meaning 'supple')
Habitat: Temperate to warm hills and mountains
Distribution: Meralgia
Mass: 36kg
Length: 150cm (includes tail)
Vocalisation: Very vocal, from mewing and humming to cough-like bark and eerie howls, particularly at night around human campsites

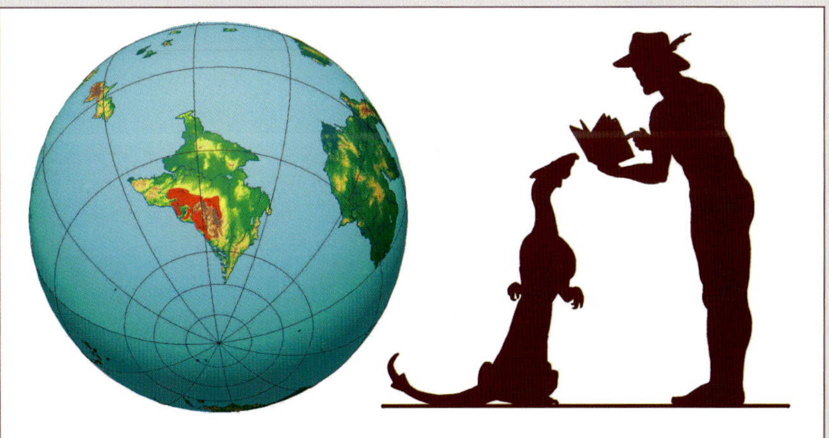

Swobbler

Popino inmodicus

Swobblers are little furfeathery animals, living in tropical woodlands. They are not highly specialised except for their feet, which have large sideways-pointing flexible toes, so they can grasp branches with ease.

A-128
50 Km
'Pointless Mountain' 1736 meters
'Nohome Hills'
'Stinking Feet Forest'
'Footrot Fields'
'Morehills Hills'

▲ Furahan geographers learned the hard way to only add names to their maps when arguments about those names had settled. The island where swobblers live goes by the indication 'A-128'. The names shown here are invented by expedition members. In this case, it seems unlikely that the names will be accepted officially.

Baignac

The large conspicuously coloured fruit is a baignac. Botanists wondered which evolutionary pressures might drive the baignac tree to invest so much energy in its fruit. The basic reason plants produce fruit is clear. Fruit entices animals to eat them, complete with embedded seeds, and to deposit the seeds, now embedded in faeces, on a suitable spot. What makes botanists think twice is why the fruits are so large – can't the tree reach the same objective at less expense?

A nice thing about purpose in evolution is that the observation usually suggests the answer. Botanists came up with several lines of thinking to explain large fruit.

'Surely,' some said, 'the efficiency of Furahan photosynthesis means that plants do not have to be frugal, pun intended, so they spend their wealth!'

'Perhaps,' others said. 'But surely the answer must take specific fruit eaters into consideration. Large fruit is wasted on small animals such as swobblers, as they cannot eat an entire baignac in one go. The fruit must be meant for large herbivores on the forest floor. Once a baignac falls, the tree may get its seeds into the guts of large animals that are superior dung-spreaders.'

'Pfah,' some newcomers said. 'Competition is the answer. Many trees bear fruit, so the large size of baignacs is simply superior advertising, well worth the price.'

Perhaps all these reasons are true to some extent. At any rate, baignacs are rather large, for which there must be a reason.

To Eat or to Sate, that is the Question

Swobblers eat almost anything. Like all animals they must balance the energy spent to acquire food with the energy yielded by that food. The best balance is to get paid without doing any work, but in a jungle that wish is rarely granted. There are many fruit trees but also many frugivores. Adding more fruit simply results in more frugivores. Predators quickly learn that fruit trees offer fruit eaters, so life is not a tropical paradise for swobblers. Still, swobblers hardly ever starve.

Do not Mess with Nature

At one point in Furahan history, humans allowed themselves to treat animals as pets. Swobblers immediately exploited human cuteness instincts and begged for free food. Unfortunately, their eating and satiety instincts are not well-matched, so pet swobblers stuffed themselves until they weighed twice as much as wild swobblers. Luckily, human culture swung back to non-interference, forcing swobblers to fend for themselves once more.

In nature, where they belong, swobblers are considered delicacies by almost all carnivores. Marblebills will kill to snap up a nice juicy specimen and usually do. If swobblers are not careful, their existence is just a short interval in the turnover of energy from fruit tree to predator.

► A swobbler enjoying a baignac.

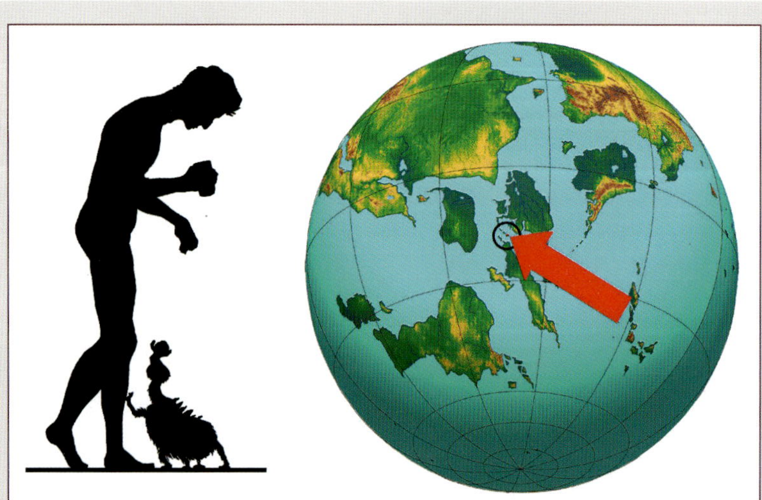

◄ **Swobbler**
Popino inmodicus
Name derivation: *Popino* (L.) (meaning 'glutton') *inmodicus* (L.) (meaning 'immoderate')
Habitat: Arboreal, tropical jungle, mid-level to high canopy
Distribution: Archipelago
Mass: 1–2kg
Length: 25cm
Remarks: The distinction between species and sub-species is unclear. The form *Purpuropennatus* shown here has only been seen on A-128.

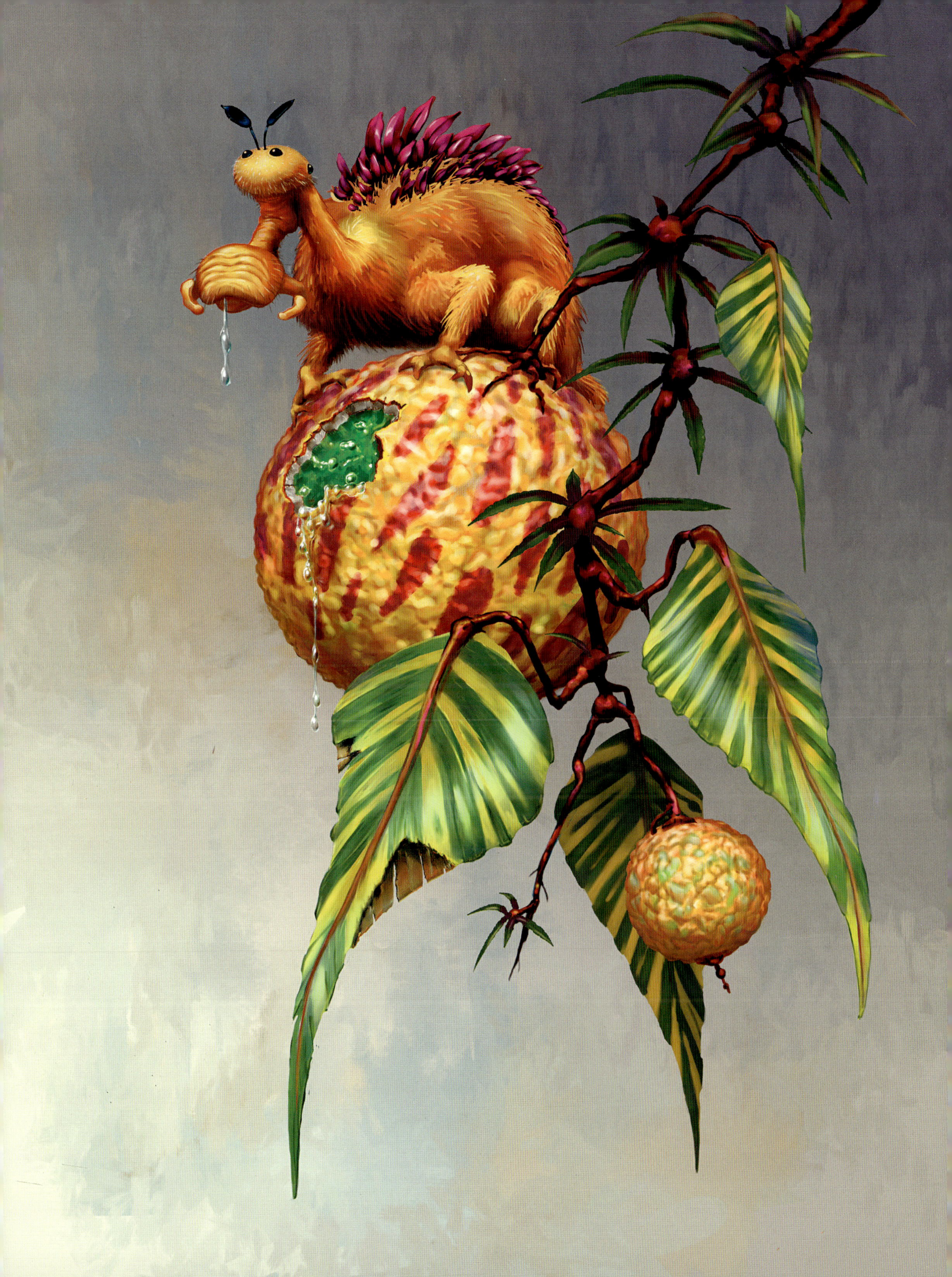

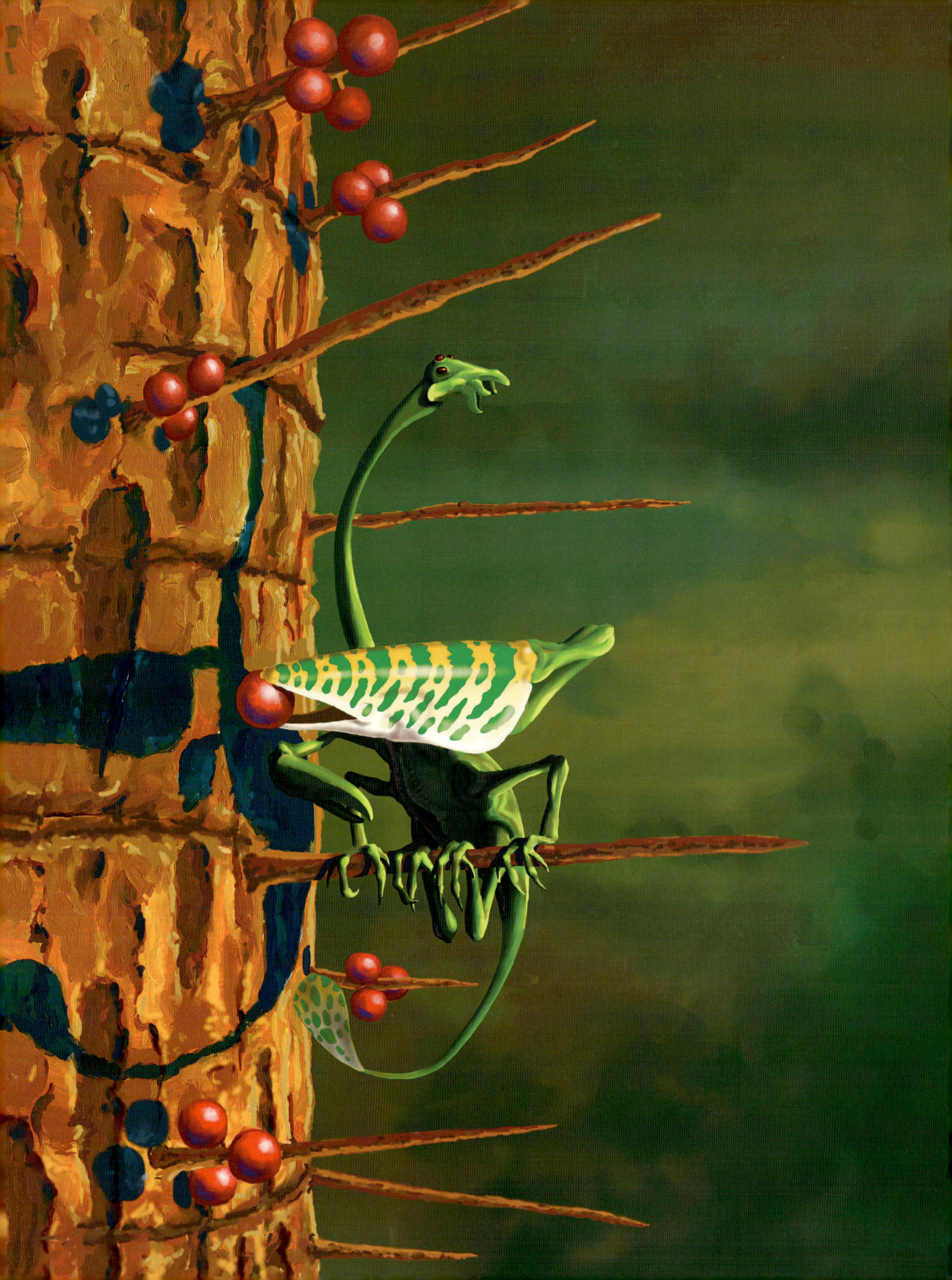

A Conspicuous Dandy

Callopistes poikilocheirus

For most of the year the dandy is an unassuming inhabitant of tropical forests, where it is found hopping on branches or clinging to tree trunks.

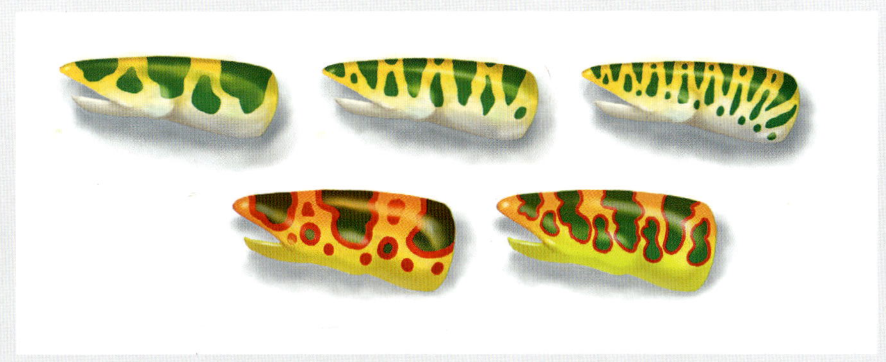

▲ Colour Pattern Formation
This range of claws, sampled from a large number of dandies, shows the variability in claw colour patterns. The variations are probably due to two gene complexes. One determines how large the spots are. If expressed early in embryonic life, the animal gets a few large spots, while later expression results in numerous small spots. The other complex determines colour itself: messenger molecules diffuse outwards from the centres of the spots and different concentrations cause different colours.

Divided Digits

Its sharp nail-like digits are very reminiscent of Earth vertebrate fingers and toes, in that they consist of linked segments. They depart from Earth designs in their branching pattern. On Earth, digits such as human fingers all branch from the wrist at the same level, but fingers do not split from fingers. This does happen on Furaha however. Some of the dandy's digits split from other digits, which is well visible in this particular image. The wrist splits into two digits and these then split into two branches each.

Claws to Grasp Attention

The two front claws have no digits but are interesting for another reason. In males, one claw grows much larger than the other, which has no mechanical advantage. After all, female dandies do perfectly well with two equally-sized claws, and males even prefer their smaller claw to grasp berries. For most of the year the giant claw is as green as the rest of the dandy, allowing the animal to remain inconspicuous in the forest. But once a year the claw becomes brightly coloured, coinciding with ripening of the berries on ruberry trees. Ruberry tree trunks sprout a large number of berries, conveniently placed on short branches. These berries provide dandies and other frugivores with an abundant food supply. Once the males' claws are fully coloured, the males leave the protection of the leaf coverage and seek out open spots where they wave their claws about, hoping to attract females. They may first have to fight off other males with the risk of being evicted. The greater risk is being spotted by a more sinister forest inhabitant, such as a marblebill or becdacier.

Paying the Price

The theory behind all this is that only the fittest of individuals can afford to be conspicuous, so standing out must indicate good health. Obviously, there are costs. Not only does growing a claw increase the male's energy bill, but the main cost is as simple as it is horrible – death. It is no wonder that males on the lookout inspect anything that moves. Is that a female, is it another male, or is it, please let it not be so, a becdacier?

◄ Dandy on the lookout.

► Dandy
Callopistes poikilocheirus
Name derivation: *Callopistes* (Gr.) (meaning 'dandy'); *poikilocheirus* (Gr.) (meaning 'many-coloured hand')
Habitat: Tree dweller in tropical forest
Distribution: South Palaeogaea
Mass: 2kg
Length: 30cm
Diet: Mostly fruit, but larvae and other small animals are readily eaten

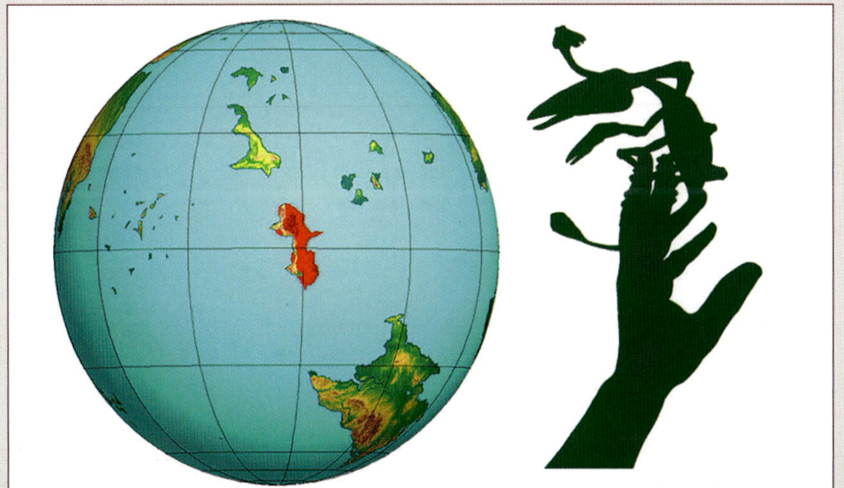

The Ornery Ochreback Thresher

Ira tarda

Threshers are solitary. This is worth celebrating, as a meeting with just one is already quite unpleasant. Some say that threshers attack without any provocation, but this is unjust. Thresher attacks are always provoked, for example by existing in their presence. The provocation need not cause an immediate response, but thresher anger builds up slowly and steadily. The illustration shows an early stage of anger, meaning a slowly simmering thresher. So far, the animal has not done more than kick dust around. Once its threshold is crossed, a thresher will methodically attempt to erase the cause of the disturbance from their vicinity in particular and from existence in general.

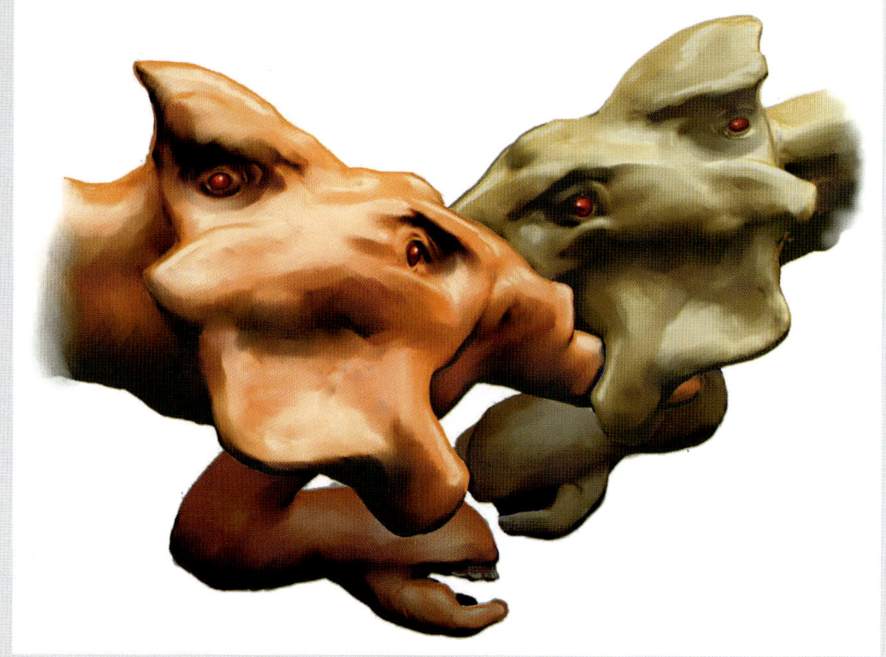

▲ **Nose Grinding**
This sketch shows how aggression shaped the horn and facial ridges. One thresher's nose horn pushes up against the side of another thresher's head, while its own shield traps the other's nose horn. Once in this mutual position, the combatants try to move their nose horns upwards, while resisting the other's efforts. The eyes are surrounded by bumps and ridges, but even so quite a few threshers lose an eye during such contests.

Wide Stance

Threshers are typical Megatheria, large hexapod herbivores with head and body adornments. In Megatheria the middle pair of legs is much stouter and has a wider stance than the other pairs. This allows the massive gut to be placed near the centre of gravity; it largely defines it, in fact. The front and hind legs can move between the middle ones, preventing threshers from stumbling.

The Jaw Apparatus

Thresher jaws finely evolved to grasp plants, with the lateral jaws pushing fodder inside, so the upper and lower jaws can cut stems. In many hexapods the upper head, housing ears and eyes, is never used for display, but thresher evolution went the other way.

Grumpiness as an Art Form

Threshers dislike field scientists, perhaps not altogether surprising, as these scientists have a habit of shooting threshers with sampling darts. Threshers even dislike one another and avoid contact. At their territories' borders, they must test one another's strength on occasion. When dust-kicking fails to settle the contest, horns lock, or rather, noses. Looking for a mate is a tricky affair, as it requires that threshers overcome their considerable grumpiness. What happens then and there is better left unsaid.

▶ A grumpy thresher.

▼ **Ochreback Thresher**
Ira tarda
Name derivation: *Ira* (L.) (meaning 'wrath'); *tarda* (L.) (meaning 'tardy or slow')
Habitat: Plains, savannah, open forest
Distribution Western coastal area of *Imparia Septentrionalis*
Mass: 3000kg
Length: Up to 5m
Diet: 95% vegetarian, but does consume animals within easy reach, such as fresh carcasses
Vocalisation: Usually silent. Grunts only when threatened
Social habits: Usually solitary (*see* main text)
Remarks: May be fooled into not attacking by confusing it with suddenly appearing colourful objects (black umbrellas are useless)

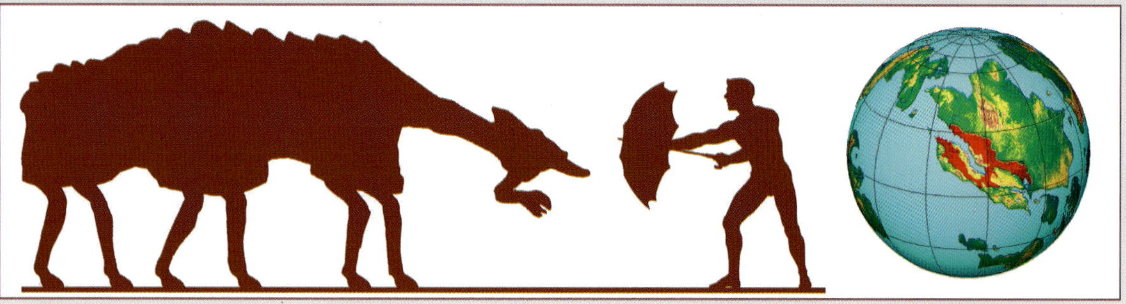

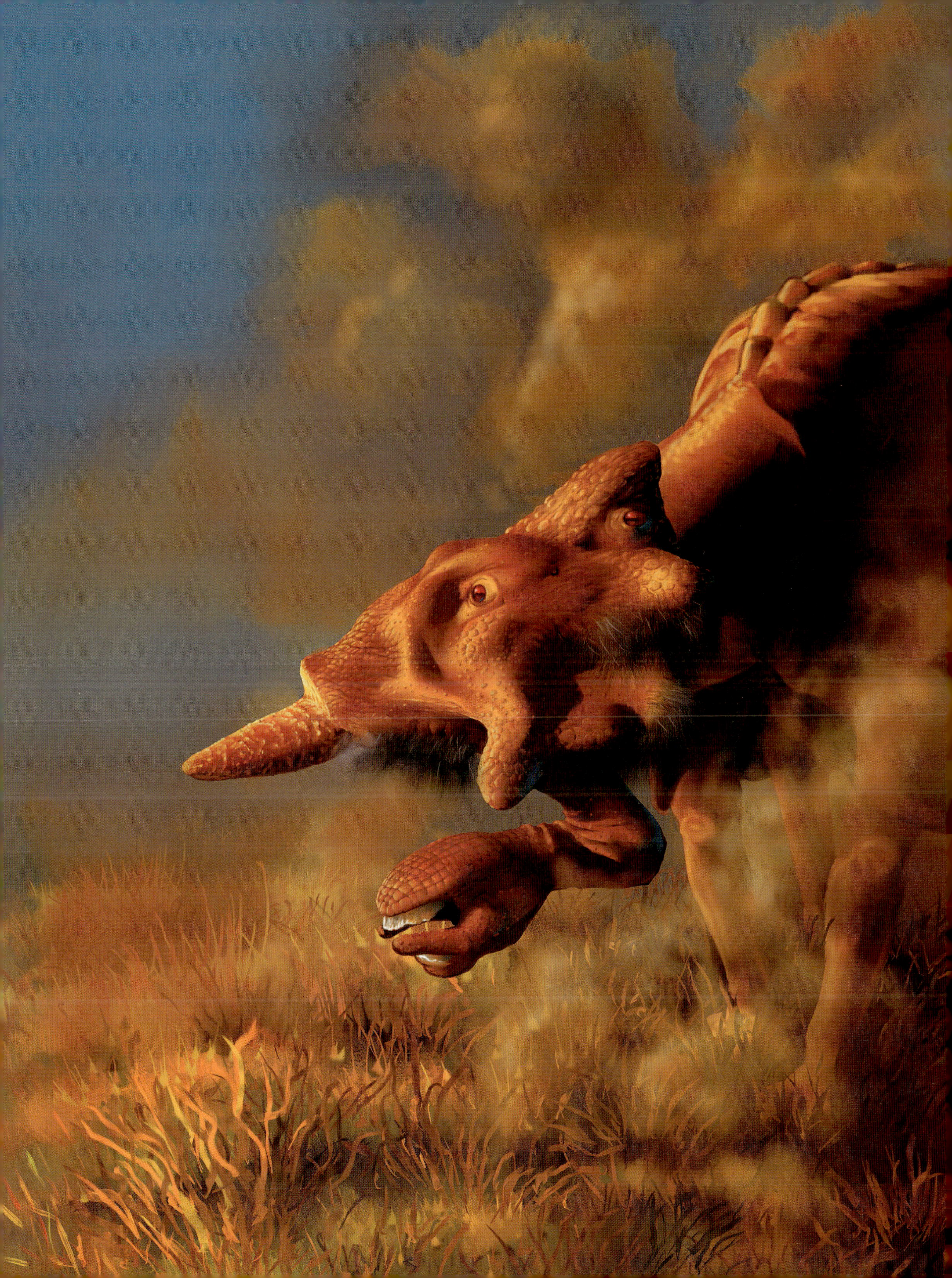

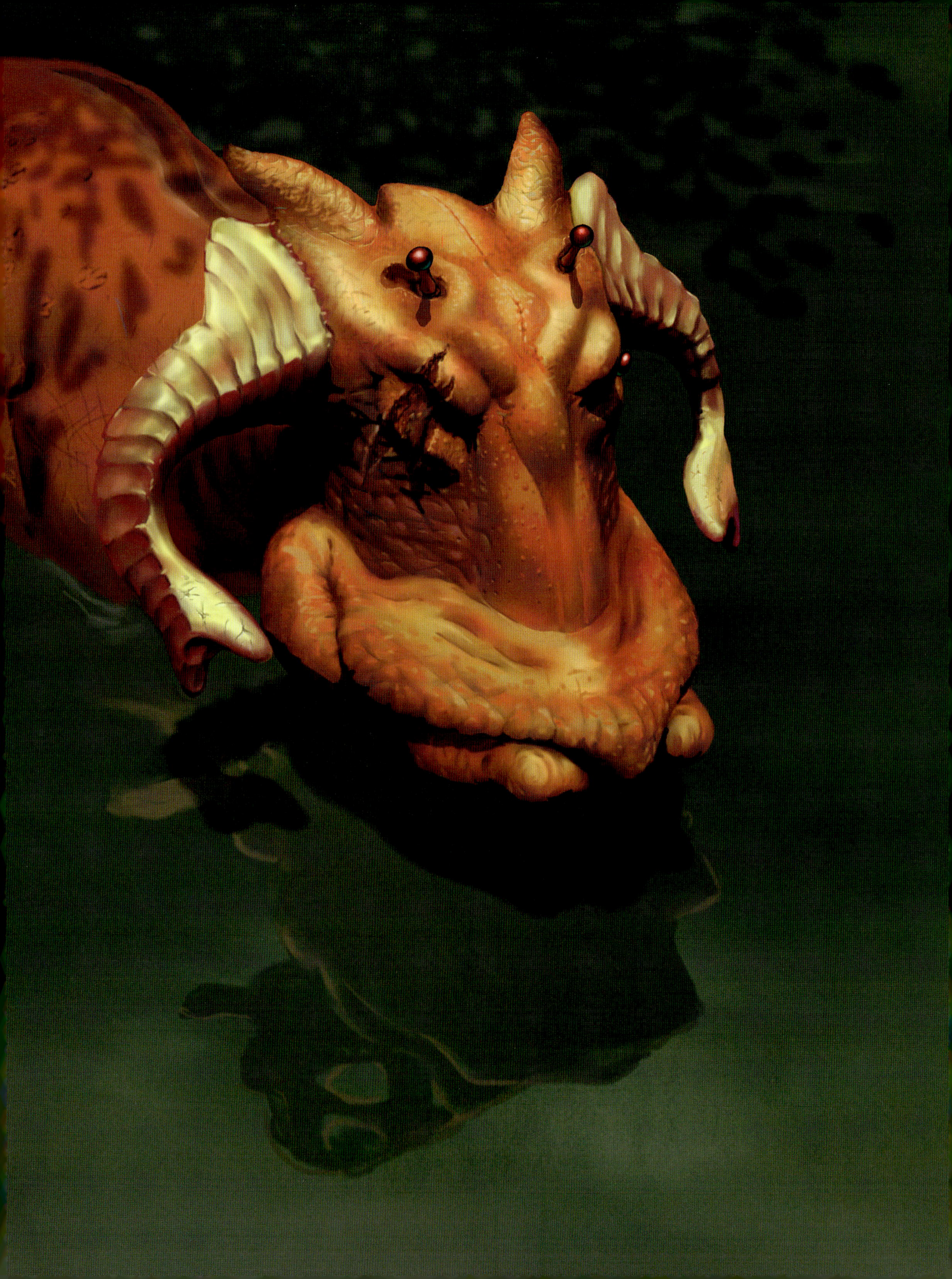

Marshwallow

Falcifer inacetum

The marshwallow is amphibious. The top of the head protrudes vertically, allowing the jaws to deal with the vegetation while the eyes keep a clear field of vision. The aft eyes take this even further, as they are placed on stalks high on the head. But in other aspects the genus *Falcifer* (meaning 'scythe bearer') does not show adaptations to an aquatic existence. The lateral jaws (the scythes) move sideways and help to pull marsh plants into the mouth. This works as well with drier vegetation. The broad feet aid swimming, but also help the animal not to sink into mud.

Marshwallows live in groups called congregations. They usually settle a bend in a stream for many years and flatten the undergrowth, creating wallow meadows that attract other herbivores because they provide a good spot to look out for predators.

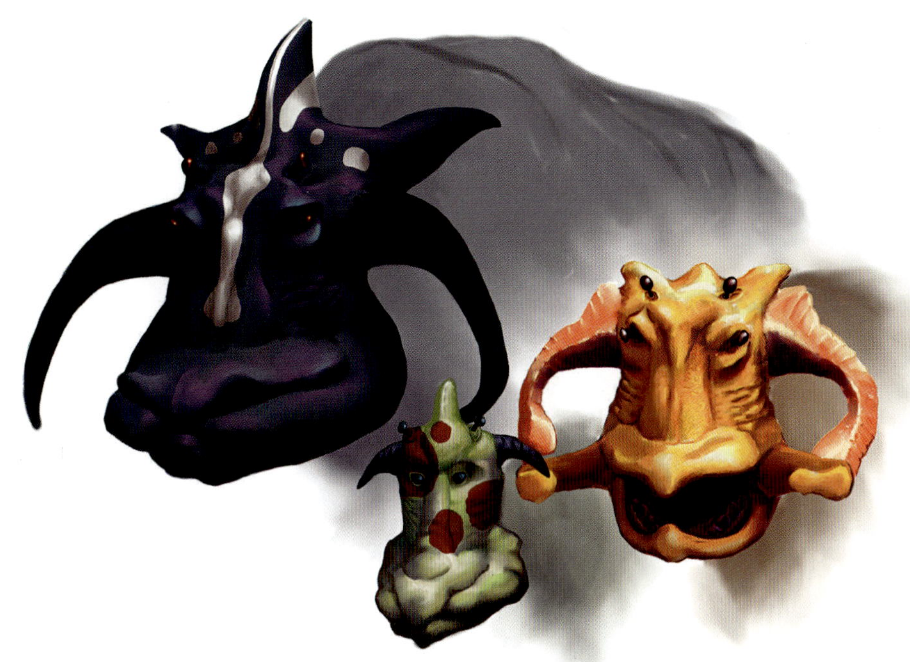

▲ From left to right: blackgammon, gna, marshwallow. The marshwallow on the image opens its mouth so the lateral and median jaws are well visible.

A Character

The marshwallow's head is impressive, with its wide mouth, fringe at the back of the head and huge lateral horns. To human eyes all this suggests a ferocious animal, but the animal is not given to extremes of emotion, making marshwallows boring as well as a bit, well, dense. Marshwallows live in an unsteady truce with sawjaws (*Serrabucca sp.*). Experienced sawjaws will never approach a congregation, but a young sawjaw may mistake a marshwallow whelp for an easy meal. The congregation converges on the sawjaw and bites and kicks it. Apart from such altercations, the marshwallow's physique serves mainly to impress other marshwallows. Their courtship involves much snorting and mudslinging.

◀ A marshwallow bathing.

Adaptive Radiation

Present-day wallow species have evolved different ways to exploit their environment, such as adaptive radiation. The hypothetical 'protowallow' must have looked much like the 'spotted gna' (*F. interstinctus*), the head of which is shown in the foreground above. It probably fed by digging for roots and tubers. The spotted gna lives near streams and lakes, but is seldom found in water for long periods.

The marshwallow in the image opens its mouth so the lateral and median jaws are well visible.

The blackgammon (*Daemon oculoscintillus*) is the largest wallow species, with the most exuberant neck shield. Blackgammons are too large to be considered suitable prey by the more common predators. Perhaps for that reason they move onto dry land more often than marshwallows. They certainly act rather less humbly than the meek spotted gna or the immobile marshwallow. Never take a blackgammon for granted.

▶ **Wallows**

Names and derivations: Blackgammon *Daemon oculoscintillus* (*Daemon* (L.) meaning 'devil' and *oculoscintillus* being a concatenation of *oculus* (L.) meaning 'eye' and *scintillus* (L.) meaning 'spark')

Marshwallow *Falcifer inacetum* (*Falcifer* (L.) meaning 'carrying a scythe' and *inacetum* (L.) being an antonym of *acetum* meaning 'sharpness of wit', so dim)

Spotted gna *Falcifer interstinctus* (*interstinctus* (L.) meaning 'spotted')

Species: At least eight, but some are fiercely disputed

Habitats: Amphibious to varying degrees, but no species is fully aquatic. Most are hippopotamoid in nature

Distribution: The globe shows combined area of the three depicted wallow species. The map shows overlapping distribution per species

Social habits: Communal, from family groups of six individuals to congregations of about 60 adults

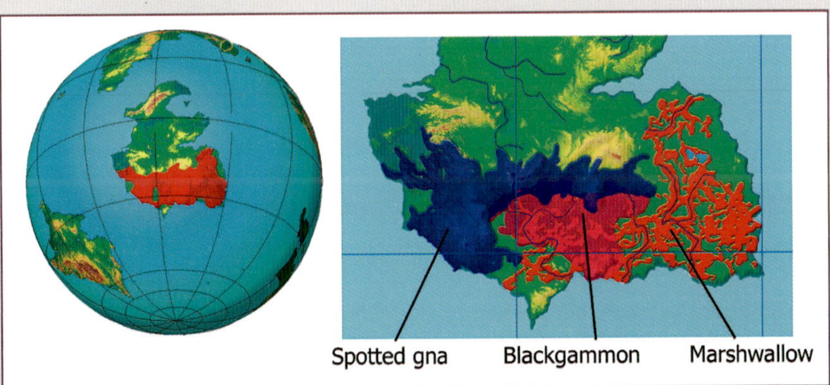

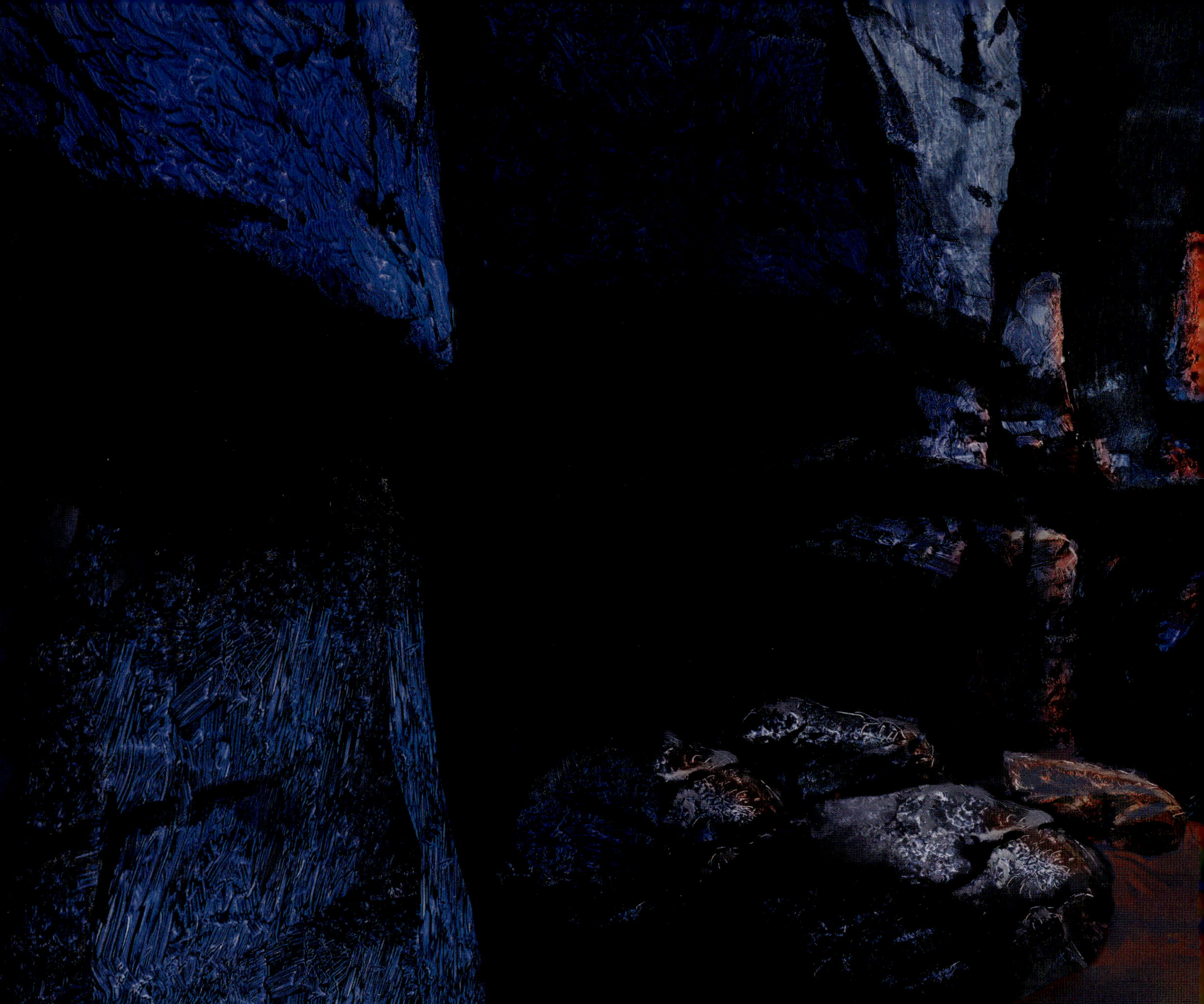

Desert Ghost

Pygargus vepallidus

Furaha's two largest deserts are each larger than Earth's Sahara. They are found on the two Imparia continents, both Septentrionalis and Orientalis. A sizable part of these continents spans a latitude where little rain falls anyway and the sheer size of these continents means that clouds, floating inland, will have evaporated or dropped their rain long before they reach the interior. The result is a succession of habitats shaped by a progressive lack of water. Life has adapted, up to a point, and includes plants enclosed in waxy cocoons. The deepest desert is virtually devoid of life.

The Habitat of the Desert Ghost

The desert ghost shown here certainly cannot survive in the dead deep desert. The presence of open water shows that this is not the inner desert at all. The scene is situated somewhere in the foothills of the eastern mountain range of *Imparia Orientalis*, where some water trickles west, percolating through the mountains and forming an occasional spring here and there.

In the scene shown, a trickle of water flows out of a sheltering rift to an open pool, surrounded by a few plants. The deep desert is still some distance away, but even so, the next open water in the west is the Intermediate Sea, some 5000km away.

Nocturnal Encounters

The desert ghost is one of the larger groups of hexapod herbivores with similarities to Earth's deer, called 'talopi', the plural of 'talopus'. Some consider this correct plural form an affectation and speak of 'talopuses'. Anyway, talopi vary in size from cat-like to camel- or hippopotamus-sized animals, and inhabit a very wide range of habitats, from polar ones through jungles to deserts.

The desert ghost is pale, active at night and walks soundlessly. Many a student on a field trip has been startled at night to find a desert ghost just a few metres away, curiously studying the student.

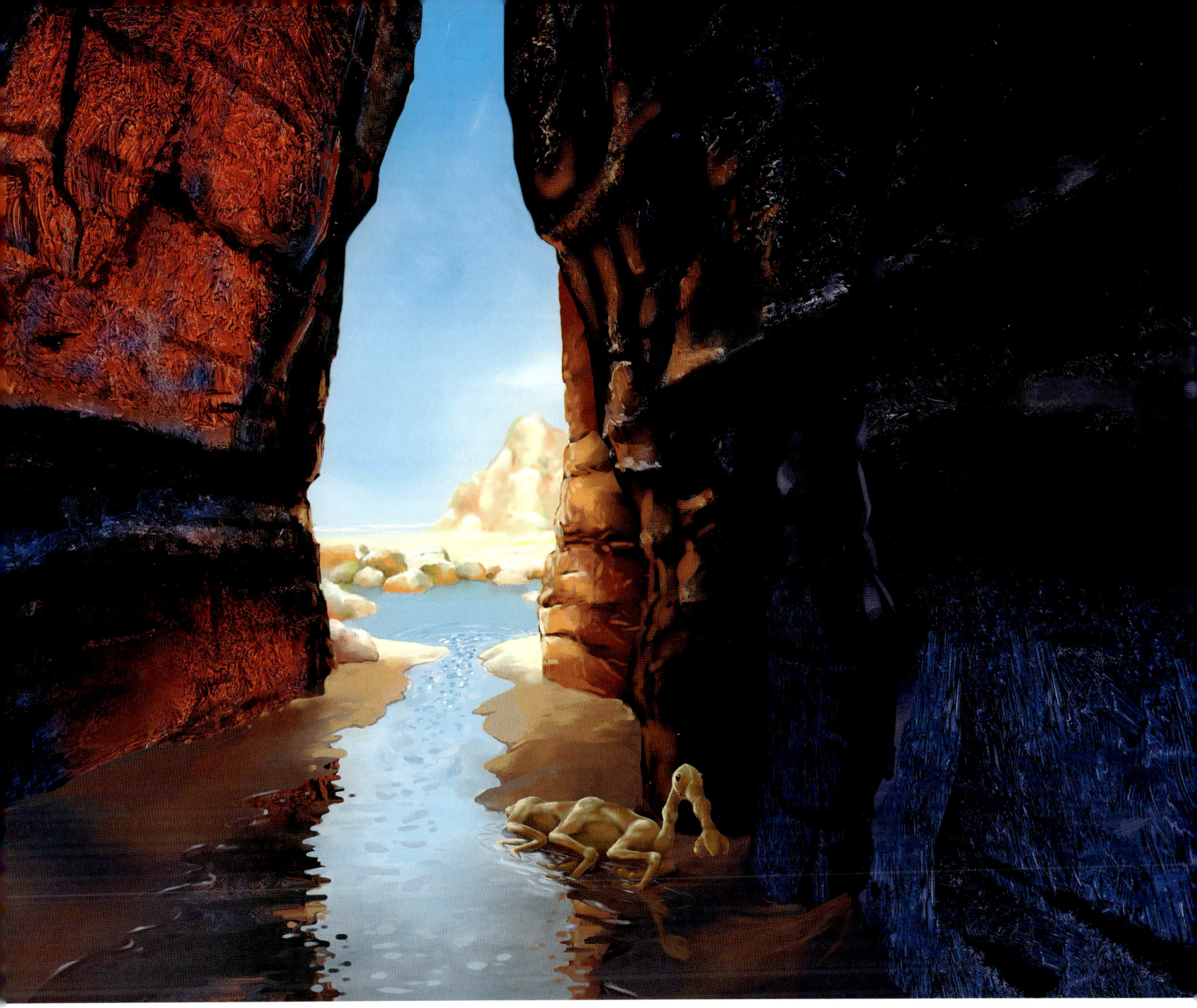

▲ Ghost cooling off.

► **Desert Ghost**

Pygargus vepallidus

Name derivation: *Pygargus* (Gr.) (meaning 'with white end of rump'); *vepallidus* (L.) (meaning 'deathly pale')
Habitat: Desert with some rainfall
Distribution: Eastern margin of *Imparia Orientalis*. Only one species in the genus *Pygargus*
Mass: 280–250kg
Length: 2.5m
Diet: Plants
Remarkable traits: Mostly nocturnal. Extremely silent thanks to cushioned feet. While the rump is as pale as the scientific name of the animal suggests, the rest of the animal is no less pale

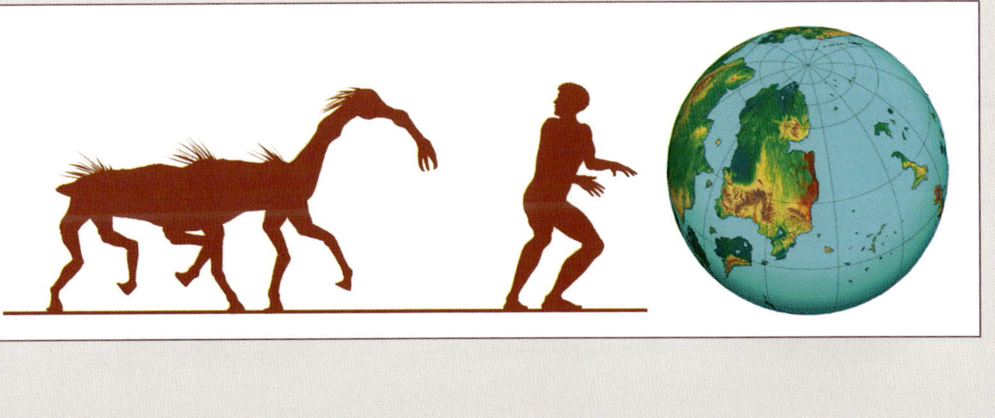

The Rufous Woolly-Haired Shuffler

Brutum rutrum

Shufflers are large herbivorous hexapods, living on the tundra around the North Polar Sea and in Northern Snow Forests. They are well adapted to this cold and bleak environment, where low temperatures and meagre food supplies pose difficulties for all organisms.

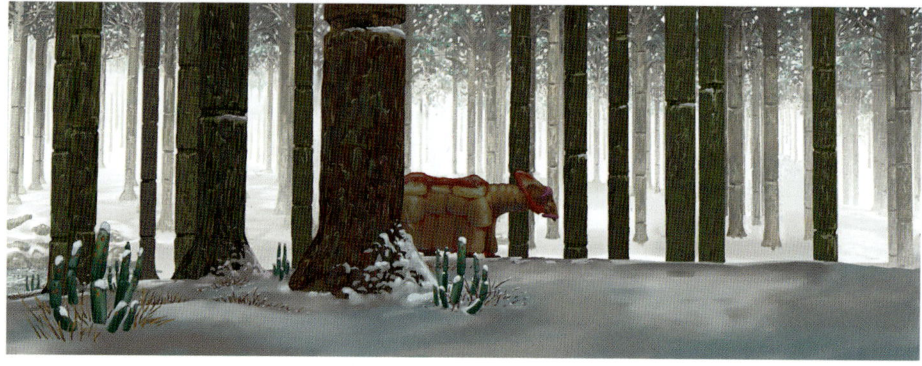

▲ The Northern Snow Forests are gloomy places, dominated by few tree species, with column trees the most common one (*Orthostatis miguelanxi*).

A Beast for All Seasons

In the summer, the tundra thaws and is transformed into marshland alternating with soggy meadows. Stony fields with dry barbgrass occur on hillsides. The shuffler's broad feet allow it to remain afoot in marshland as well as in the snows of winter. Its lateral jaws are almost immobile and team up with the maxilla to form a functional shovel. During winter, the animal uses its shovel to search for plants under the snow and in the summer the shovel helps the animal to uproot tubers from tundra soil and uproot water plants from lakes.

Splendid Insulation

Shufflers have a unique furry covering to help them withstand the cold of winter. In the autumn, their skin grows large leathery flaps, hanging down over the rest of the skin. These flaps overlap one another and cover the hair on the skin underneath the flap. The outside of the flaps remains naked and hairless. When the flaps have reached their full length, their blood supply dwindles, leaving the skin to dry out and form a thin leathery cover. When winter comes, the animal is equipped with a layer of fur, protected from rain, hail and snow by the dead surface, itself needing no protection. The covered fur functions much as the coating in a human-made coat. In spring, the flaps of skin drop off, but before they do, shufflers walk about with the remainder of the flaps sticking out, looking a bit like gigantic pinecones.

A Parasite Paradise

The disadvantage of this well-insulated fur coat is that it forms a very welcoming environment for parasites. Some parasites, called *trichophages* (meaning 'hair eaters'), live off skin flakes and hairs. They form a nice ecological community, but the story does not end there. Recent studies showed that shuffler skin glands secrete *trichophagocides*, meaning substances poisonous to trichophages. Trichophages responded to this raising of the evolutionary stakes by developing a partial insensitivity. Apparently, the situation has lasted long enough for new arrivals to evolve into tiny predators eating the hair eaters. Accordingly, these tiny monsters are called *trichophagophages*.

You can tell populations of shufflers apart by their parasite populations. Studying these tiny ecosystems is not as hard as you might think, because shufflers are usually not aggressive. However, do not sample their parasite populations in the pinecone stage, as shufflers are then rather irritable – probably because of an itchy skin.

▶ Shuffler drinking.

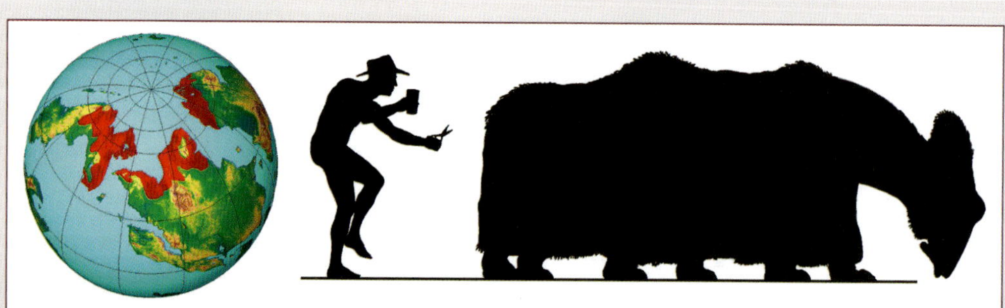

◀ **Woolly-Haired Shuffler**
Brutum rutrum
Name derivation: *Brutum* (L.) (meaning 'beast'); *rutrum* (L.) (meaning 'shovel')
Habitat: Tundra, marshes, ventures into Northern Snow Forest
Distribution: Northern regions of continents circling Northern polar sea. There are four species of shuffler (some think the 'rufous' and 'ruddy' species ought to be combined into a new species, the 'rufescent shuffler')
Mass: 1800–2200kg
Length: 3.5m
Diet: Plants
Vocalisation: Booming calls in autumn and spring

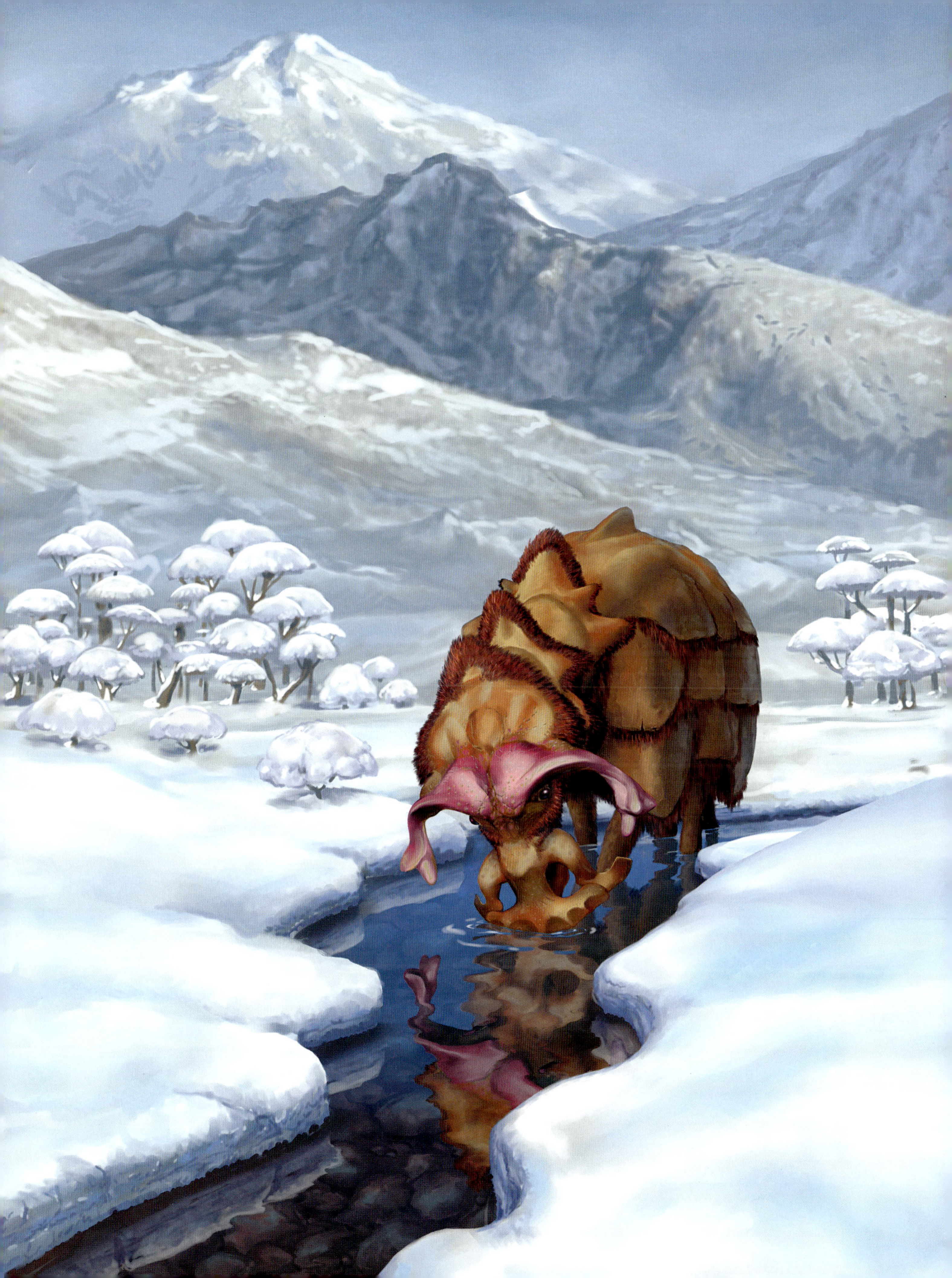

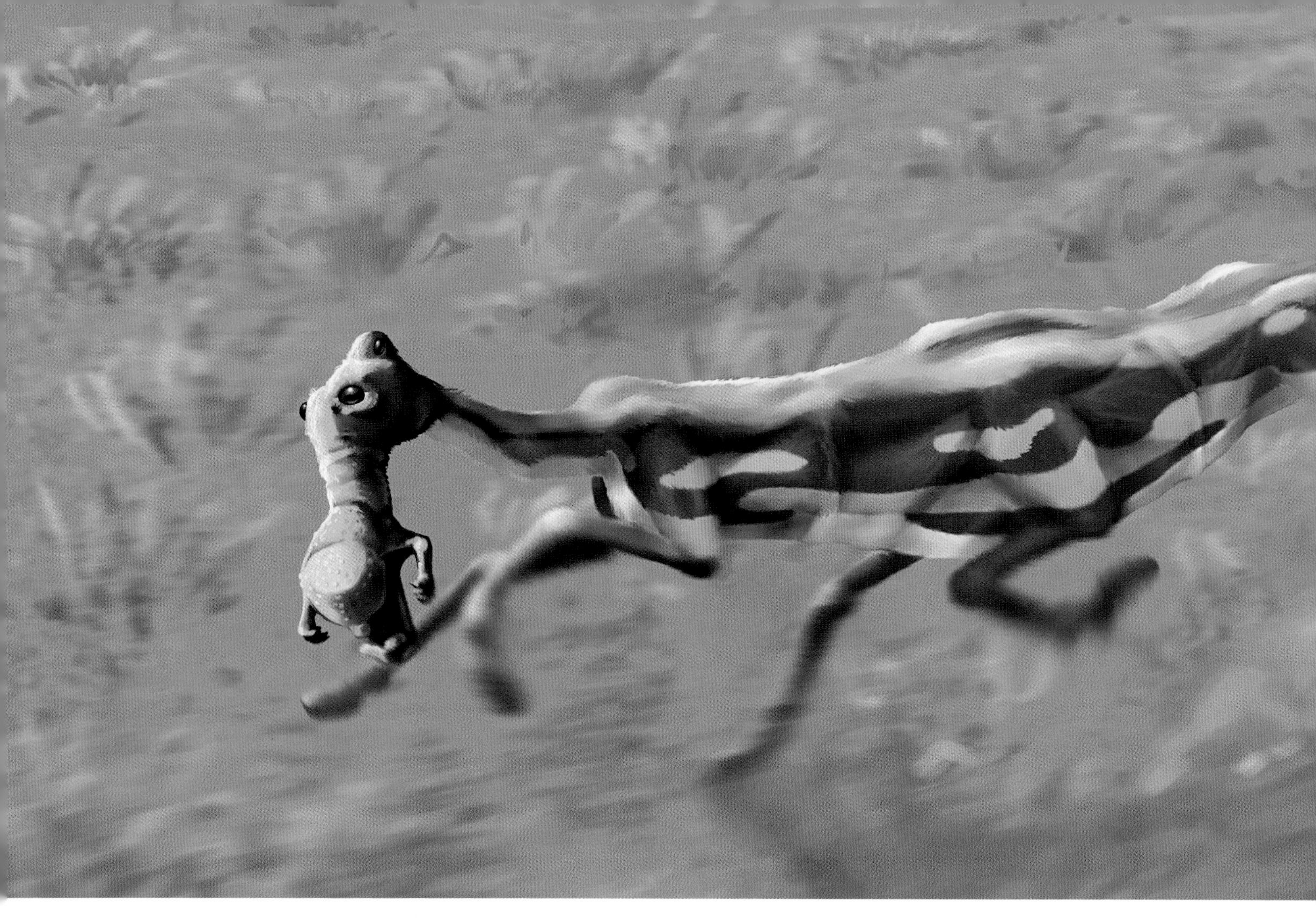

Prober and Bobbuck, Racing

Acerbus acutus and *Swala perceler*

This blurred action scene captures one of the most dramatic events on the Auralgian plains: two of the fastest terrestrial animals of the planet, a prober and a bobbuck, are running a life-or-death race.

Centaurism and Clavigers

The predator, the prober, is a neocarnivore, meaning it hunts with its modified front limbs. These are no longer used for locomotion, a process known in Furahan biology as 'centaurism'. Once freed from weight-bearing, these limbs underwent quick evolution. Neocarnivore front limbs can look like spears, clubs or other weapons. The prober is of the 'claviger' type, meaning it uses its front legs like clubs. These clubs swing sideways and easily bring down a running bobbuck. Once down, the bobbuck's legs are broken to immobilise it. The prober's relatively weak beak is never used for the kill, but only to reach into a carcass and to dislodge lumps of organs that after swallowing are masticated by the internal gnathes (jaws).

Individuals Racing for Survival

The prober is faster than the bobbuck, but not by much, and the bobbuck has more endurance than the prober, but not much more. These differences lie at the root of prober and bobbuck tactics. If a prober starts a full-out attack from too far away, it will not overtake the bobbuck quickly and will have to give up the hunt due to exhaustion. Bobbucks must always keep their distance from approaching probers. Still, no bobbuck can afford to spend all its energy running from probers at first sight. Its best strategy is therefore to trot away while the prober is also moving slowly and to start an all-or-nothing escape run slightly before the prober starts its full-speed attack. Starting a run is therefore a calculated and exhausting affair for both animals. Probers cannot afford to fail too often, as they will then have insufficient energy to succeed ever again. A bobbuck, having escaped, must take great care not to be pursued again shortly afterwards, because it will then easily be overtaken.

An Evolutionary Race

Their evolutionary race has driven the capabilities of both animals to extremes and the differences between success and failure are small. From an evolutionary point of view, a situation like this is not stable. A small difference in performance or behaviour may shift the balance. Both species must adapt quickly or face extinction. 'Quickly' in this respect still takes many generations, but for the animals themselves survival depends on just a few seconds.

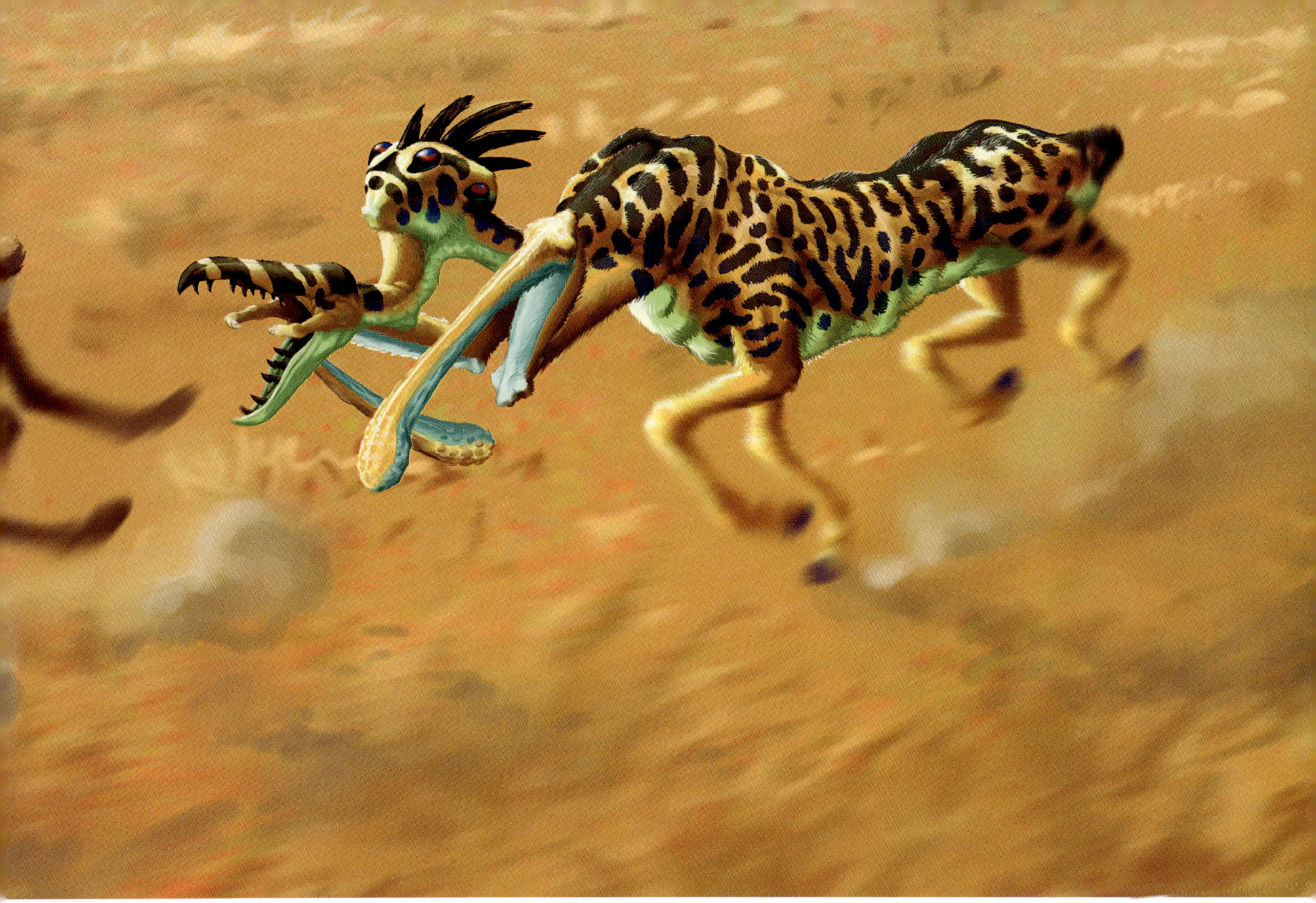

▲ A fast hunt.

▼ **Prober**
Acerbus acutus
Name derivation: *Acerbus* (L.) (meaning 'sourness'); *acutus* (L.) (meaning 'sharp')
Habitat: Plains, open savannah
Distribution: Northern Auralgia. Four sub-species
Mass: 40–60kg
Length: 1.3–1.6m
Diet: Bobbucks! If there are none, any suitably sized animal will do
Vocalisation: Territorial calls at dawn. This is a characteristic high-pitched 'yee-ap' sound, rising in volume at the end of the call and repeated often

▼ **Bobbuck**
Swala perceler
Name derivation: *Swala* (Sw.) (meaning 'antelope or buck'); *perceler* (L.) (meaning 'very fast')
Habitat: Plains, savannah
Distribution: Northern Auralgia
Mass: 60–70kg
Length: 1.3m
Diet: Preferably high-energy shoots and leaves

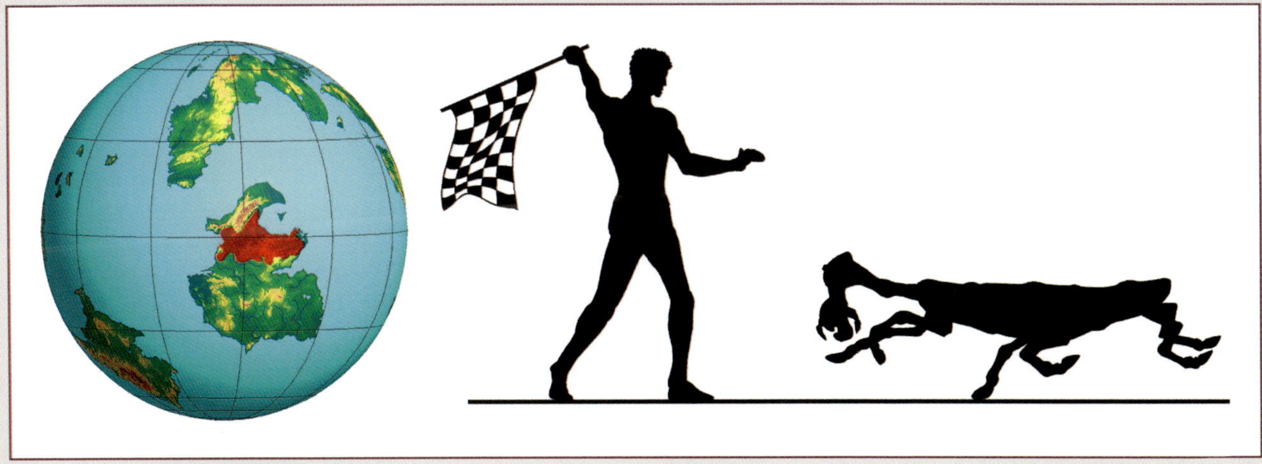

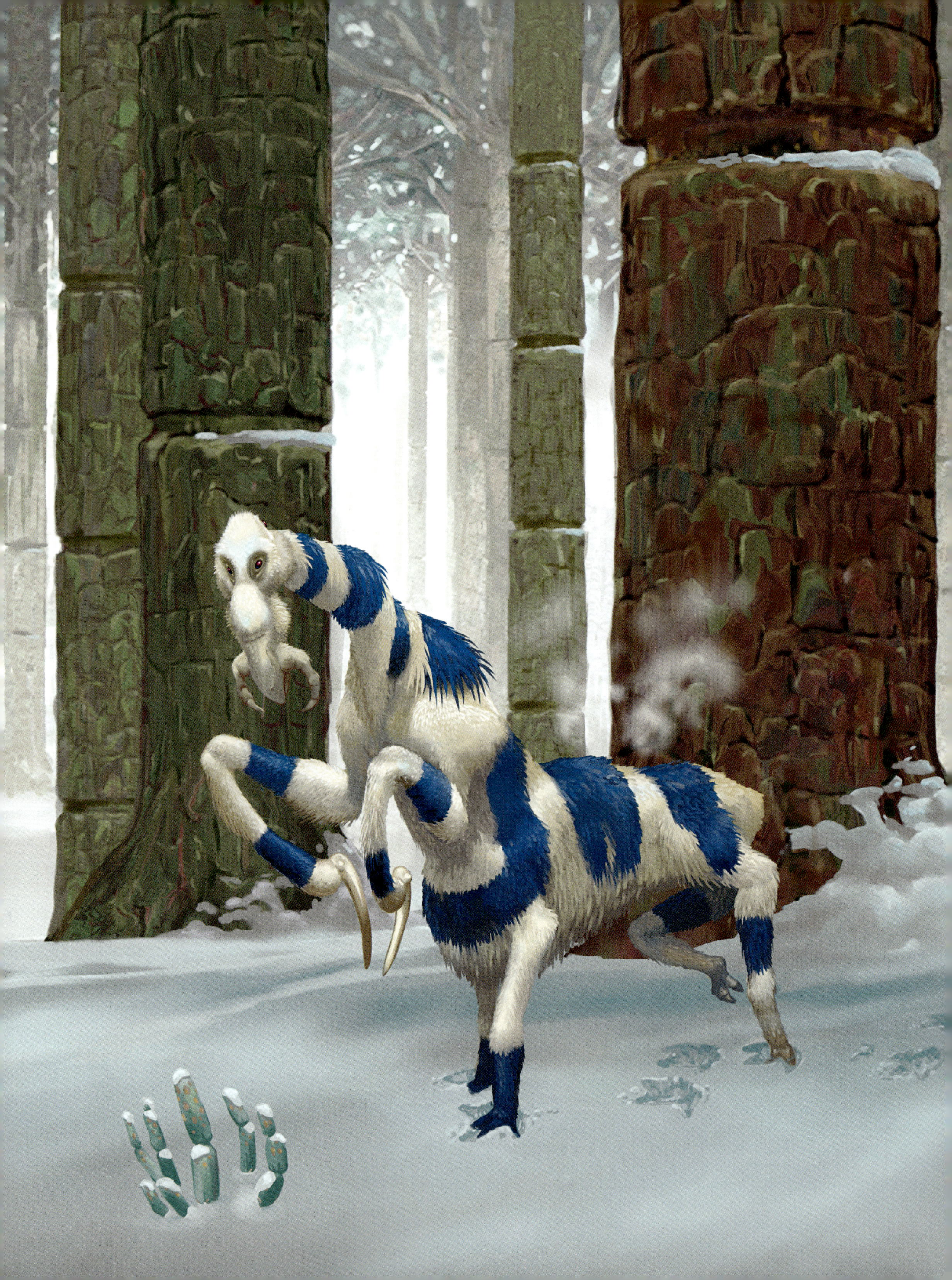

A Curious Pied Stickler

Perfixor artifex

Katarzyna Altanero, who devoted her life to the study of hexapod carnivores, described sticklers as 'prototypical gregarious piluferentic centauraptors'. This statement shows that doctissimus Altanero was one of those people who, once they themselves master something, think that all other people are born with that knowledge already in place. The second realisation, one that first requires her cumbersome prose to be digested, is that she was correct.

Centauraptors

Sticklers are obviously centauric hexapods that turned their front legs into weapons. Centaurism does not only involve quick evolution of the front limbs, but also results in drastic changes in an animal's balance and locomotion. The trunk's front part became shorter and was tilted upwards, while the former middle legs increased in size and moved forwards.

These changes must have evolved hand in hand with specialisation to a specific way of catching prey. Predators relying on speed should not carry massive clubs, while those bringing down armoured prey must carry large clubs but do not need to be fast. Scavengers are slow and carry some of the heaviest weaponry of all.

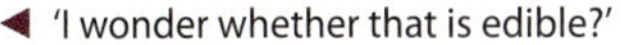

◀ 'I wonder whether that is edible?'

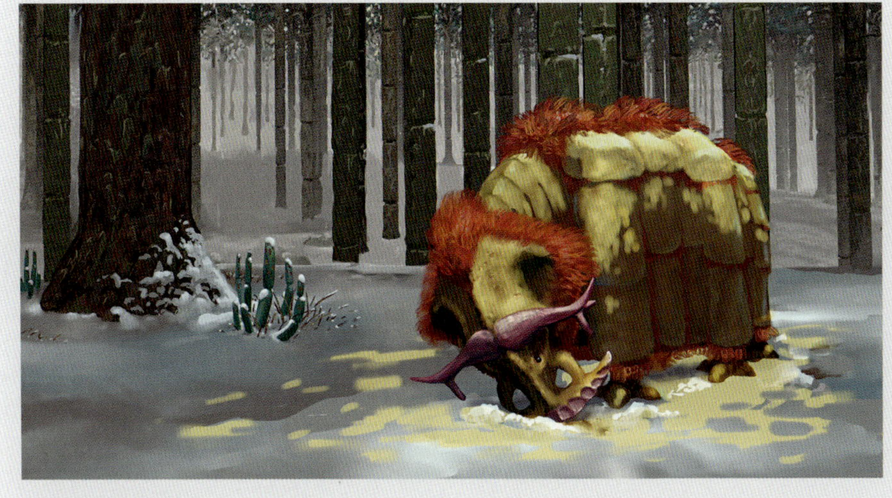

▲ **Stickler Habitat**
There are animals in the Northern Auralgian forests, but many stay hidden in winter. You may encounter various sub-species of shuffler, such as *B. rutrum silvestris* or *B. rutrum scourum*; the latter variant is shown here.

Piluferentic

Centauraptor weapons betray an interesting array of forms. Some are clubs and others are 'pointy end' weapons. Limbs known as 'axes' have a sharp edge on the underside, useful to hack open a carcass, but the same weapon usually has a heavy bulge too, allowing it to double as a club to knock prey off their feet or to bludgeon the prey's head, blinding it.

The stickler's weapon is no exception to this 'Swiss army knife' nature. While its weapon is primarily shaped like a club, there is a sharp protrusion, the erstwhile foot that can be swung into action. This explains the word 'piluferens', meaning 'lance carrying'.

Gregarious

As the word suggests, sticklers live in groups. In this species' case, tasks differ between group members, allowing a relatively complex society that must run on advanced cognitive capacity.

Do not underestimate sticklers. One may appear to study you with open curiosity, but its pack members may well be on their way to encircle you.

▶ **Stickler**
Perfixor artifex
Name derivation: *Perfixor* (L.) (noun from the verb *perfigo*, meaning 'to impale or pierce'); *artifex* (L.) (meaning 'cunning or skilled')
Habitat: Forest, prairie, marshes
Distribution: North Auralgia
Mass: 120kg
Length: 2.1m
Diet: Animals (not fussy)
Peculiarities: Colour bands blue and dirty white in winter, shifts to grey-green and ochre in summer

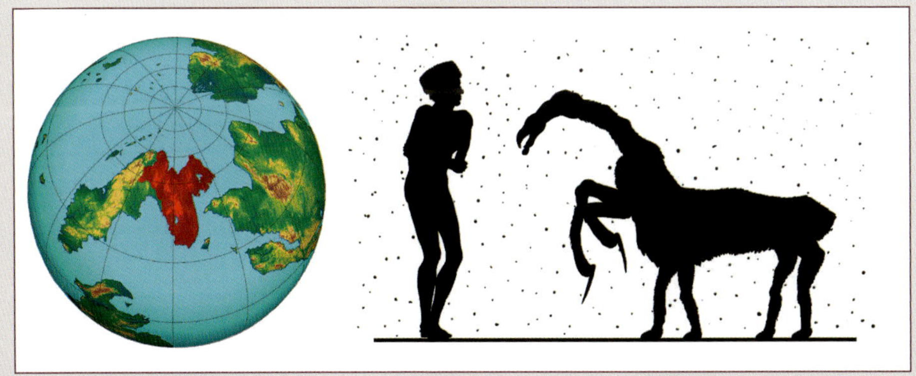

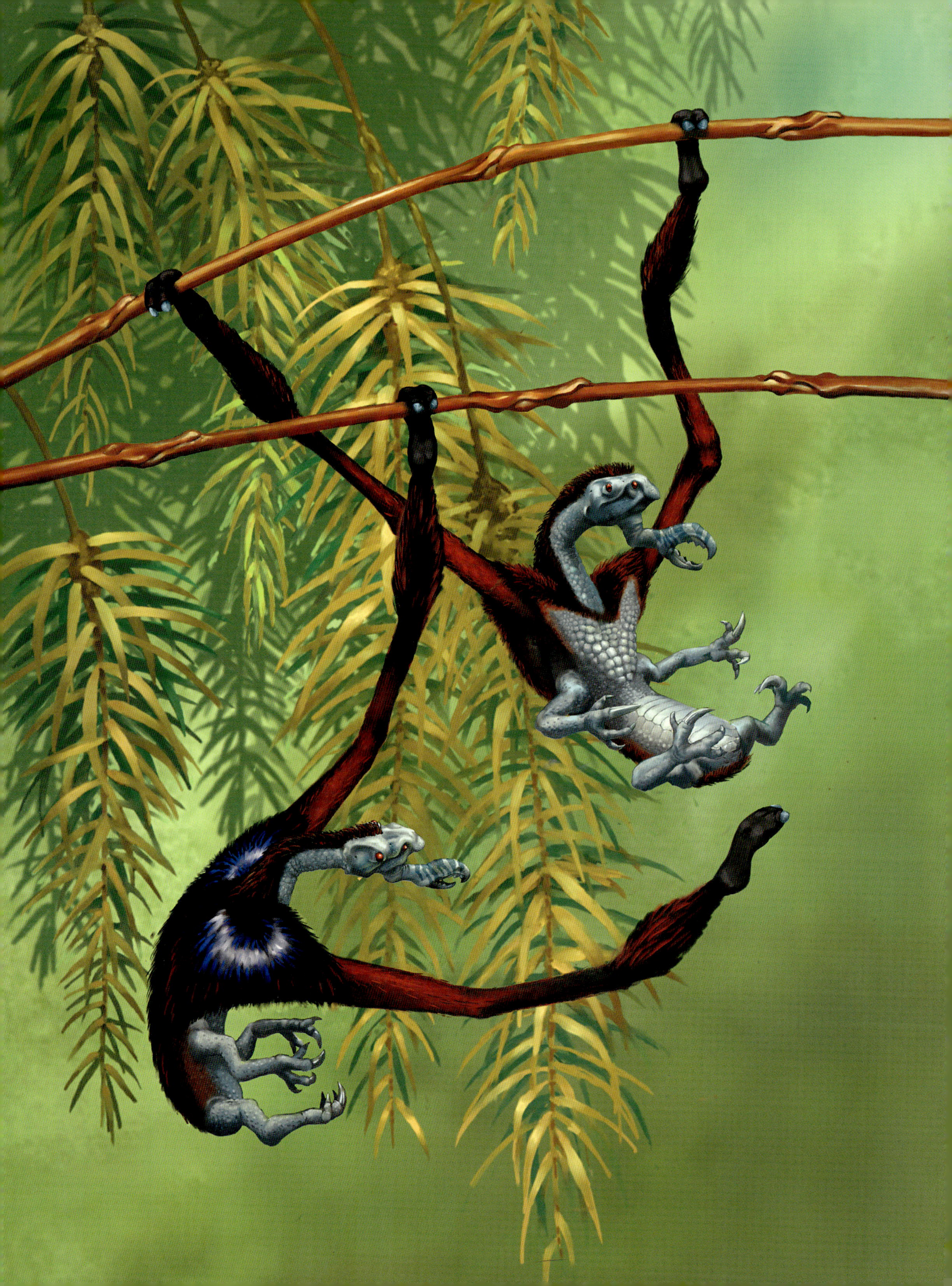

Marblebill on The Prowl

Iaculator weismuelleri

The marblebill is an arboreal carnivore with interesting specialisations. A quick glance reveals that it is a brachiator, moving by swinging from its arms. These, the front legs, are about three times as long as the second and third pairs of limbs and end in simply built but very powerful hooks.

Suspended from these hooks the animal swings from branch to branch. A firm grasp is literally a matter of life and death for a brachiating animal. You might therefore expect the marblebill to have something like a hand with an opposable thumb, instead of a simple hook. But thumbs curl around a branch and if the timing is the least bit off, the thumb hits the branch head-on and suffers injury. A simple hook is much less prone to such injuries and serves just as well.

The marblebill's other legs are very short, which makes sense for a brachiating animal as a large dangling mass would ruin the pendulum effect of the swings. The animal uses these legs to crawl up tree trunks or around branches and occasionally to walk on the ground. A better word would be stumbling, as they are rather clumsy out of trees. Mind you, they are still dangerous on the ground.

▲ **Marblebill Visual Fields**
The spheres around these models' heads show the marblebill's visual fields. Visual fields of the hind eyes are shown on the left and those of front eyes on the right. Some parts of the field can be seen with both eyes (not coloured); light blue areas can be seen with one eye, while dark grey areas cannot be seen at all. Both front and hind eyes have a blind area behind the animal, smaller for hind eyes than for front eyes.

Carnivorous

The lower four limbs also serve to grasp a victim. The marblebill's favourite prey is the 'berbie' (*Berbex ululator*, shown overleaf), a tree-dwelling herbivore. Once caught, the victim's feeble attempts at defence have little chance of success against the marblebill's armoured chest and abdomen. There is little time for resistance anyway, as marblebills usually incapacitate their victims quickly.

The Eyes Have it

Marblebills have the common hexapod pattern of four eyes placed as front and hind pairs. In ancestral hexapods, what are now front eyes were previously top ones, and what are now hind eyes were previously bottom ones. The former top eyes now face forward to help judge distances and fixate prey. The size of the ommatidia, the individual eyelets in a compound eye, varies according to their place in the eye. The ommatidia facing forwards are small, providing good acuity but limited sensitivity to light.

The marblebill's hind eyes protrude from its head, providing a wide field of vision in the horizontal and vertical planes, which makes sense for an animal moving in three dimensions. The visual acuity of the hind eyes is not good, but they are sensitive to low light.

Screeching at Dawn

Marblebills live in pairs and are fiercely territorial. At dawn, pairs spend their time announcing their presence with ear-splitting screeches. But they hunt with great stealth. A gaggle of berbies, never the brightest of beasts, may notice a branch swaying and a baignac falling. Only when they hear the marblebill's triumphant howl does it dawn upon them that one of their comrades had just now been munching that baignac.

It is this post-hunt 'Aa-ia-iaaaaa' howl that explains the species' name. There are various related species, all with their own calls, such as *I. greystokeii*.

◀ Threatening marblebills.

▶ **Marblebills**
Iaculator weismuelleri
Name derivation: *Iaculator* (L.) (meaning 'swinger'); *weismuelleri* (after Johnny Weismueller, twentieth-century Tarzan actor)
Habitat: Middle to high levels of dense tropical forest
Distribution: Naivasha in The Archipelago
Mass: 60kg
Length: 80cm
Diet: Strictly carnivorous; staple food is the Berbie (*Berbex ululator*)

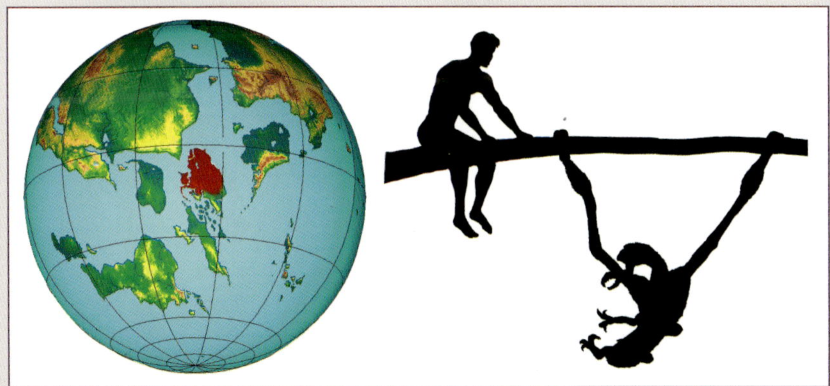

Vigilant Guardian Berbie

Berbex ululator

Poor berbies. In most books they feature only as clueless prey for the more spectacular 'trapeze terrors'. That is what Furaha's brachiating predators are often called by publishers catering for the uneducated. But even more serious works devote most attention to marblebills, tenterhooks, nuntsjuks and becdaciers. In kill scenes, the unlucky prey is usually suggested to be dispatched quickly, without undue suffering. 'Nature red in tooth and claw' is what the voice-over says, but the gore is never shown (hexapod blood is not red, but a greenish blue anyway).

Hold and Kill

Do predators kill their prey quickly and, if so, why? Pity plays no role in their chilling calculations, which only deal with whether a struggling prey can still harm the predator. If the prey is small, unarmed and mostly harmless, predators simply start eating the prey before it is dead. But if the prey can still cause injury, it must be rendered immobile. Apart from this general strategy, marblebills have another reason to kill their prey quickly. If they drop their prey, it will fall to the forest floor, where other predators may snap it up quickly. This is why marblebills must catch their prey, prevent it from falling and immobilise it, all as quickly as possible.

Berbies: Built to Digest

Berbies are herbivores living in large groups. They are brachiators, as are their predators. The predatory and herbivorous brachiators represent different evolutionary lines that took to the trees at different times. The zoologists studying them, brachiologues, think that predatory brachiators first evolved the ability to attack smaller and slower arboreal animals and only later took to attacking larger animals such as berbies.

Meanwhile, berbies and their relatives adapted to life in the trees as well as their predators. Berbies eat fruit but are not exclusively dependent on such high-energy food. In this they differ from the much smaller zoomers. Zoomers are specialised fructivores and behave as if they live on sugar bombs, which is what the fruit they eat amounts to.

Poor Berbies

Berbies mostly eat leaves and stems that provide little energy, probably because this food is always available, unlike fruit. The downside is that this low-quality food requires a considerable gut to allow leaves to be slowly processed into something more nutritious.

Berbies adapted well to this limitation. Their mass is concentrated in a rotund belly, allowing them more agility than you might think.

Defence

Berbies defend themselves by flailing their arms at marblebills or plunging to the forest floor, where marblebills are not welcome. There are of course predators down there, but the risk is worth it when pursued by a marblebill. The best defence is to prevent marblebills coming near. While individual berbies are not overly bright, a gaggle of berbies is relatively smart, as there are always some sentinels keeping guard. Such a sentinel is shown here sitting on a Voronoi tree. Sentinels stay motionless for an hour or two, quietly observing anything moving.

An Aside on Voronoi Trees

Voronoi trees are named for the pattern of their barks. Like many Furahan trees, they are hollow and provide access to their interiors with holes here and there. Many Voronoi trees house spidrid colonies; whether the bond is commensal or mutualist is unknown.

Noise Makers

Poor berbies. Sentinel duty is dangerous. If a pair of marblebills manages to sneak close enough to begin their famous final two-pronged attack run, the outcome is often clear. But when a sentinel spots its enemy in time, it howls its loud predator warning, alerting the entire gaggle. Other berbies take up the horrible screech, agitating the entire forest. Berbies may even start mobbing the marblebills from a safe distance, ensuring that the day is lost to the marblebill couple.

All in all, berbies deserve to be known for more than being marblebill fare.

▶ Sentinel on duty.

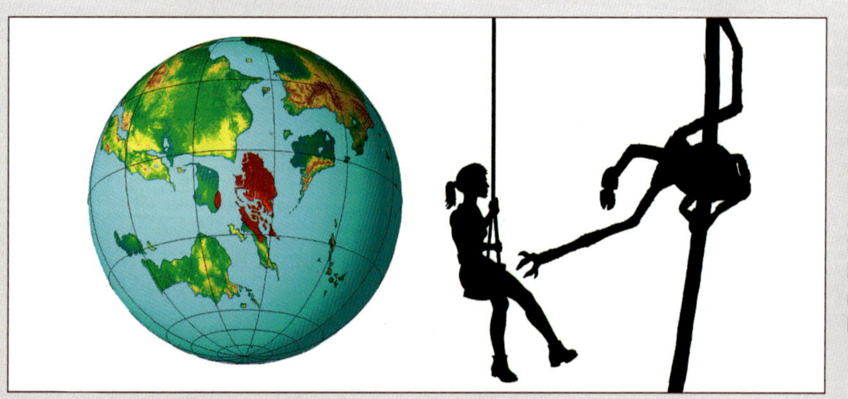

◀ **Berbie**
Berbex ululator
Name derivation: *Berbex* (L.) (meaning 'stupid/sluggish person'); *ululator* (L.) (from the verb *ululare*, meaning 'to celebrate or proclaim with howling')
Habitat: Middle to high levels of dense tropical forest
Distribution: The Archipelago, mostly Naivasha
Mass: 60kg
Length: 80cm
Diet: Herbivorous. Can digest leaves with low nutritional value
Remarks: Three species in the genus *Berbex*

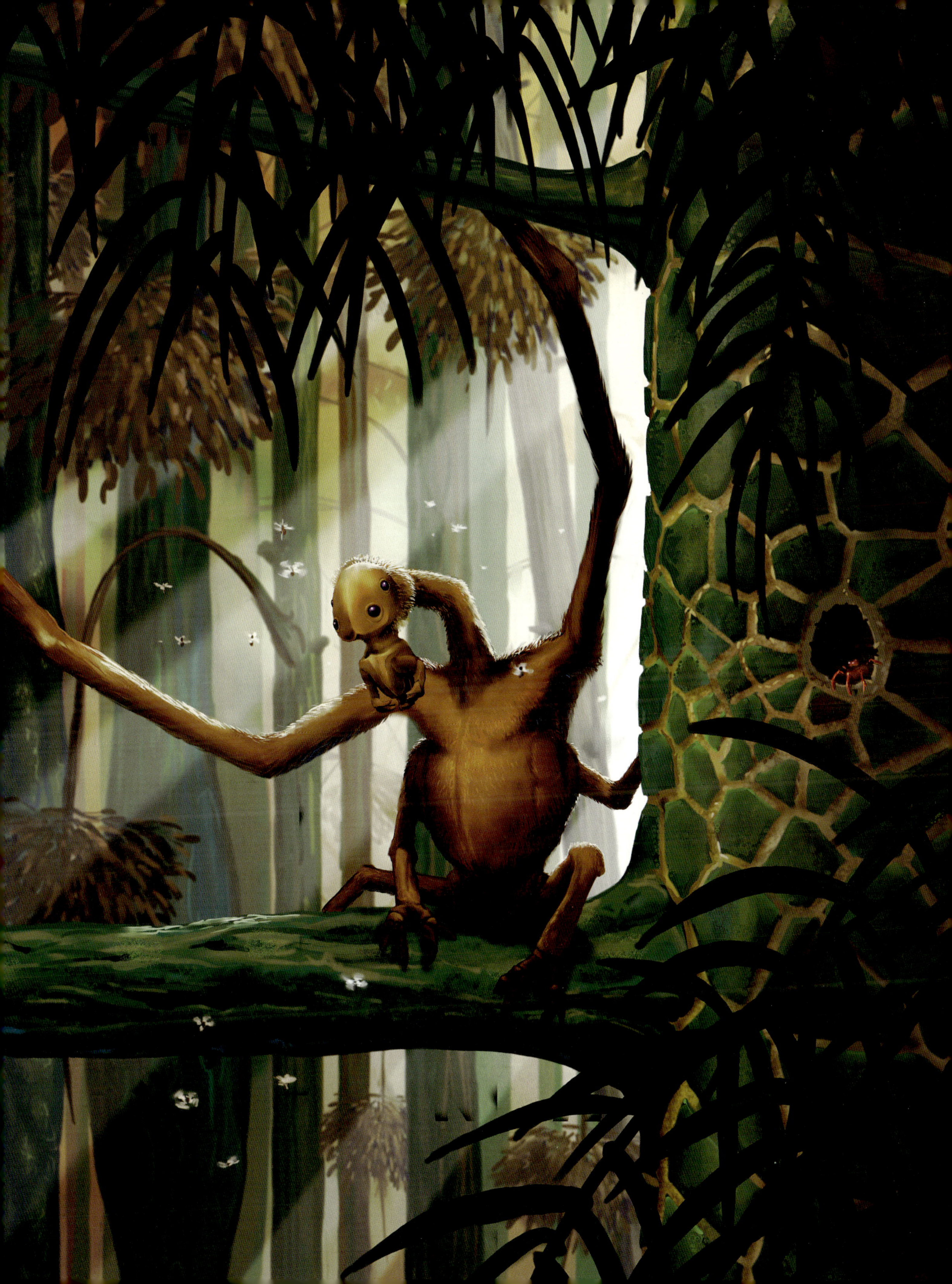

Shadowfisher

Draco umbraferens

On Earth, vertebrates gave rise to three different groups of flying animals: pterosaurs, birds and bats. On Furaha, evolution produced just two large groups of hexapod flyers, the Quadrialata and Dialata. Their names give away their main difference: the first group has four (quadri-) wings (-alata) and the second group just two (di-). Shadowfishers are dialate.

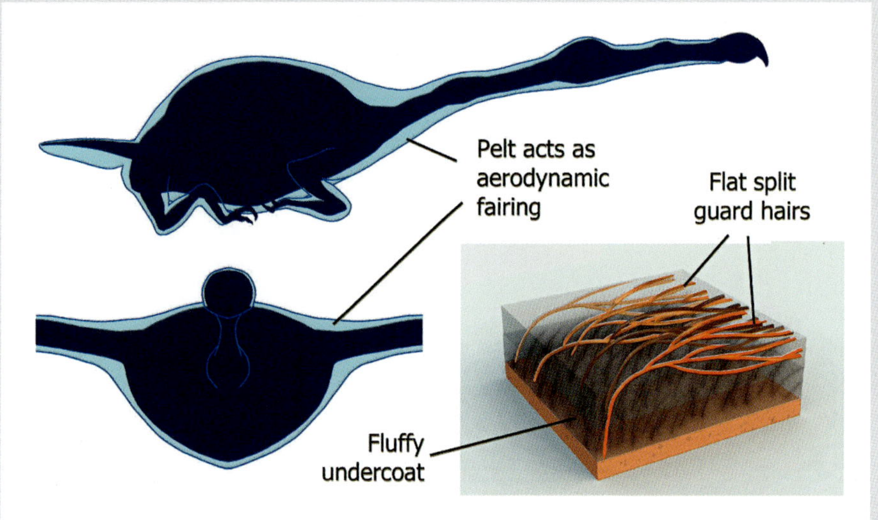

▲ **Scientific Notes**
The places where the limbs and neck join the body are smoothed with hair, which makes them aerodynamically more efficient. The pelt consists of an undercoat of split hairs, overlaid with long flat split hairs. These guard hairs, of which only a few are shown, resemble both feathers and hair, and provide a smooth surface.

Greek to Me

At the very start of Furahan biology, even major groups were named almost daily. Even so, the process was not straightforward even then.

The tempers of biologists studying and naming Furahan 'avians' were tested almost immediately after the ship Ngonjera landed, as some animals literally flew into sight within one minute. The onboard biologists learned about avians at the same time as everyone else did. When the non-biologists saw that some of the flying animals had four legs and two slightly bat-like wings, they shouted 'Dragons!'. As a rule, professionals do not like lay people interfering with what they think is their business. The biologists explained that the animals in question were not dragons, could not be dragons and that there was no such thing as a dragon anyway. But their resistance was futile. These animals are now known to one and all as 'dragons'. Never ask a Furahan biologist whether their dragons breathe fire. They have heard the joke before and were not amused the first time.

Fishing in the Dark

The shadowfisher lives in marshlands, where it uses its four legs to cling to reeds while looking for small amphibods, such as kermitoids and the like. They use a clever trick, shown in the image. The animals find a brightly lit spot, position themselves there and then spread one or two wings over the water. The shadow this casts is mistaken by some water creatures as a safe haven to shelter from aerial predators, which it is not.

Shadowfishers often put their heads in the shade as well. This is not always wise, as doing so means shadowfishers do not see those who prey on them.

Gaudy

Shadowfishers are not all as gaudy as the one shown here, or at least they are not as bright as this all the time. As usual, animals are usually conspicuous for reasons having to do with securing a mate and so it is for shadowfishers. Also, as usual, there is a price to pay, in terms of becoming prey. But shadowfishers are quick and agile flyers, so they get away with it, most of the time.

▶ A shadowfisher shadow fishing.

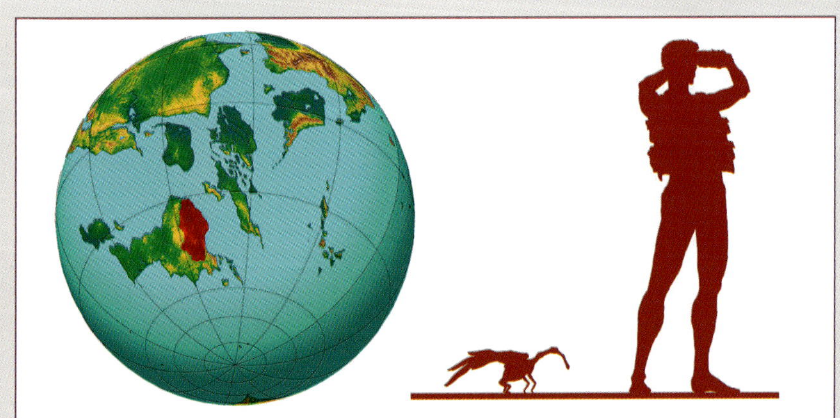

◀ **Shadowfisher**
Draco umbraferens
Name derivation: *Draco* (L.) (meaning 'serpent' or 'dragon'); *umbraferens* (L.) (from the noun *umbra*, meaning 'shadow' and the present participle of *fero* meaning 'bearing or carrying'; together: 'shadow bearing')
Habitat: Open forests and plains, in areas with open water
Distribution: Eastern Meralgia
Mass: 0.4kg
Length: 15–20cm
Diet: Mostly small grassland or water animals, but also fruit and seeds
Vocalisation: A soft Kiii-oh! Kiii-oh!
Remarks: Coat colours vary according to season, sex, age and colour morph

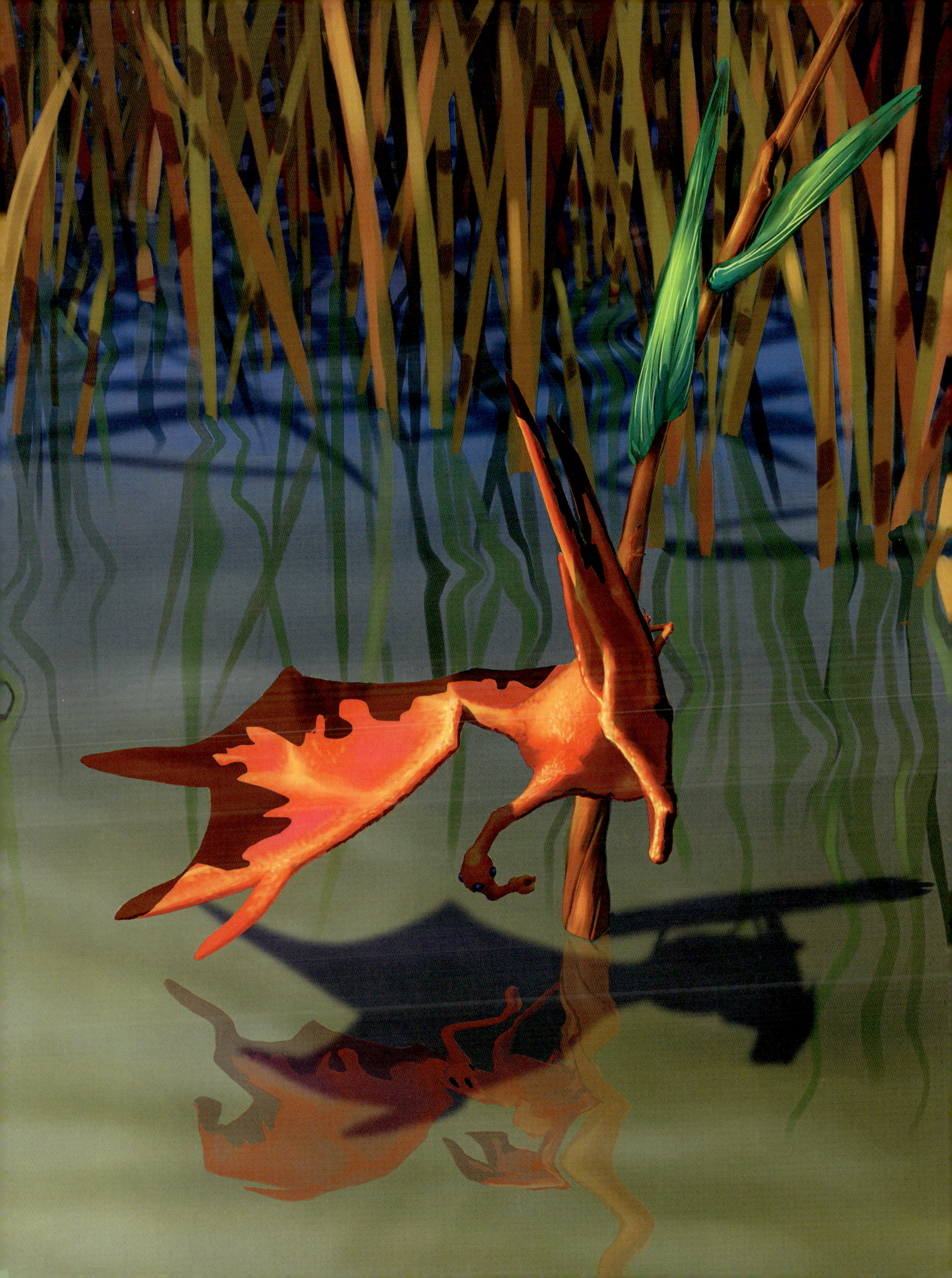

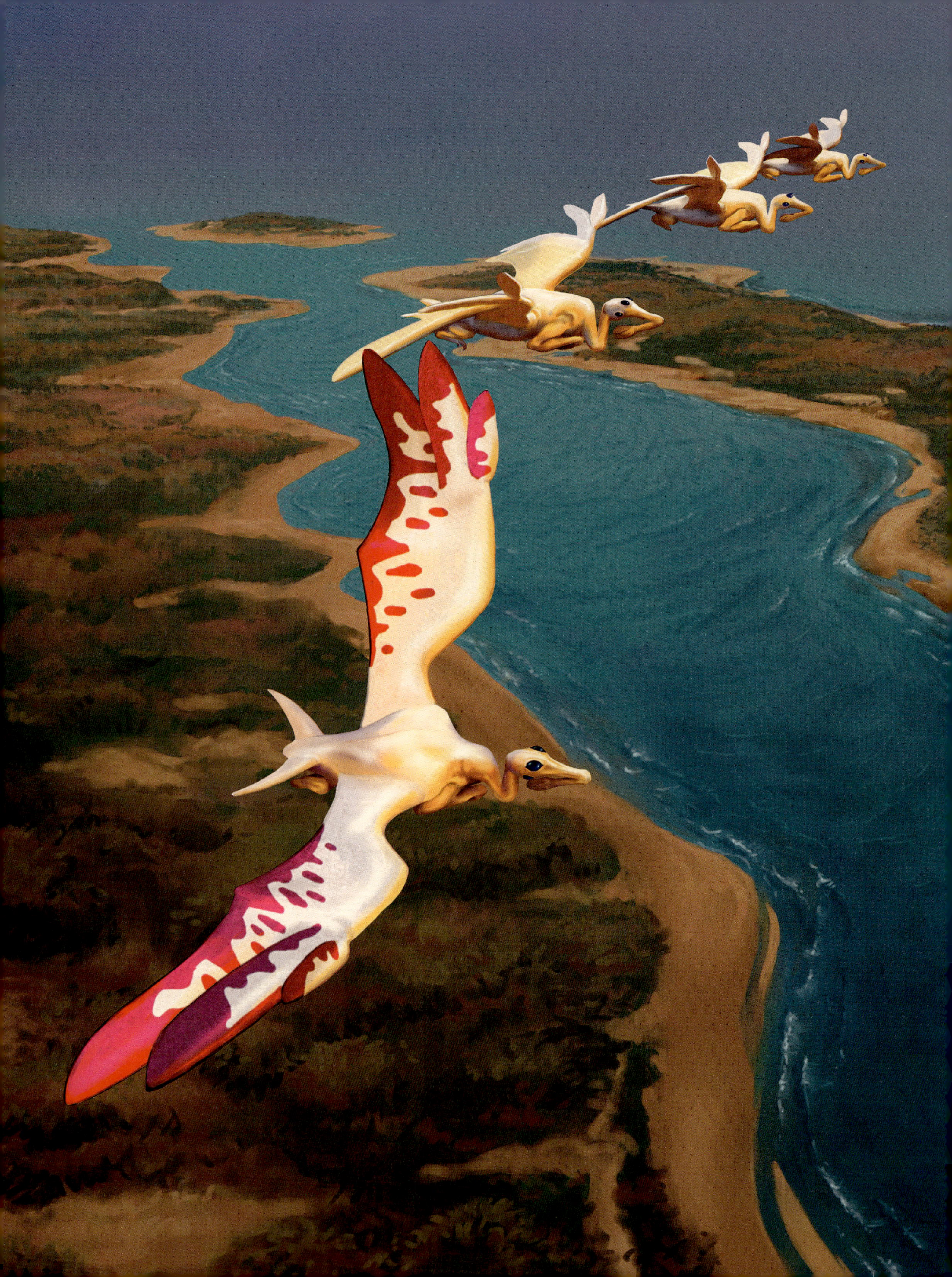

A Korongo Dawn Patrol

Dromodraco drungus

Korongoes are large dialate avians, found in troupes of five to twelve individuals. They sometimes fly in formation in a line abreast, as shown here, but more often in a diagonal staggered formation, with each body behind its predecessor's wing tip. Walking on the ground, they also patrol their territory in staggered formation. The foremost korongo may frighten prey into breaking cover, giving those behind it a good chance of catching it.

Colours

The only life stage in which this dragon species shows sexual dimorphism is when one member of a troupe reaches senior male status. This 'squadron leader' develops bright colours and is tasked with dominance displays. When he grows and matures to the mature female position, he, by then 'she', will lose the bright colours again.

Wing Size, Air Speed and Taking Off

As avians become bigger, physics dictates that their wings need to grow relatively more than the body. There are limits to this, so large avians generally have somewhat small wings relative to their mass. This disadvantage can be offset by high speed, but this in turn makes landing and taking off difficult. Taking off then presents the challenge of either running very fast or gaining enough height to start flapping. This is where the dragon's four legs become an advantage. Korongoes use them to catapult themselves into the air.

This pattern looks like the four-limbed starting jump that was proposed for Earth pterosaurs. Of course, pterosaurs being extinct, that behaviour remains conjectural, but the 'jump-clap-fling-and-flap' of Furahan korongoes can be observed as fact.

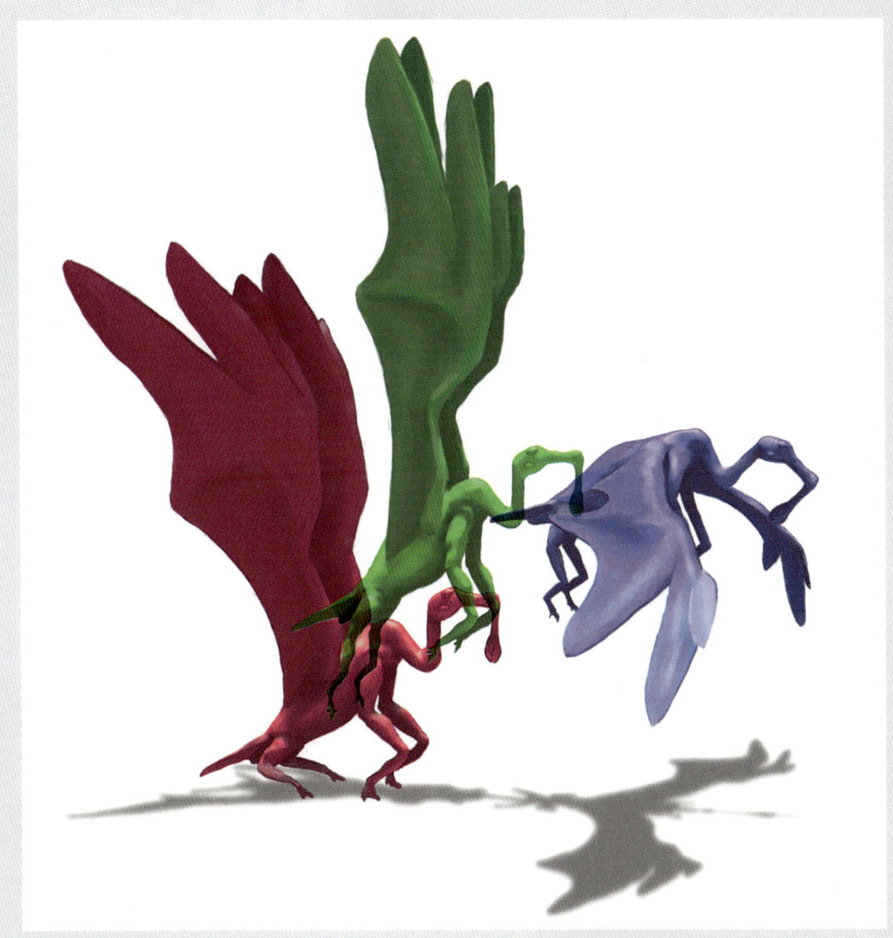

▲ **Korongo Take-off**
To take off, the animal makes a few hops, ends in a deep crouch (shown in red) and then jumps up, wings raised, powered by all four legs (green). It then forcefully beats the wings down (blue), providing 'clap-and-fling' lift, followed by fast and strong beats to send it on its way.

◄ Korongo patrol.

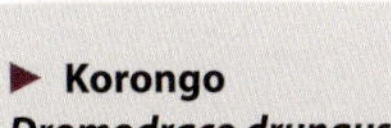

► **Korongo**
Dromodraco drungus
Name derivation: *Korongo* (Sw.) (meaning 'stork'). *Dromodraco* from *dromos* (Gr.) (meaning 'a run') and *draco* (Gr./L.) (meaning 'dragon'); *drungus* (L.) from *drouggos* (Gr.) (meaning 'a troupe of infantry')
Habitat: Open ground; steppe, savannah, beaches
Distribution: Southern reaches of Imparia
Mass: 80–100kg
Length: 1.5m and wingspan up to 5m
Diet: Opportunistic predator and scavenger. Typically found in formation, advancing slowly across the ground
Remarks: Exuberant colours apply to usually only one specimen per troupe (the 'squadron leader')

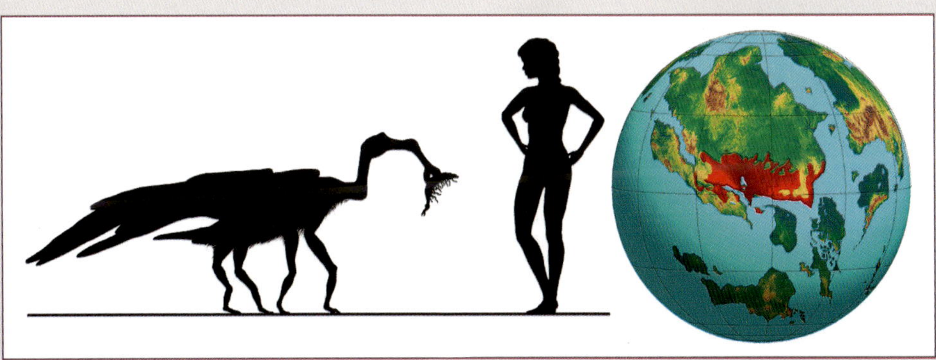

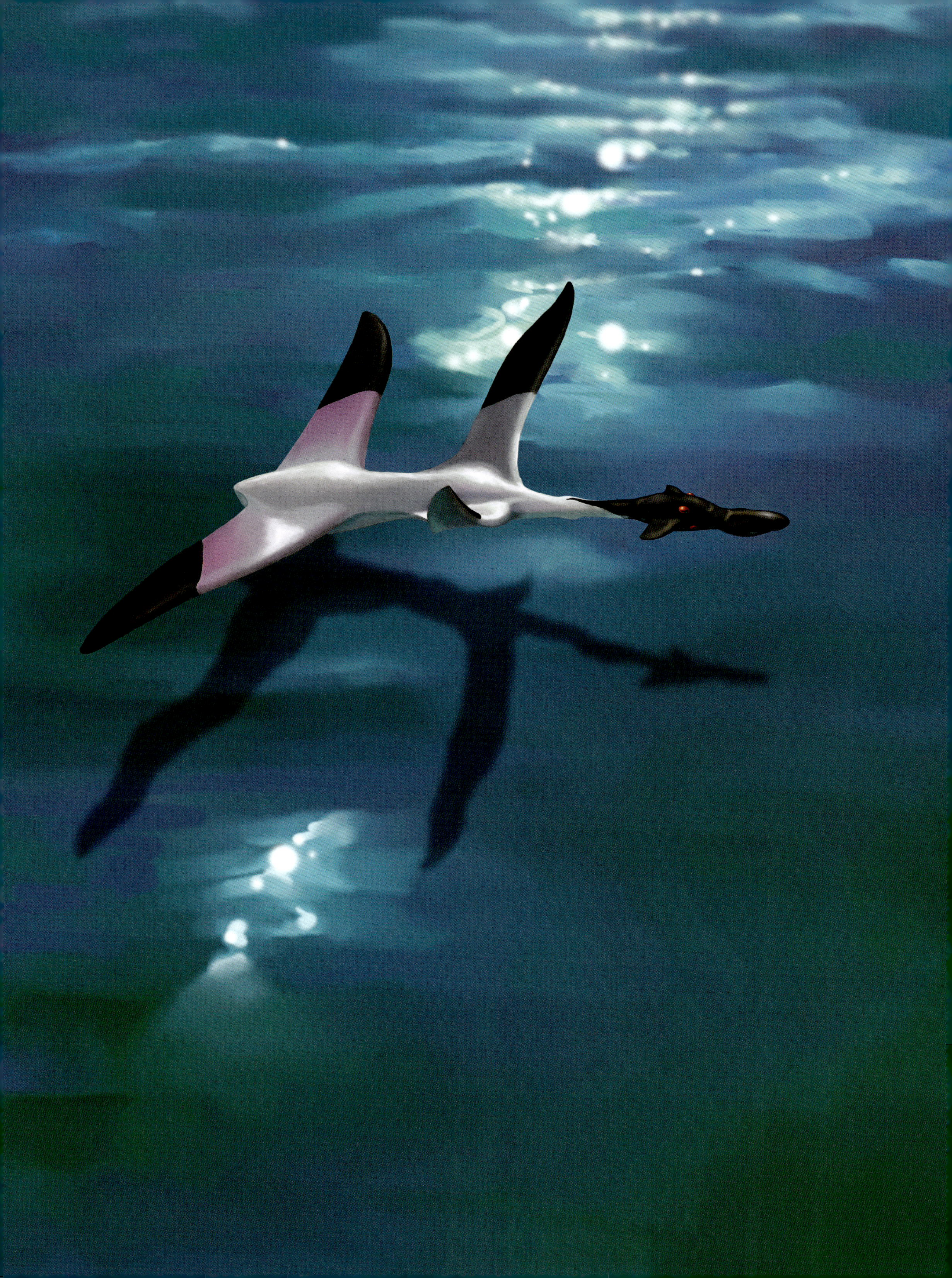

Porcelain Seasoar

Nimbus undasocius

Seasoars are intriguing quadrialate ('four winged') flyers that have puzzled researchers since their discovery, first for their flight plan and later, when more became known, for their procreational habits.

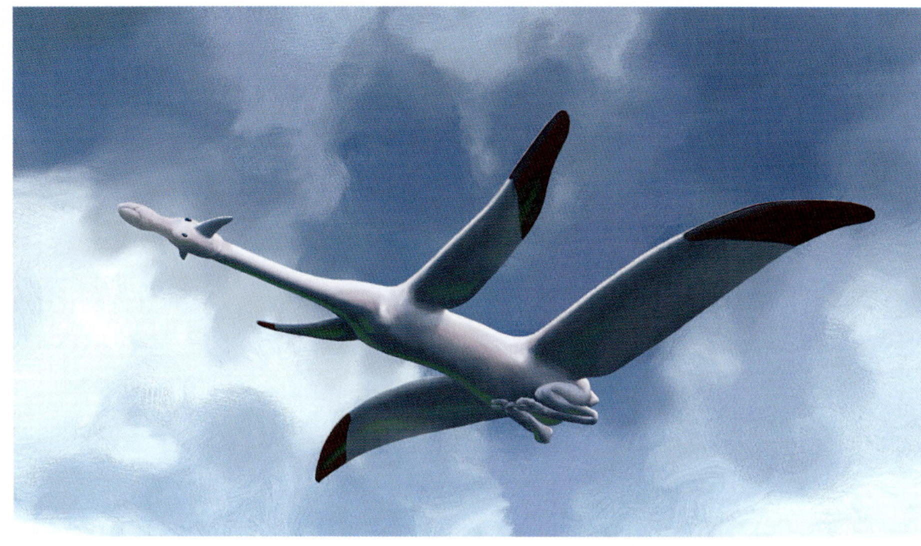

▲ This image of the white-headed seasoar shows how the hind legs fold up diagonally, allowing the leg segments to lie partly next to another in flight, reducing drag.

The Seasoar Flight Plan

The wings of the seasoar are placed differently than those of most Furahan four-winged flyers. The usual arrangement is for the front pair to overlie the larger main wings. This 'canard' arrangement is thought to increase lift and to allow precise manoeuvrability. Instead, seasoars' two pairs of wings are positioned very far from one another. This pattern provides a considerable amount of lift at the expense of manoeuvrability, but this is not a handicap for the seasoar.

Porcelain Seasoar

The species shown here is the 'porcelain seasoar', so named for the almost translucent aspect of its white coat. Seasoars live on open seas, where they skim the waves looking for any fish or cloakfish that can be snapped up. It is very adept at this type of flight and spends hardly any energy in keeping aloft. The merest twinge or resetting of a wing angle is enough to make the most of the ever-changing winds. Still, do not expect seasoars to turn instantly, as they do so only slowly – majestically, some would say.

Sex is Rarely Simple

Seasoars used to be seen as excellent examples of hermaphroditism. They are not males who only deliver sperm, nor are they females who only carry the metabolic burden of producing yolk or nurturing their young. The stereotypical view of hermaphroditism is that both sexual partners share the reproductive burden equally, so they both fertilise one another and both spend the same amount of energy in forming a young each.

Large adult seasoars indeed follow this textbook path. But small individuals cannot easily afford the cost of growing a young. Their best chance to reproduce is to follow the male route only where they try to impregnate a larger individual without growing a young themselves. Mating with such a rogue is bad for a large individual, as only one young is produced instead of the usual two. Small individuals are therefore not popular partners, so small seasoars generally wind up with other small ones. They may try to become a rogue by seeking a large mate, for some reason unable to secure an equally large partner. Of course, when such small individuals grow large, they scorn the attentions of small suitors and will only consider partners or equal of larger size.

Sex is rarely simple, hermaphroditic or otherwise.

◄ Gliding over the waves.

► **Seasoar**
Nimbus undasocius
Name derivation: *Nimbus* (L.) (meaning 'cloud'); *undasocius* (L.) (meaning 'wave follower')

There are at least fourteen species in four genera. All are adapted to soaring flight.
Habitat: Open ocean, can be found in coastal waters in the breeding season
Distribution: Temperate zone of the entire Southern hemisphere
Mass: 40kg
Length: Wingspan is 2m for front wings and 2.5–3m for the hind wings
Diet: Any surface-dwelling sea-animal of adequate size
Vocalisation: Usually silent under all circumstances. Surprisingly weak 'Prwwww' during courtship

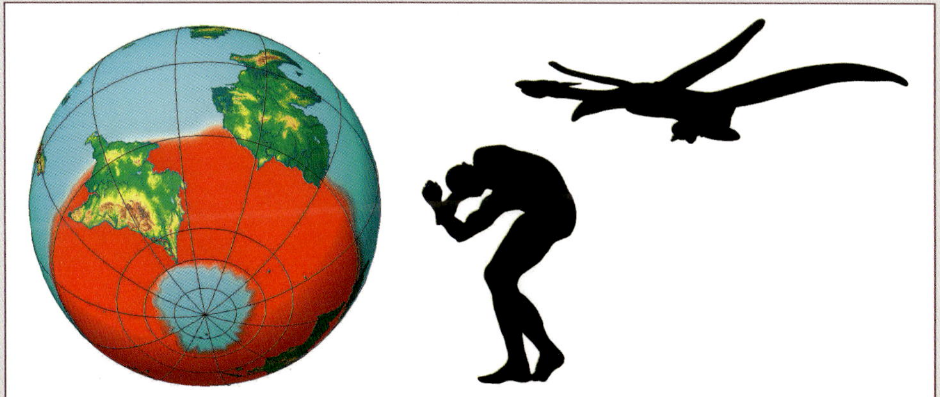

DIPEAU

VELIVOLATOR MOEBII

Food and Travel

This arid landscape shows the edge of the deep desert, providing a difficult environment for any animal. The spot is close to the one shown on the 'desert ghost' pages. Food is scarce and the next meal may be a long way off, something that holds for herbivores as well as carnivores. All animals here must be able to travel long distances surveying the land without using much energy or water. If they are unable to do so, the amount of energy spent looking for food will outweigh that derived from it. If the search is unsuccessful too often or is too long, such an imbalance can end in only one way where the unlucky animal finds its carcass helping someone else's energy balance.

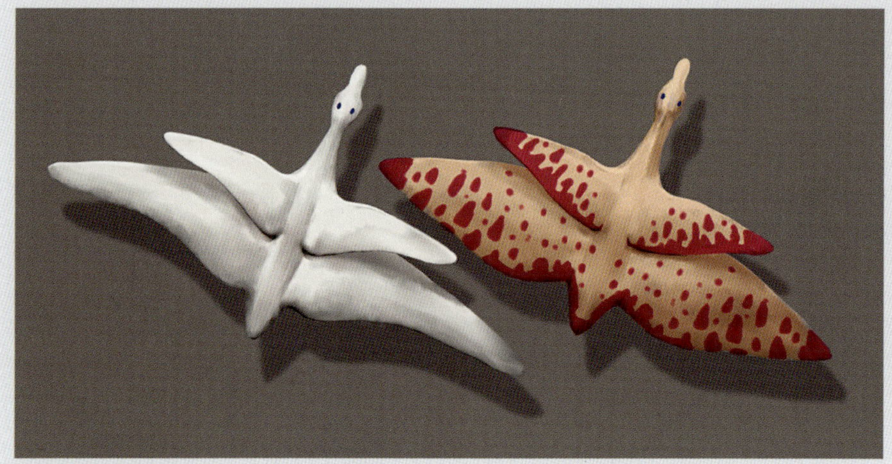

▲ **Dipeau Variability**
The variant at the left, also shown on the main illustration, is how dipeaus look at the Northern and Western parts of their range, as deep into the desert as they manage. This variant is called *V. moebii arzachii*. Note its overall pale colour and the relatively long and slender wings. Next to it the variant at the Southern and Eastern dipeau range. This variant is *V. moebii gruberti* and it has stubbier and broader wings. The variants are not separate species. One may find every intermediate shape but never intermediate colours.

Surveying the Land

Flying animals have a definite edge as far as long-distance surveying is concerned, especially those that have mastered the art of gliding, which takes much less energy than active flapping flight. Good gliders use columns of rising air to gain height almost effortlessly. Gliding is helped by having an aerofoil shape well-suited to this mode of flight. The dipeau is reasonably well adapted in this respect. Still, its wings are not as long and slender as might be expected. Of course, the presence of two pairs of wings is a great advantage, as the front canard wings direct air over the main hind wings and augment the amount of lift considerably. The dipeau also has the brain capacity necessary for subtle but efficient control, meaning it uses minuscule movements of the position and shape of its wings, making the most of changing winds, vortices and air currents.

Adaptation Limits

Dipeaus are not restricted to the deserts shown here, but also live in friendlier biomes, where they typically fly lower and need to be more manoeuvrable. In such surroundings, broad and short wings serve best. In view of the wide range of the dipeau it should come as no surprise that not just the animals' colour, but their wing shape also vary according to their habitat.

The dipeau in the image is probably looking out for food. Dipeaus will relish the occasional fruit, but mainly thrive on animal remains.

▶ Dipeau over Moebian landscape.

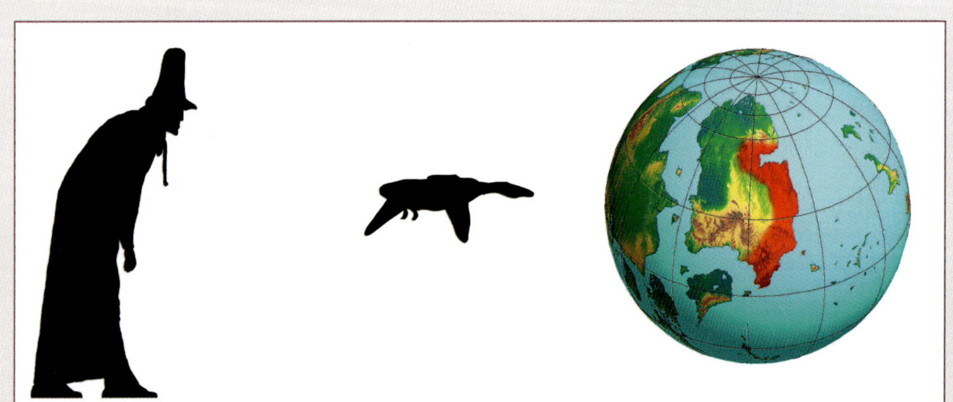

◀ **Dipeau**
Velivolator moebii
Name derivation: *Velivolator* (L.) (meaning 'flyer with sails'); *moebii* (L.) (after Moebius, famous twentieth-century Earth artist)
Habitat: Air, over desert to open savannah
Distribution: South-east of *Imparia Orientalis*
Mass: 10kg
Length: Wingspan up to 2m for hind wings
Diet: Omnivorous, mostly carrion
Vocalisation: Usually silent

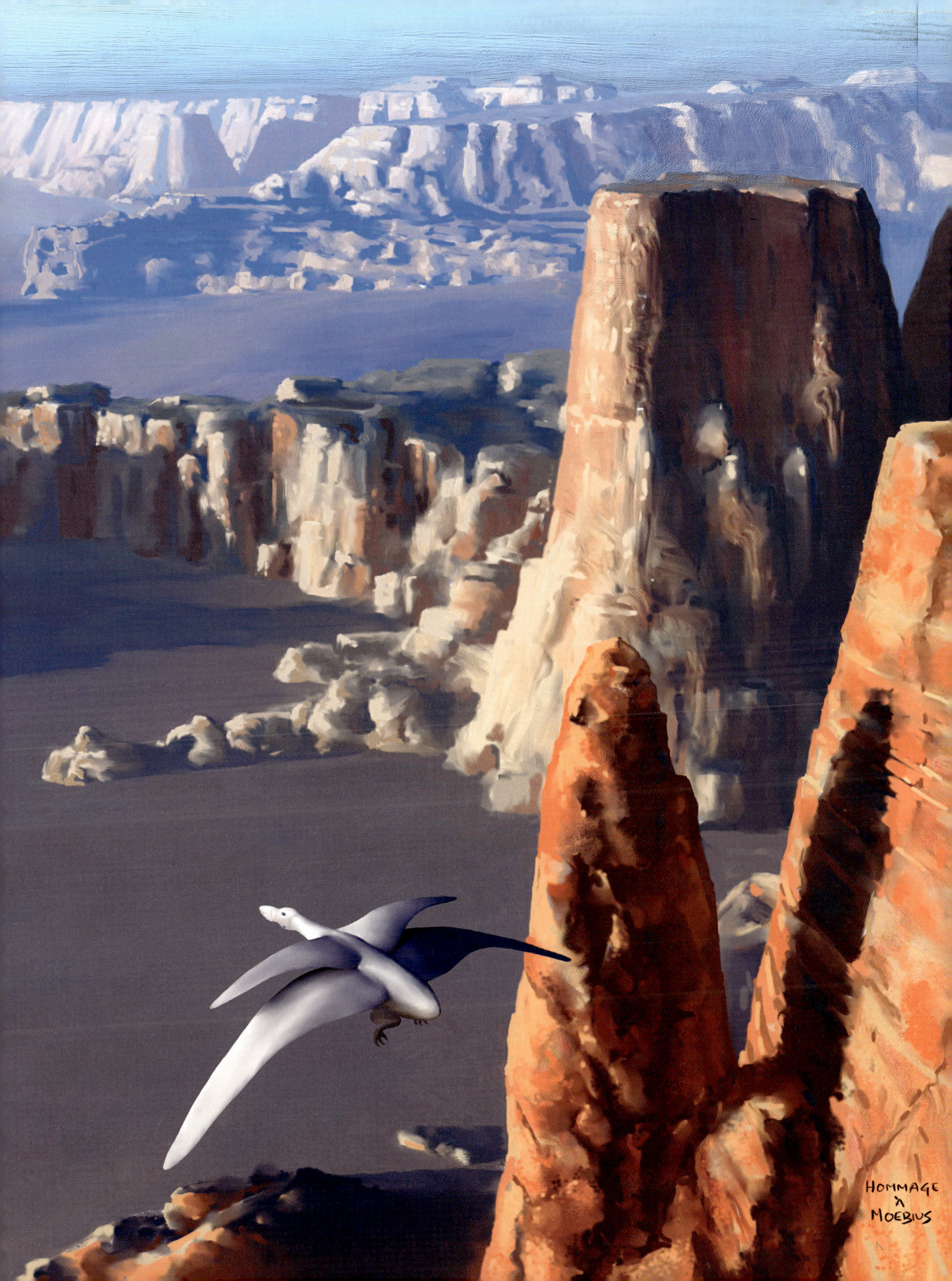
HOMMAGE
A
MOEBIUS

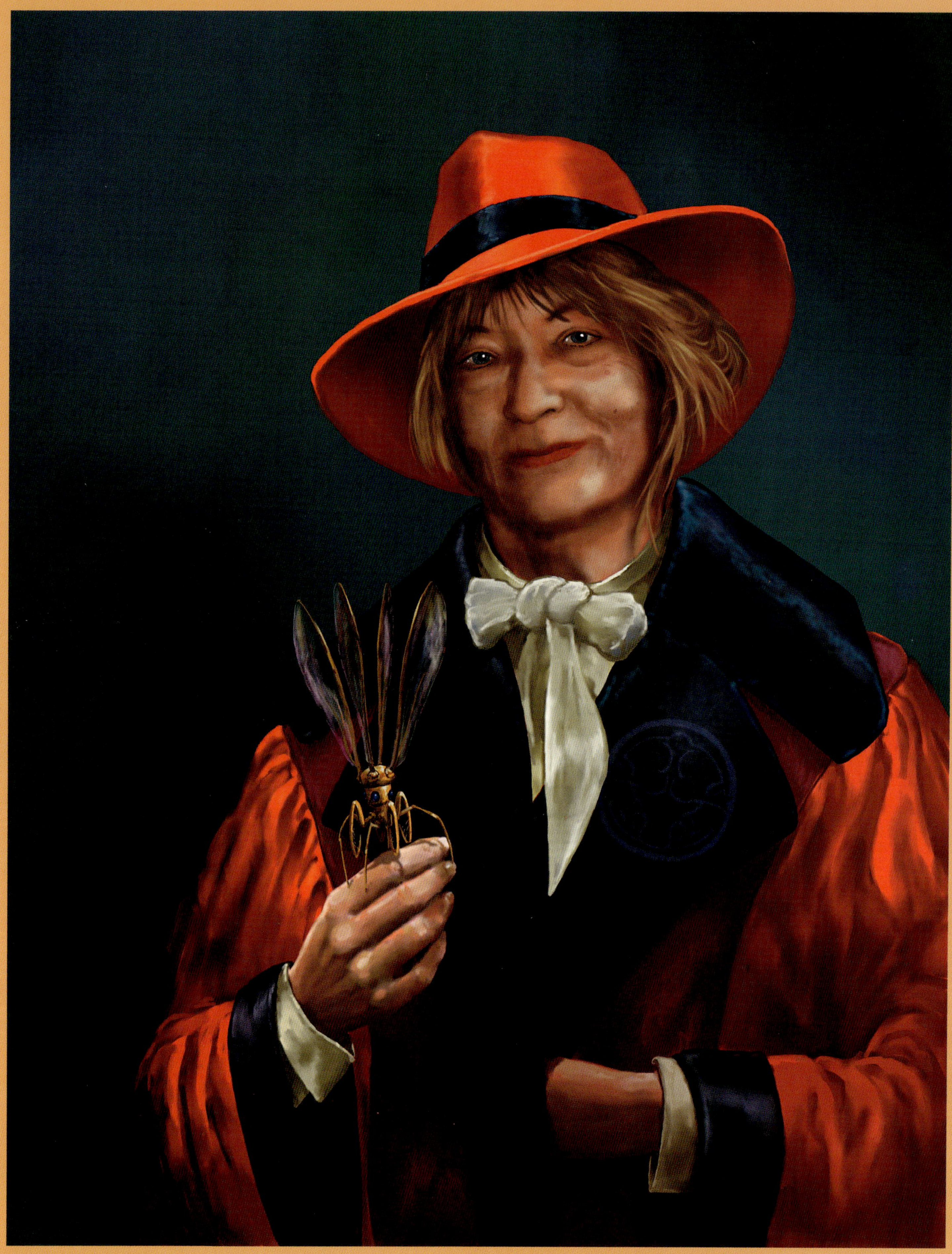

Chapter 7

Rusps

'Some people expected life forms on other planets to be nightmarish monsters, without reason or design to their bodies and anatomies. Some wanted alien life to differ from Earth forms so badly that they allowed their fantasies to cross physiological or physical boundaries, conjuring up animals with X-ray vision or one-tonne spiders.

But evolution simply forces life forms to provide answers to the questions their environment poses them. The same question is likely to yield the same answer. If the environment asks how best to shape an animal leg bone or a plant stem, blind evolution is likely to come up with a tube to combine strength with low weight. Countless other examples show that evolution always attempts to solve mechanical problems while balancing energy demands. It is this evolutionary energy sieve that warrants the axiom 'Darwinian evolution honours Newtonian mechanics'.'

From *Worlds Apart: Natural Histories of Furaha and Earth* by Souren Nyoroge

◀ This painting of *Profissima* Tartufa Salomé Rulyinka was one of the few displayed in the Institute Hall. This was an honour bestowed only on those with extraordinary Academic skills. Her thesis, finished while she was still a *Studiosa*, on jaw movements of woolly-haired shufflers (*Gigatheron latifrons*), earned her a *Doctior* instead of the common *Doctor* degree. She later wrote extensively on biomechanics, reflected in the *epitheton laborans* chosen to accompany her *Profissima* title, 'Locomotorologica'. As Chair of the Biomechanics Department, she expanded its influence considerably and she later became *Decana Augusta* of the Zoology Faculty.

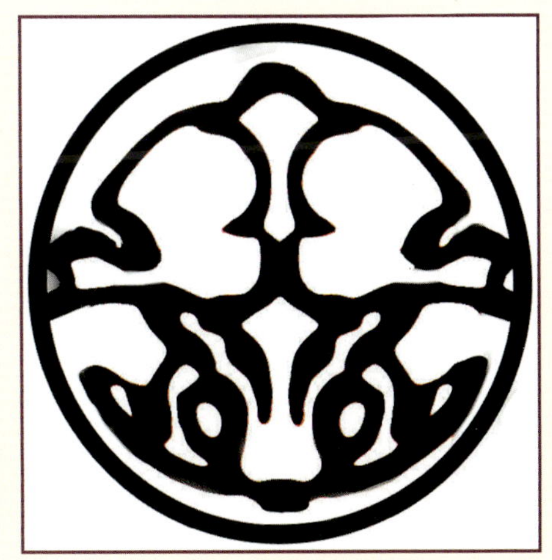

▶ Rulyinka's cartouche, a stylised Shuffler head, honours her Clan, as her grandfather had introduced that cartouche when he elevated Biomechanics to a full Department. The wide-brimmed hat provides an unexpected contrast with the formal black beret expected from *Profissimi*. This, as well as the informal coiffure, must concern a conscious choice to emphasise her predilection for Field Studies. Contemporary viewers would have deciphered these clues at a glance, but yet may have missed one element. The tetrapter on her hand is often interpreted to refer to her own work, but it may hide a more personal message. Prof. Rulyinka may have added the tetrapter to honour her partner, the Zoomath Grover L. Nastrarruzzo, with whom she collaborated on tetrapter flight.

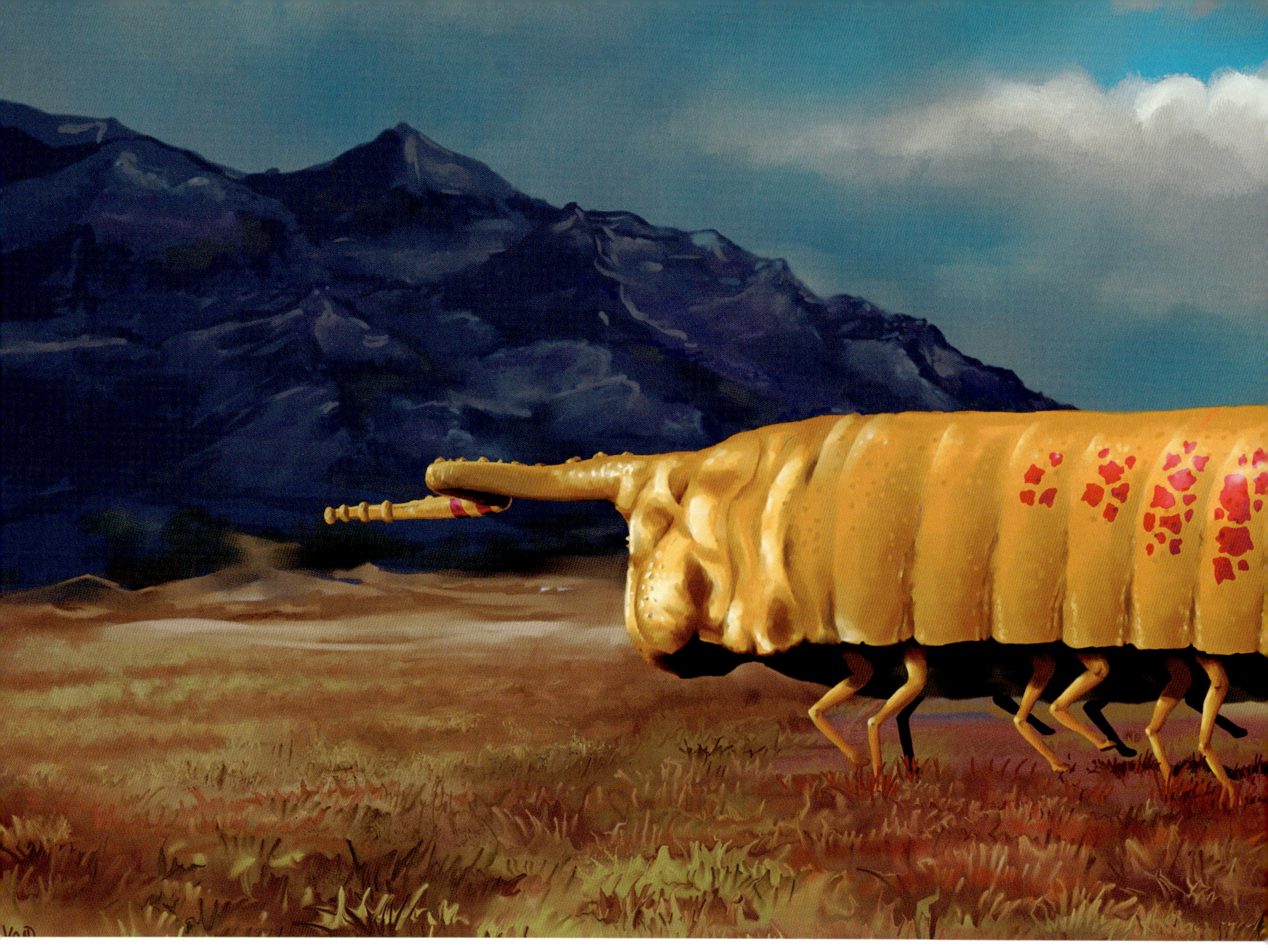

MEGARUSPS

BRONTORUSP

BRONTOCRAMBIS BRUCUS

Although there is a wide variety of body schemes on Earth, large animals are almost always vertebrates: this is most often true in the sea and always true on land. No one group has a monopoly on large size on Furaha. Hexapods, the most common large animals, have to make room for animals with a different plan. In the case of megarusps, making room has to be taken quite literally.

The 'Janus Effect'

Rusps are full of anatomical oddities. Their most conspicuous feature is probably that they seem to have two independent heads. This needs more explanation. The term 'cephalisation' indicates that the mouth, major senses and the brain tend to evolve near one another at the front of an animal. Rusps definitely have a head performing all these functions at its front. There is a strong whip there too, with which the animal defends itself. The brain is ring-shaped and is therefore called the 'neurannulus'. At the rear of the animal is a secondary head. There is no mouth there, and the brain is much smaller, for which reason it is called the 'neurannulinus' (meaning 'little ring brain'). Although there is a thick cord between the two ring brains, the neurannulus seems to have overall control. This situation resembles the cooperation between the many brains of Earth's octopuses.

Present thinking holds the very long bodies of rusps responsible for their double-headedness, something called the 'Janus effect'. These long bodies would be vulnerable if sense organs and defence mechanisms would only be located at the front. There is evidence that eyes featured on each segment very early in rusp evolution, so the hind head seems to have been present in rusps from their earliest start. In fact, cephalisation in rusps may have acted on both ends of the animal.

Whips and Defence

Adult rusps are covered by an extremely tough carapax and defend themselves with whips, so they are almost immune to predation. Adult rusps protect young rusps. The whips are rarely completely at rest. Even when grazing quietly, a group of rusps offers the mesmerising sight of their whips swaying slowly. This probably means something to other rusps, but this whip language has so far not been deciphered. The whips are equipped with bony rings at their ends and the skin there is even thicker than elsewhere. The whips can be swung with great velocity, so one hit is usually enough to maim or kill a predator.

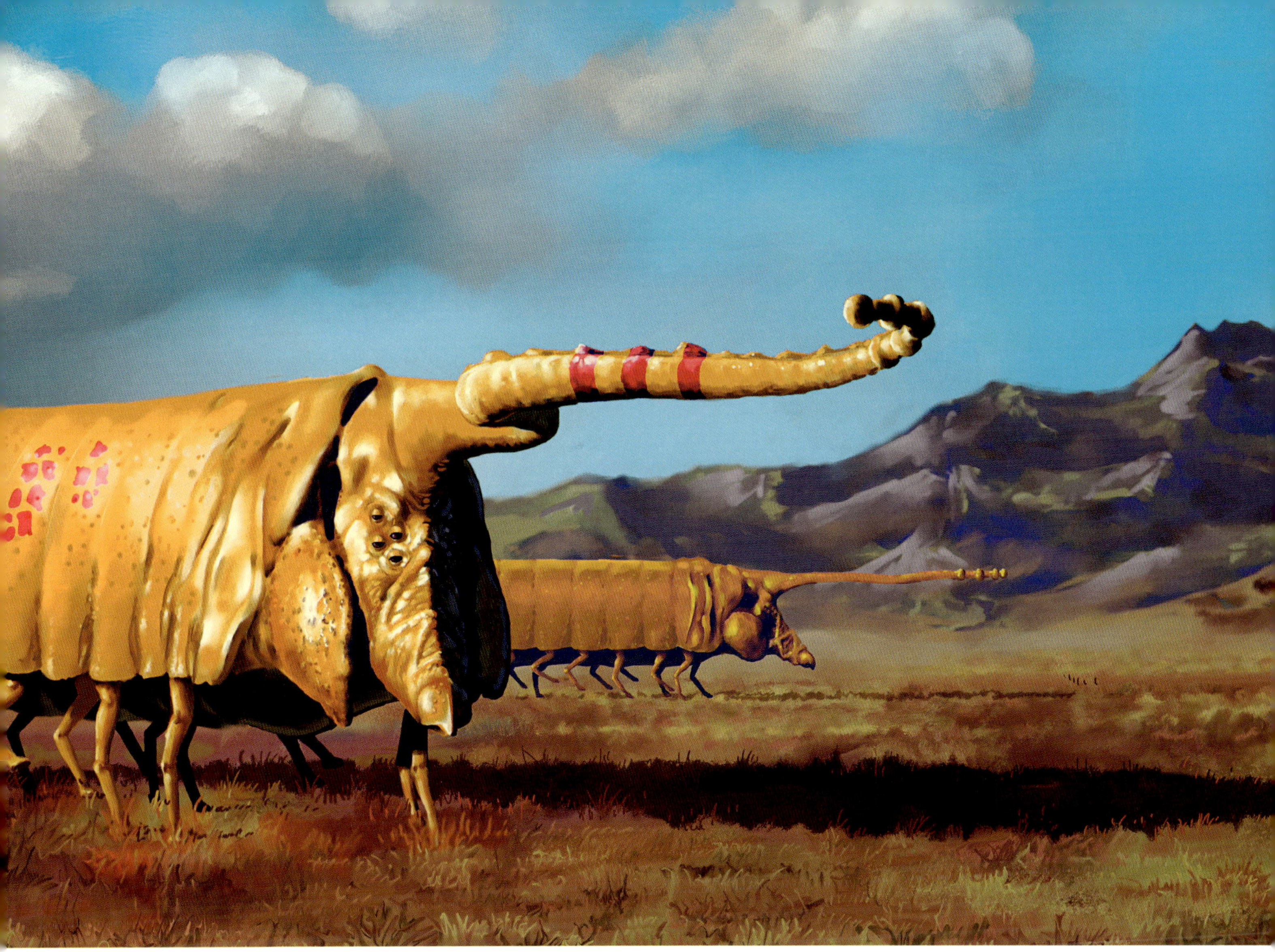

▲ 'What is it you want?'

At present no predator can tackle an adult rusp, but there is some tantalising evidence that this was different a few million years ago.

At present, predators will have to wait for a brontorusp to die of natural causes, so they can scavenge the carcass. That is no easy task, as the carapace becomes even tougher after death.

▼ **Brontorusp**
Brontocrambis brucus
Name derivation: *Brontocrambis* (Gr.) (concatenation of 'thunder' and 'cabbage-caterpillar'); *brucus* (L.) (meaning 'caterpillar')

Habitat: Steppe, savannah, open woodland ('dotted forests')
Distribution: *Imparia Septentrionalis*
Mass: Unknown
Length: 25m without whips

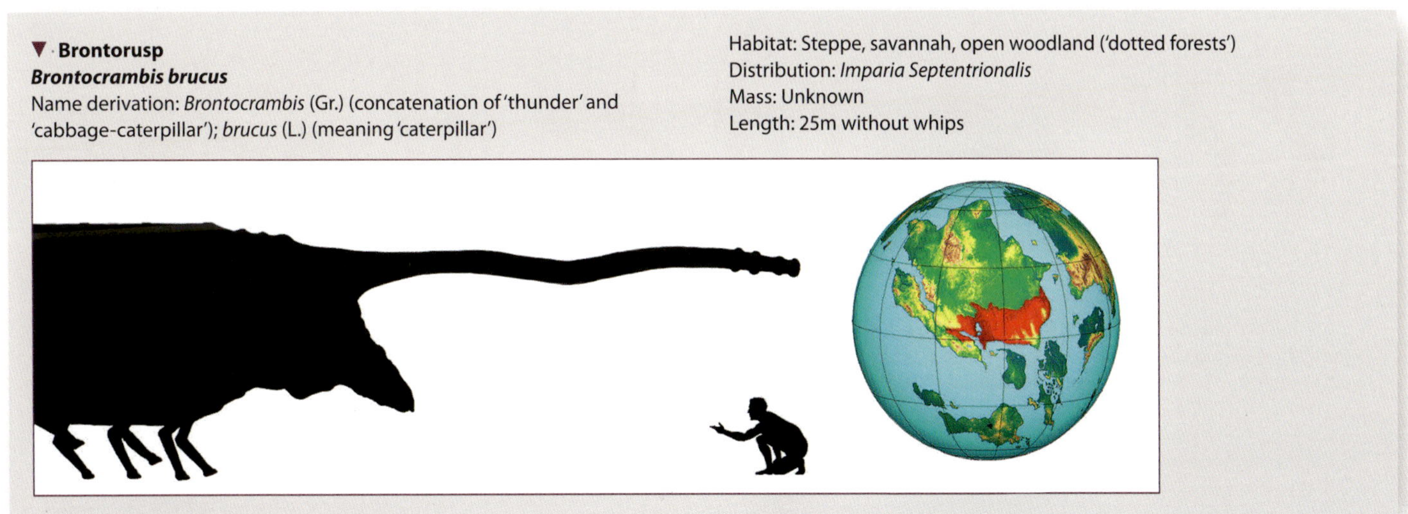

Megarusp Anatomy and Locomotion

The external armour or carapax of rusps looks a bit like a segmented exoskeleton, but it does not function as scaffolding at all, only as defence. Rusps have a proper endoskeleton, with each rusp body segment having a ring of bone.

There is no vertebral column though, so rusps are not vertebrates. Please do not offend Furahan crambologists (rusp experts) by calling their ring bones vertebrae. Two bones join the rings, forming a V-shape at the sides, bottom and top. This ensemble allows a accordion-like movement with the odd consequence that the entire animal can become shorter or longer. Still, the joints between the accordion bones at the top are quite stiff, so they allow little movement.

A favourite joke of anatomy professors at the Institute is to ask junior students about rusp length; students retaliate by consistently talking about 'rusp vertebrae'.

Whip Anatomy

The whips, held horizontally, share an anatomical trick with sauropods, Earth's long-necked dinosaurs. Bones at the underside of the whip resist compression and strong ligaments at the top withstand tensile forces. In effect, rusp whips are built like suspension bridges.

Legs to Stand On

Having many legs offers possibilities that animals with four or six legs do not have, but also poses unique difficulties. One disadvantage is that most rusps are slow and cannot jump at all, rather like elephants. Another disadvantage is that the combined mass of all these legs is higher than the combined mass of just six, four or two legs, even if each these is sturdier. If legs would only have to support the animal's weight, rusps could save much mass by reducing their number of legs. But there is value in redundancy and one expression of that is resistance to injury. An animal with two or four legs is usually doomed when it breaks a leg. A hexapod is already much more likely to survive that injury. If, however, you break one leg of a rusp, the animal will prove that it is not crippled by coming after you with a vengeance.

Leg Anatomy

The three major leg segments, the coxa, femur and tibia, point forwards, backwards and forwards again. This zigzag folding resembles that of mammal legs and helps to keep all the joints and the foot in close to a vertical line from the hip joint down. This position minimises bending forces due to gravity.

The legs mostly move forwards and backwards, and are not splayed outwards. Placing a foot far away from the point directly underneath the hip requires strong forces to keep the leg in that position. The necessary muscle force increases considerably as an animal increases in size. This is why only very small animals such as insects can afford to place their feet far away from the point just underneath the hip. In contrast, large animals had better keep their feet directly below the hips, which is just what megarusps do.

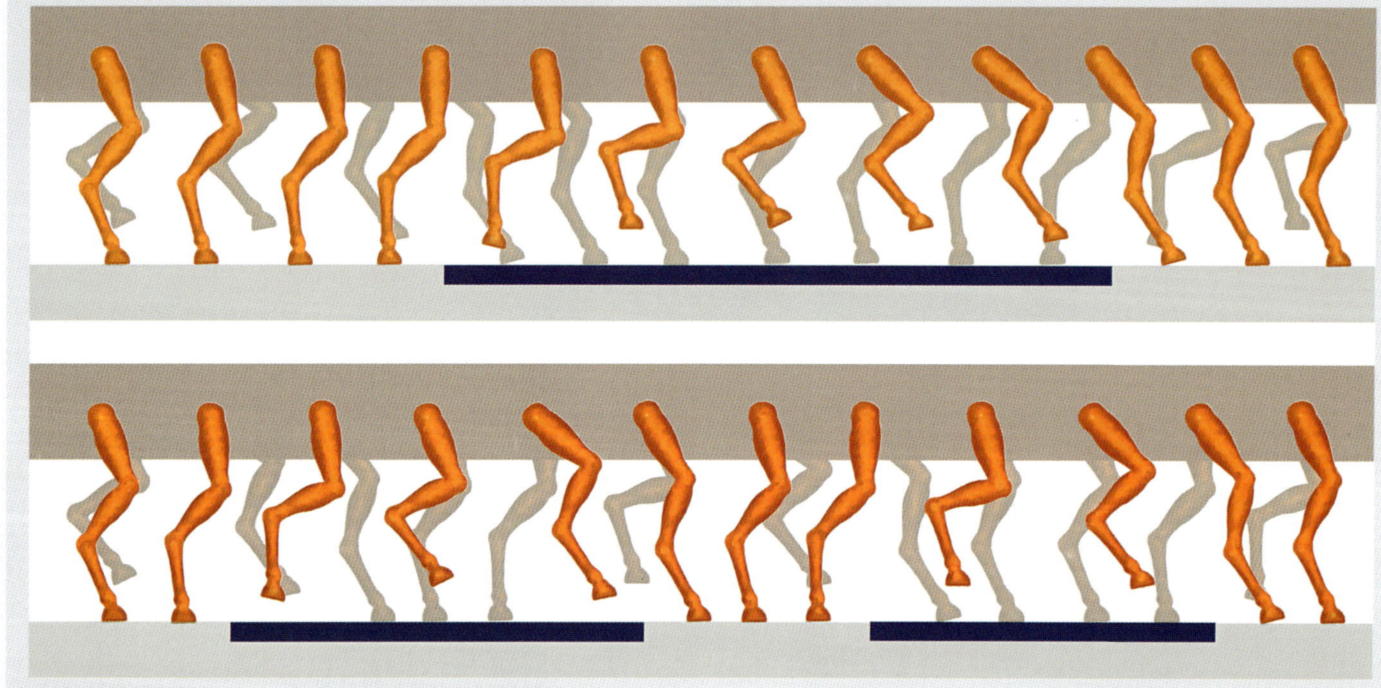

▲ Gait Phase Diagrams
These images show all 24 brontorusp legs, with the front head at the right side of the image. The top image shows a walking pattern, in which each leg is one twelfth of a cycle out of phase with the next one. As there are 12 legs to a side, that is exactly one cycle per side. Legs are off the ground for half a cycle, so six segments in a row are not supported by legs on the ground (the dark grey bar).This stresses the body skeleton, but the advantage is that the legs do not bump into one another.

The lower image shows what happens when the phase differences are doubled: now, only three successive segments are unsupported. There is less stress on the skeleton, but more chance of legs bumping into one another.

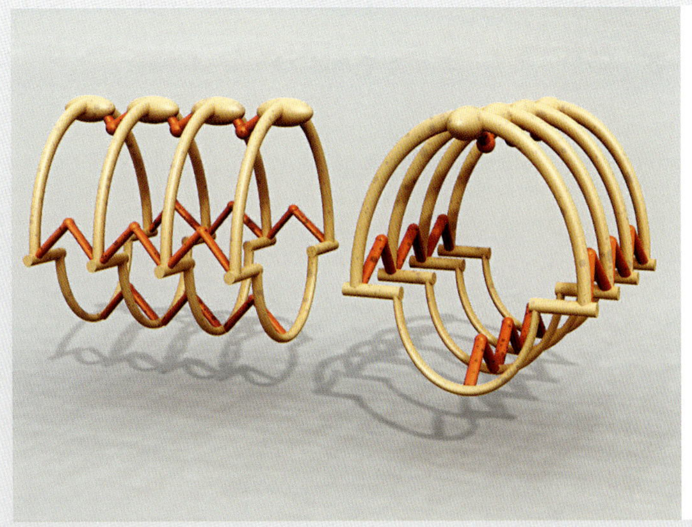

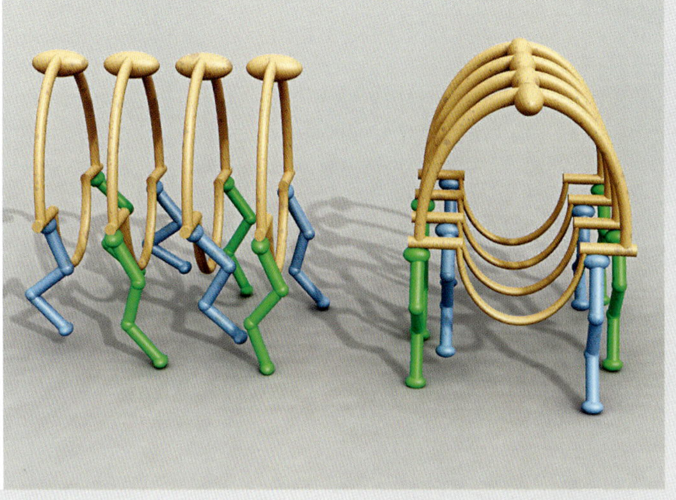

▲ **Rings of Bone**
The left image shows four schematic segments of the rusp skeleton, with a ring in each segment. The rings are connected at the top, bottom and sides by two 'harmonica bones' that allow the rings to move, bend and even twist a bit.

The right image shows how the legs are attached to the rings. On successive segments the legs are displaced to the right or the left, so each ring carries an outer and an inner leg. Outer legs are coloured green, inner ones blue. The offset prevents the legs kicking one another.

► **The Rostrum**
The rostrum (snout) is equipped with an odd skeleton. Each portion is made up of an upper and a lower V-shape of two bones each. Each ensemble of two V's can rotate around an axis through the two starting points. The brown and blue axes show such axes of rotation, so the snout as a whole is quite mobile. If the bones within the two V's move towards one another, the whole shape becomes longer and narrower.

The names of rostrum bones can be inferred, if it is kept in mind that for any inferior (lower) there is a superior (upper), and for every dexter (right) there is a sinister (left).

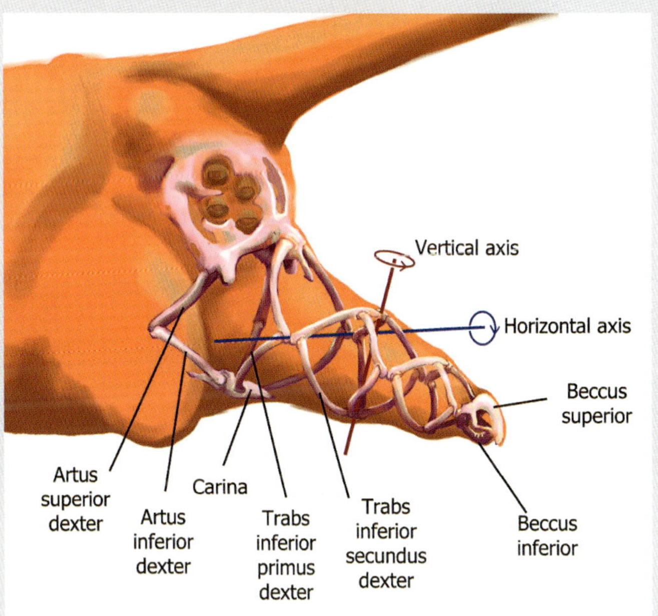

▼ The mobile snout allows rusps to feed on plants without having to move the entire body.

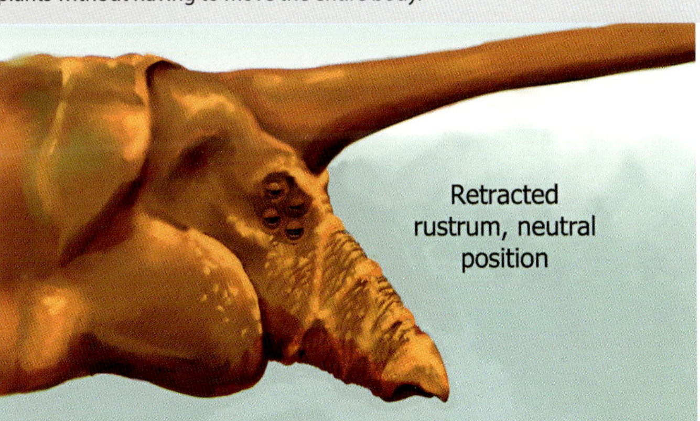

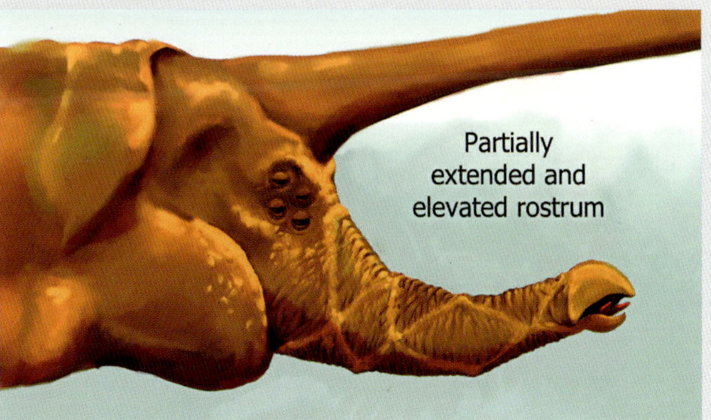

Other Megarusps

Any crambologist will tell you that you should take an interest in other rusps beside the brontorusp. On their behalf, this page is devoted to other megarusps. One major rusp characteristic has not yet been discussed, which is their number of eyes. They have four on each side. Their 'camera design' is an obvious pointer that rusps are not related to hexapods. The eyes are directed in diverse directions, which provides a large field of vision. Still, the same field could easily have been reached with fewer eyes. The number is probably a remnant from the head's formation process as it represents four original segments, each with its own pair of eyes.

The Dax

The 'copper rusp' or 'dax' is the rusp in the foreground on the left. Its legs point sideways to some extent, probably yet another rusp solution to avoid leg entanglement. Its name betrays a peculiar characteristic of the clade *Hyperchromata*, sporting interesting colour effects. The colour of the copper rusp varies with the light between ochre and blue green, which indeed looks a bit like old copper. Mind you, the effect wears off after death, so there is no point in killing one to obtain its hide to use for interior decoration.

Shimmerusp

The 'shimmerusp' or 'brilleur' belongs to the same group and exhibits even brighter colours. The physical nature of these effects will be the subject of a series of field studies, starting with the pragmatic question as to whether shimmerusps tolerate skin studies well. Most crambologists think not.

Giraffatitan

The giraffatitan is best described as a brontorusp that was squeezed and stretched to allow its rostrum to reach tree height. The modifications involve the front of the animal only. What is not apparent in this side view is the pronounced flattening of the animal. The head's vertical position means that the front whip is not very efficient at horizontal swipes. The result of all this is that giraffatitans are more fragile than brontorusps, but that is rather relative. They are still formidable animals.

▼ Various megarusps.

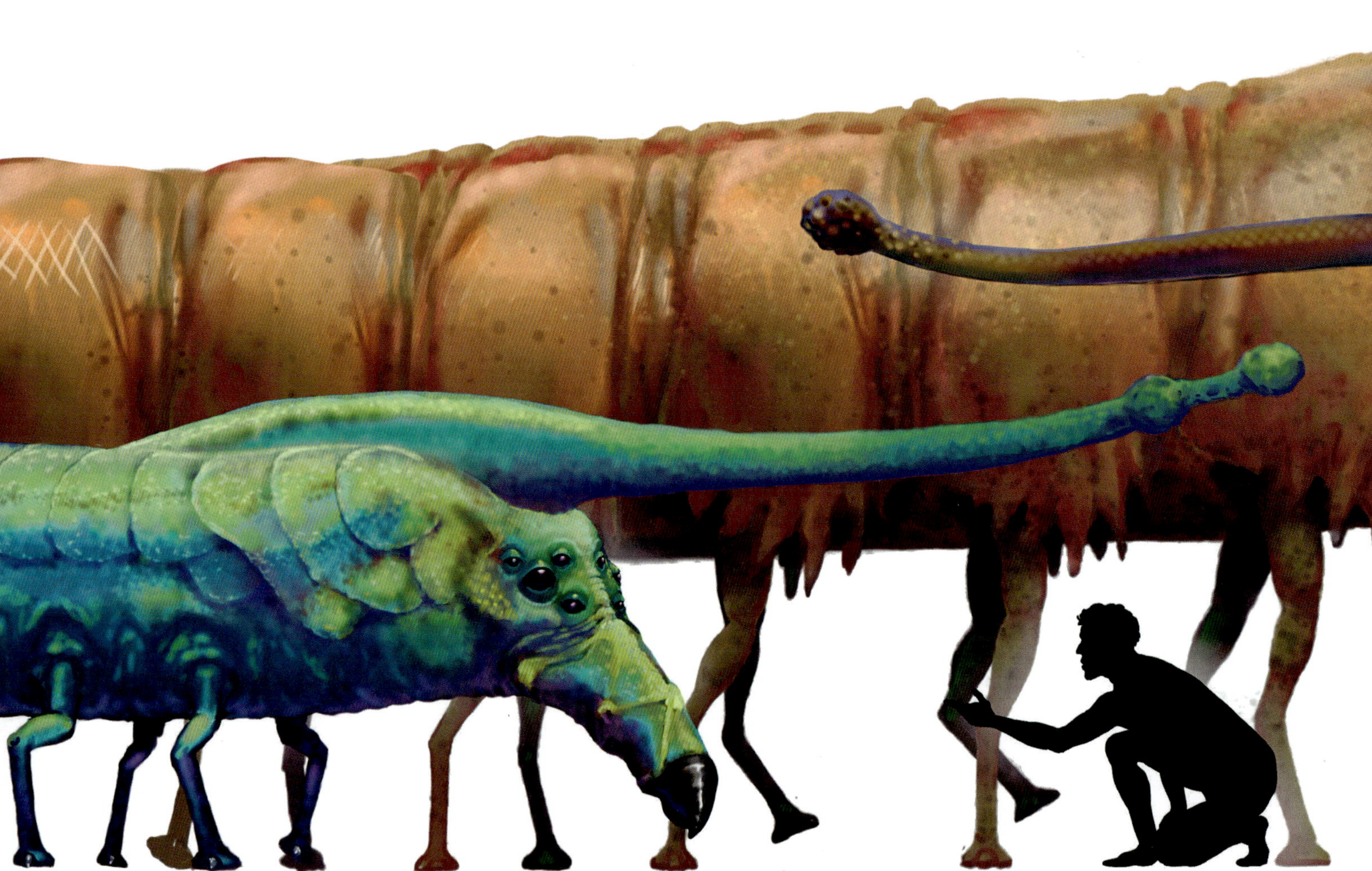

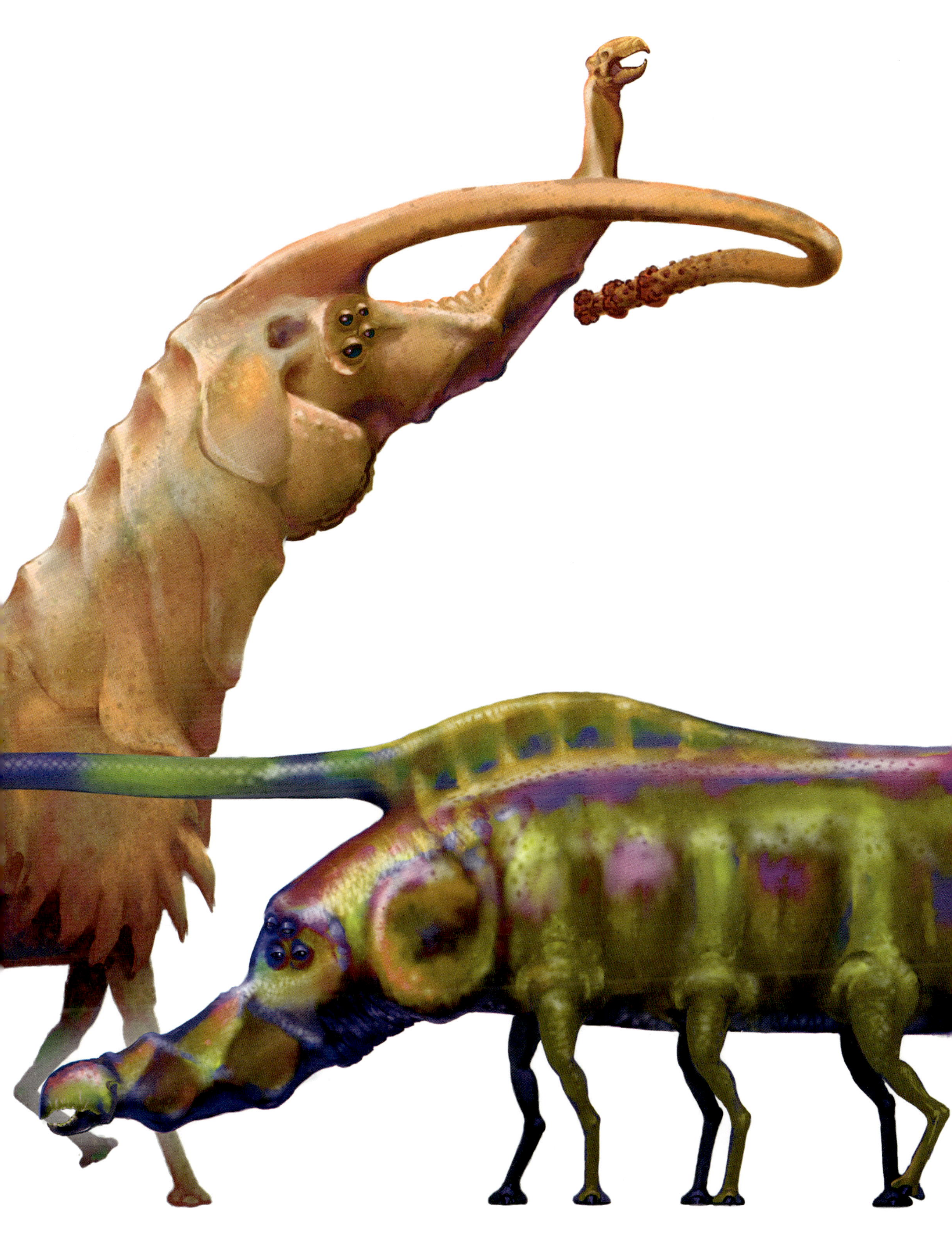

Other Rusps

Metriorusps

Megarusps, as big and conspicuous animals, usually get much more attention than their smaller cousins. Still, the medium-sized forms ('metrios' being the Greek word for medium) also merit attention. They have the same mass as Earth's horses or big antelopes, but their overall shape is quite different.

The number of pairs of legs varies between species and can in fact even vary within a species. Crambologists first thought that this was due to a genetic mishap, but studies failed to show anything wrong with individuals with a lower-than-expected number of legs (oligopods) compared to those with a higher number (polypods). The number of legs can in fact easily change up or down in successive generations. Biologists had taken it for granted that something like the number of legs would be an immutable feature of an animal's Bauplan. Unexpectedly, this was not the case for rusps. To appreciate how odd this is, just suppose that humans might have anything between four and seven fingers on each hand.

Metriorusps (and microrusps) face other problems than megarusps. The most obvious difference is that they are not protected by sheer size and must therefore rely on other protection methods. These pages show some interesting adaptations.

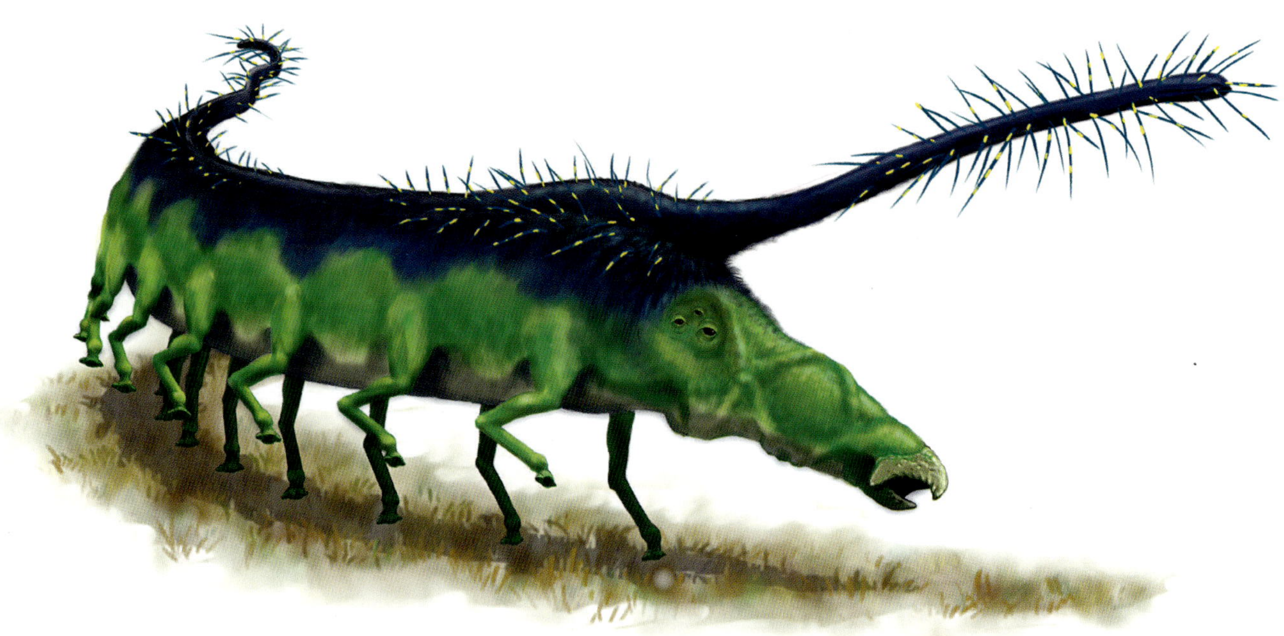

▲ Runrusps
Megarusps cannot run at all. Some blame their large number of legs for this. When large animals on Earth move fast, there is usually a phase during the running cycle in which no leg touches the ground, so the animal is in fact jumping. On Earth, many animals find it easier to run by not using all their legs while running. Some cockroaches and crabs even run on just two legs. But runrusps found another solution: all legs on one side move in unison. The image of the 'spiny runrusp' shows the animal running in this way, leaning into the bend. The spiny runrusp is not only protected by speed, but has poisonous quills as well, that dislodge easily and, thanks to tiny hooks, tend to work their way deeper into any animal unlucky enough to be stung.

▼ Spiny Runrusp
Dromeus picrus
Name derivation: *Dromeus* (Gr.) (meaning 'runner' or 'racer'); *picrus* (Gr.) (meaning 'stinging')
Habitat: steppe and savannah, also open woodland
Distribution: North half of the Peninsula
Mass: Up to 400kg
Length: 4m
Diet: Herbivorous

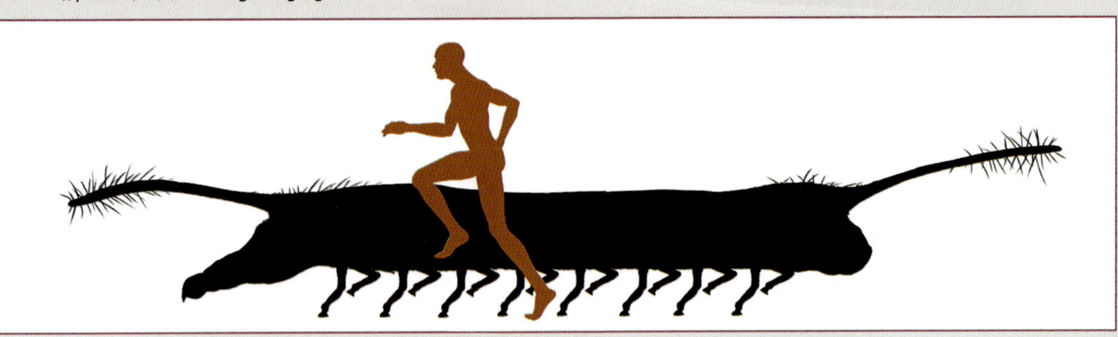

▶ Cuirassier

Running away is out of the question for these heavy and slow plodders. Instead, they are equipped with an armoured hide to withstand attacks by predators wielding clubs as well as lances.

Cuirassiers legs are not placed in a straight line but follow the animal's oval contour. The legs in the animal's middle are noticeably sturdier than those at the ends. This may represent a further adaptation for defence and helps to distribute weight.

The skin consists of tough fibre strands running in all directions, transmitting forces horizontally in the skin. In between lie ceramic-like coin-shaped bits of bone that effectively stop stabs but are too small to crack when beat by a club. The legs are protected in part by skirts. The spaces in between the skirts are supposed to entangle clubs, but that is uncertain. When threatened, the animal sinks down on these skirts, further removing the legs from danger. Their high-mounted whips are equipped with very functional 'thagomizers' to strike back at its enemies.

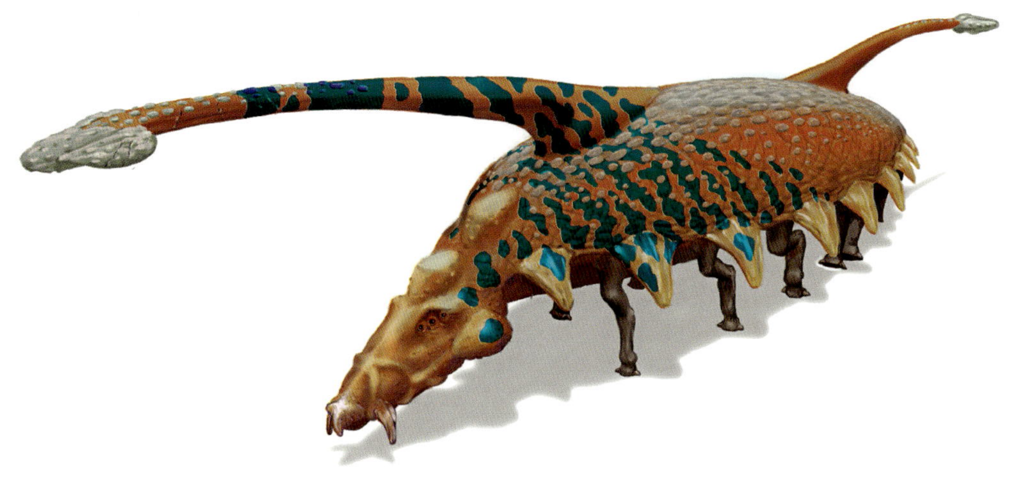

▼ Smart Armour

Cuirassiers have three to four layers of bone 'coins' in their skin. The coins are larger and thicker in deeper layers. The coins within and between layers are connected by strong fibrous bands, not drawn here.

Without coins the hide would be vulnerable to stabbing predators. If the coins formed a continuous layer, it might shatter under the impact of clubbing predators. Together, the strands and coins protect against both modes of attack.

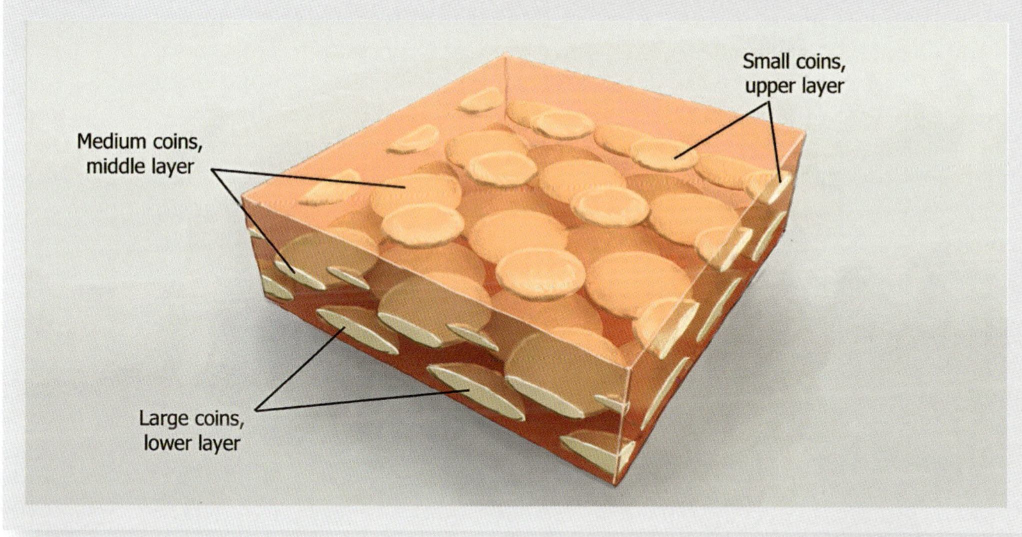

▼ Cuirassier

Sceptrum chrysopanoplium

Name derivation: *Sceptrum* (Gr.) (meaning 'mace'); *chrysopanoplium* (Gr.) (meaning 'with a golden suit of armour')

Habitat: Steppe, savannah, open woodland, forests
Distribution: Isthmus of the Peninsula, adjacent parts of *Imparia Septentrionalis*

Mass: 450kg
Length: 3.75m without whips

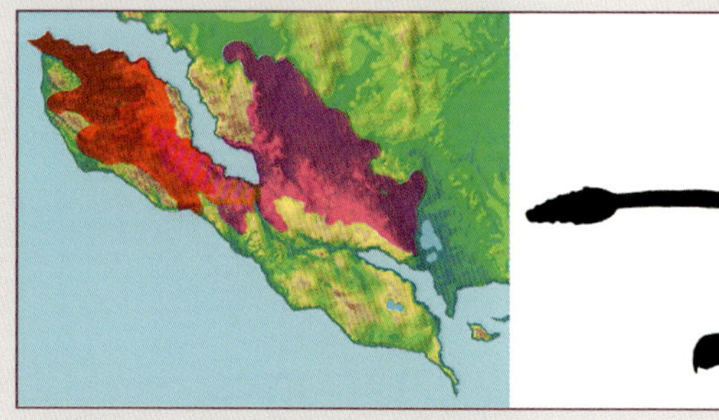

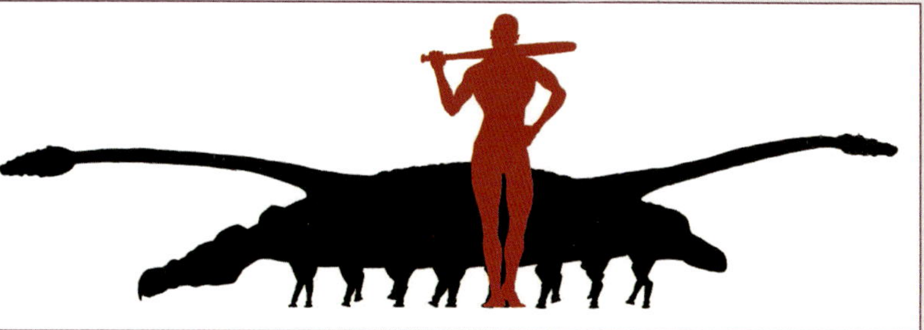

Balore or Baloar

Opipteuta episcopa

Some might say that including two microrusps in one book overestimates their importance compared to the many other life forms that could be shown instead. This is true – there are enough Furahan animals to fill many volumes. But choices had to be made, and there are more species in the microrusp group than in all other rusp groups combined. All mega- and metriorusps together do not even account for one quarter of all microrusp species.

The balore is the general public's favourite, no doubt because it featured in several popular Natural History programs. It has several things going for it to explain its popularity. One is that its surroundings provide an interesting background for balore antics.

Take the example shown here. Surely no-one can deny that early dawn in the arid Ommelandic Hills offers a unique spectacle. The 'dedobrujas' or 'Hexenfinger' (or 'witch fingers') are reminiscent of large similarly shaped plants or mixotrophs on other worlds. Earth's saguaro cactuses or Nowyswiat's prostak trees come to mind. But even in a much blander environment balores would still steal the show with their lively antics.

Balore Anatomy

The Baloridae exhibit a peculiar trait, 'limb omission'. The hypothetical protorusp from which all rusps are descended probably had one pair of legs per body segment. The Baloridae depart from this pattern as a central area of their bodies, a few segments wide, is legless. This may represent just another example of the rusp tendency to add or lose legs with apparent ease. Still, all other examples of rusp leg loss concern anterior or posterior segments, while the balore's legless area is situated right in the middle of the body. This area now separates two legged areas with five or six pairs of legs each. It would seem that this allows the animals additional suppleness.

Multiocularity

Microrusps usually have only four eyes, in contrast to larger rusps. This genetic fluidity implies a very modular connection between eyes and corresponding brain tissue. Anyway, one eye pair is placed high, so the animals can keep an eye on the horizon and the sky, whereas the other pair are pointed towards the ground in front of the balore, where the next snack is likely to be found.

Behaviour

Balores are communal and live in family groups dominated by a group of two to five sibs. When a sib group becomes old enough to irritate their elders, the sibs are kicked out of their native commune. Groups of wandering adolescents may join one another, invariably followed by a struggle for domination. The struggles involve showing off, threats, some actual fights and a sprinkling of mating, all of which probably explain why balores attracted much popular attention.

Vigilance

It is not easy to surprise a balore commune, because there are usually several members on guard duty. With four eyes each, it is not easy to sneak up on one balore, let alone on a group.

But balores being what they are, some doze off or only pretend to be on the lookout. Only one or two balores on guard duty may actually keep watch. If a lookout is caught off guard by a member of the 'oligarchy', they nip it a few times to remind it of its duties.

► Baleful balores.

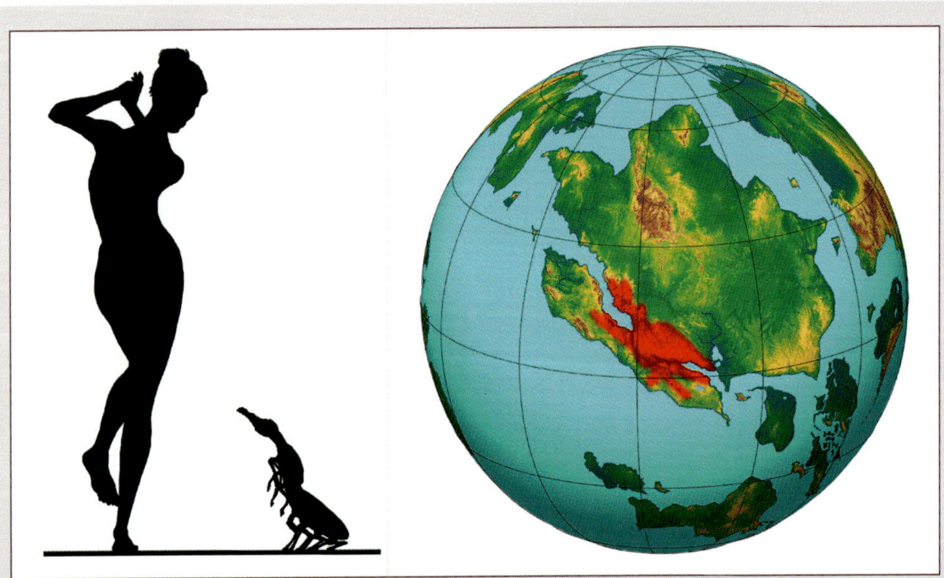

◄ **Baloar**
Opipteuta episcopa
Name derivation: *Opipteuta* (Gr.) (meaning 'looking around'); *episcopa* (Gr.) (meaning 'watcher')
Habitat: Dry steppe to desert; digs burrows
Distribution: Western central areas of Imparia Septentrionalis
Mass: 1–4kg
Length: Up to 50cm
Diet: Omnivorous
Vocalisation: Frequent vocalisations with wide variety of sounds

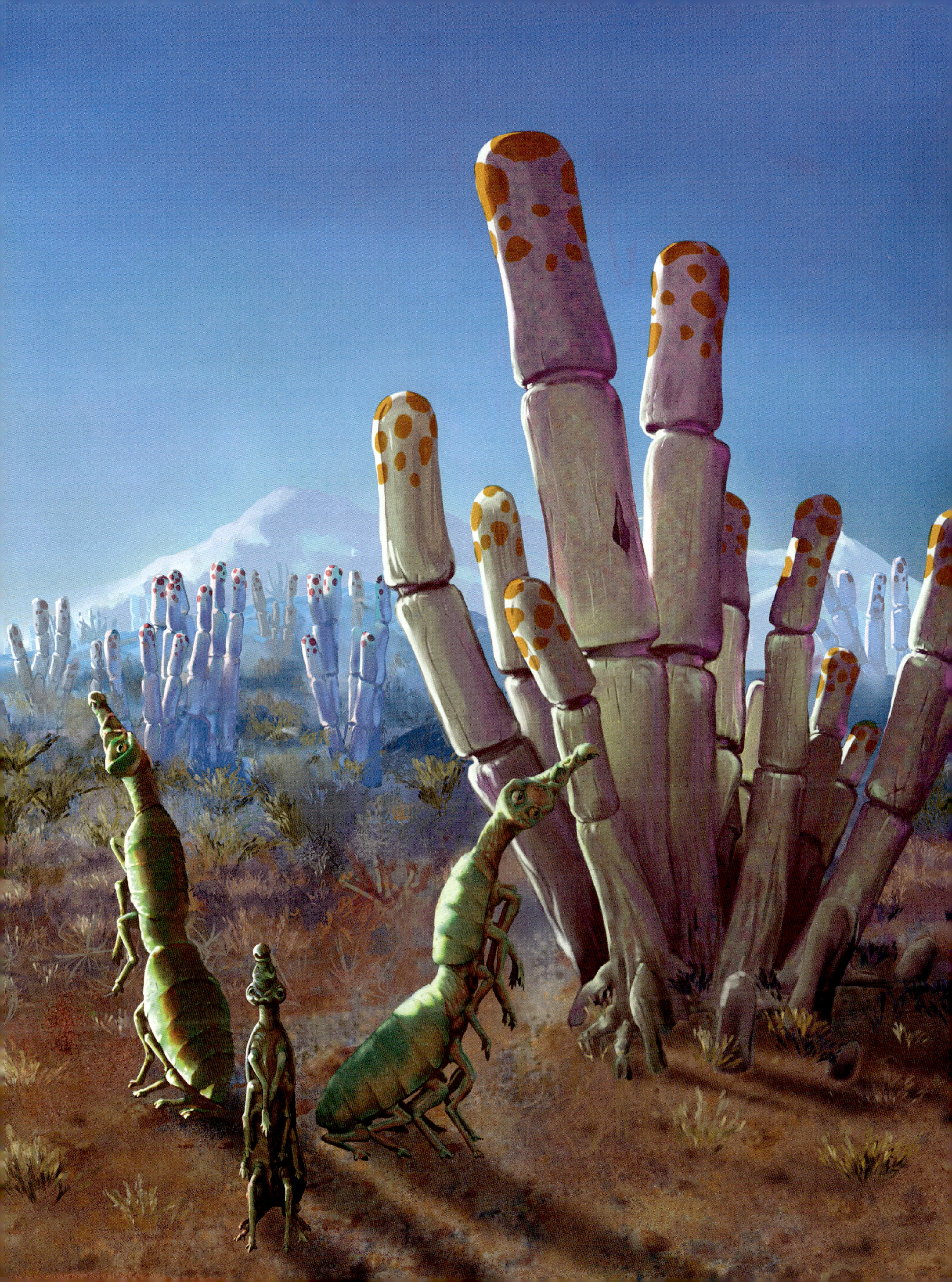

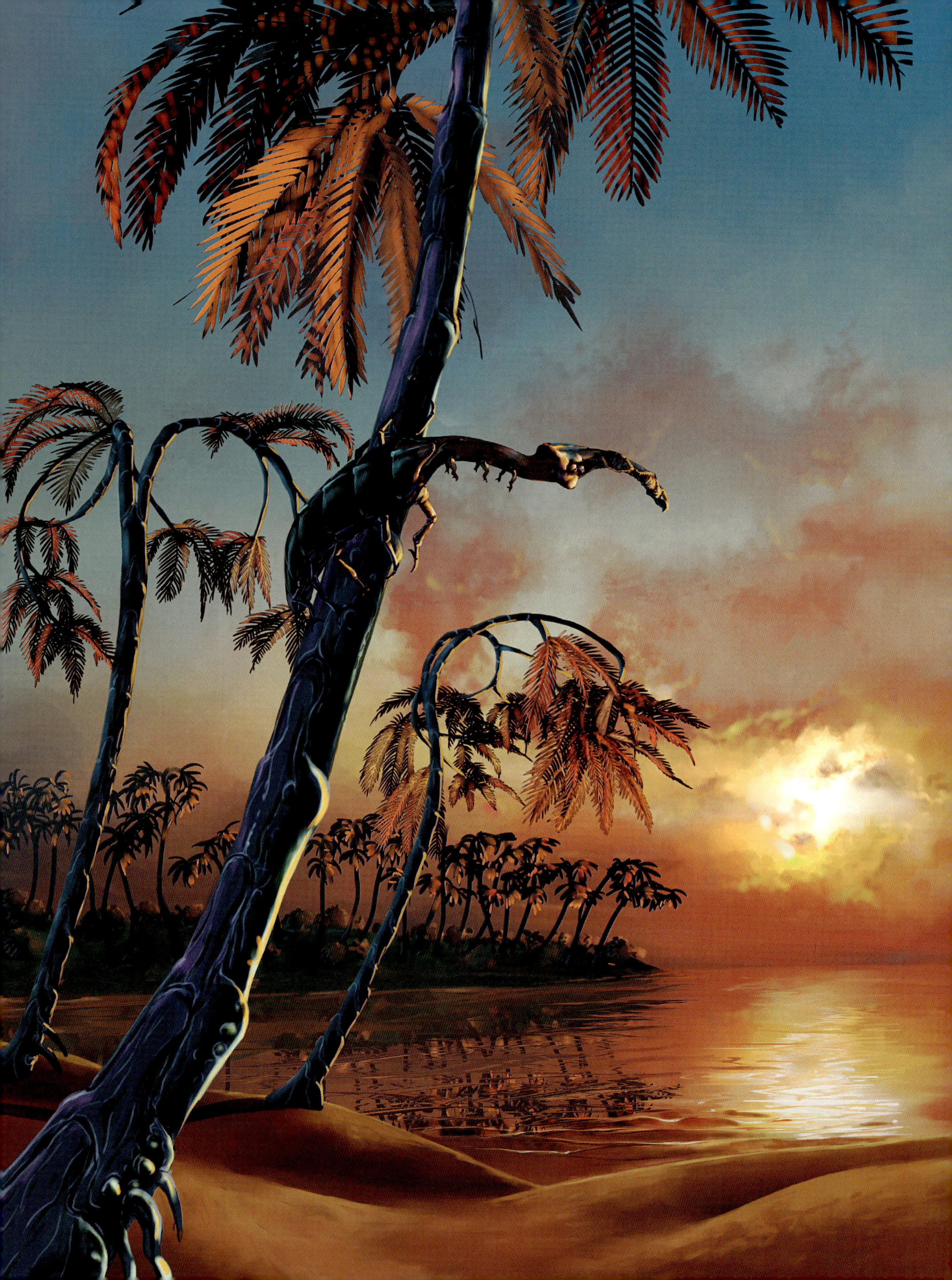

Treehugger

Amnesialata blansjarii

It comes as a surprise to many people that treehuggers are related to brontorusps, but their typical rusp rostrum (the snout) shows they are. Many microrusps such as the treehugger are adapted to a life in the trees. You might think that tree-living animals would be able to jump or fly, but treehuggers cannot do either. Their grasp is very secure, thanks to the strong pincer-like nails on their feet. In a pinch, they can grasp a branch with just the hindmost pair of legs and can use their beak to secure a new, well, beak hold. They have the most extensible rostrum of all rusps but have added another trick to elongate their reach. The rusp body plan does not normally contain a neck, but microrusps have freed several body segments from locomotion. The corresponding legs are atrophied and those segments now constitute a functional neck.

Honey Lovers

The bobberpalms, shown here, are called so because their clumps of leaves bob up and down from elastic branches. Small societal luminescent tetrapters called honeybuns make nests in the hollow segments of bobberpalms. They secrete 'honeybun syrup', a nutritious secretion containing much C_2H_5OH to feed their larvae. However, treehuggers adore the syrup and will ruin a nest to get some. Honeybuns in turn try to scare treehuggers away by spraying them with a disgusting concoction called 'honeybun glop'. A saying on Furaha sums it up, 'He who wants honey too much gets glop'.

◄ Treehugger enjoying the sunset.

Home is Where the Heart is

As is well-known, the famous writer and naturalist Sigismunda Felsacker left Furaha after some very dramatic events and never returned. Her classic book *Palaeo Days* made it back to Furaha though. These are its closing words.

And that is why I left Furaha forever. To this day, I do not know whether it could all have gone differently. People said to me that the combination of Sébastien's pride and Ramonda's ambition would have led to a collision, no matter what. But they did not look me in the eye when they said that. At the time I acted as who I was then, not who I am now. So did Ramonda and so did Sébastien. Had we been wiser, they might still be alive.

After I decided to leave, there was not much more to do than to say my farewells. I already knew I would never come back. I never dealt well with time dilation, when people come back after 20 to 30 years and have only aged half that amount. I did not want to be one of those who come back unnaturally young. I would not be able to bear it that people 30 years from now would not know who the smiling man on that image in the Academy's gallery was, standing between Ramonda and myself.

I planned my trip to the Nexus spaceport, said my farewells, packed my possessions and started the journey. I took a slow route and stopped over at Souren Station in the Archipelago, where I wanted to say goodbye to some friends. I came to a café at the beach where we had agreed to meet, and was informed that my friends would be late, and would I mind sitting on the porch in the meantime? I have never been patient, so I did mind.

But I knew in my heart that the time for haste was over. I sat down on the porch overlooking the beach and ordered a drink. Sipping it, I started taking in my surroundings. The sun was low in the sky, and everything was bathing in that yellow-gold glow you only get in the late afternoon.

There were stands of bobberpalms on the beach, gracefully moving in the soft breeze. Their leaves rustled in that odd dry way they have. Spidrids chicked in the bushes and under the porch. Bobberpalm perfume and the scent of honeybun syrup grew stronger as the sun set. I looked at the honeybun swarms, zooming in complicated patterns around their palms. I admit that a lifetime of science made me ponder the principles underlying swarming, but soon I just sat there, living the moment.

Then, suddenly, the past misery, my life on Furaha and its beauty all melded together and overcame me. I was there and then absolutely certain that I would long to be back on that porch, or at least on Furaha, for the rest of my life.

And I was right. I have been homesick ever since.

► **Treehugger**
Amnesialata blansjarii
Name derivation: *Amnesialata* (L.) (poor Latin meaning 'bringer of forgetfulness'); *blansjarii* (L.) (after Ben Blansjaar, psycharch)
Habitat: Trees, the denser the woodland the better
Distribution: Archipelago
Mass: 1–4kg
Length: up to 50cm
Diet: Omnivorous. Has strong penchant for Honeybun syrup.
Vocalisation: Occasional 'Glou glou glou' call with rising pitch

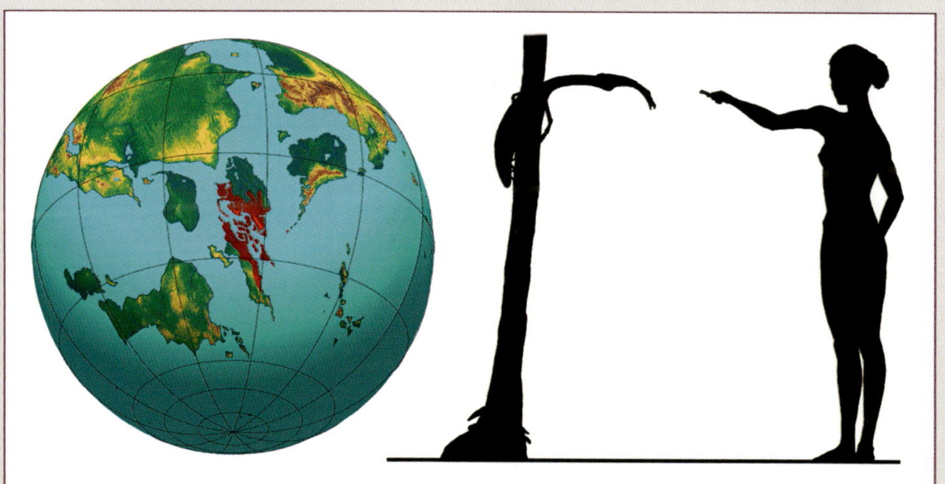

Appendix: The Author's 'Prehistory' of the Furaha Project

The Furaha project started with oil paintings on panel. When I switched to digital painting, some designs were kept, but even then evolving thoughts on the biology of the planet induced many changes. Many paintings were simply archived as fossils of the prehistory of the project. Here are some of these fossils.

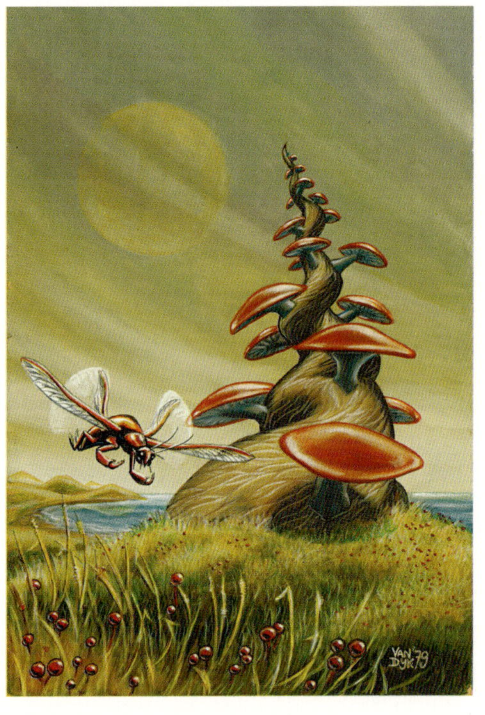

◀ A long time ago, bored with painting exploding spacecraft for others, I started painting alien creatures for my own pleasure. The flying insectoid already shows my biomechanical interest, with separate wings for lift and for propulsion. The grass really called out for a painting of herbivores, so this painting started the project.

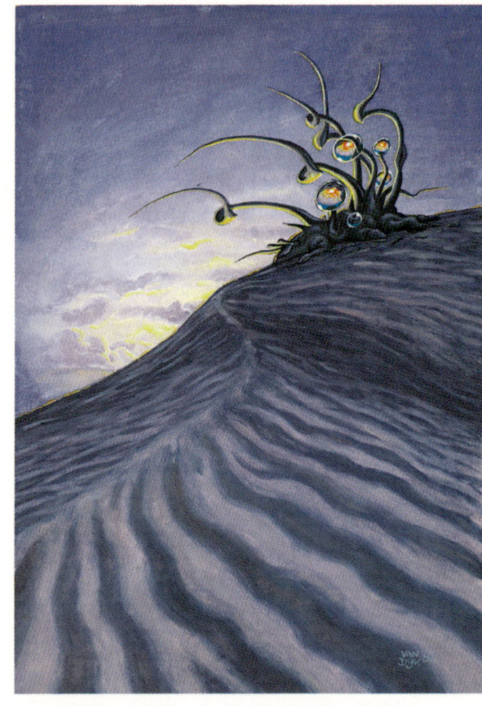

▶ This painting mostly aimed to achieve a sense of wonder, without being bothered by a biologically plausible back story. It would work as a book cover, I think.

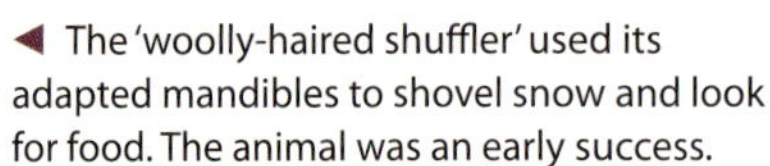

◀ The 'woolly-haired shuffler' used its adapted mandibles to shovel snow and look for food. The animal was an early success.

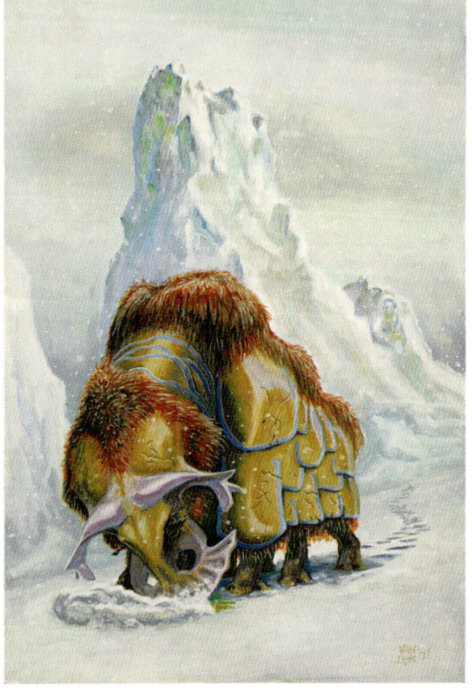

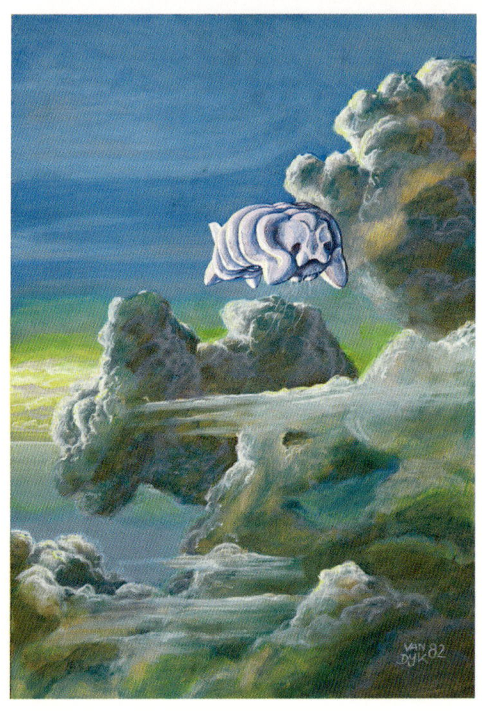

▶ Furahan skies used to be filled with 'ballonts', animals using the lighter-then-air principle to take to the air. Unfortunately, physics said no to ballonts on an Earth-like planet, so all ballont paintings ended up in the archives.

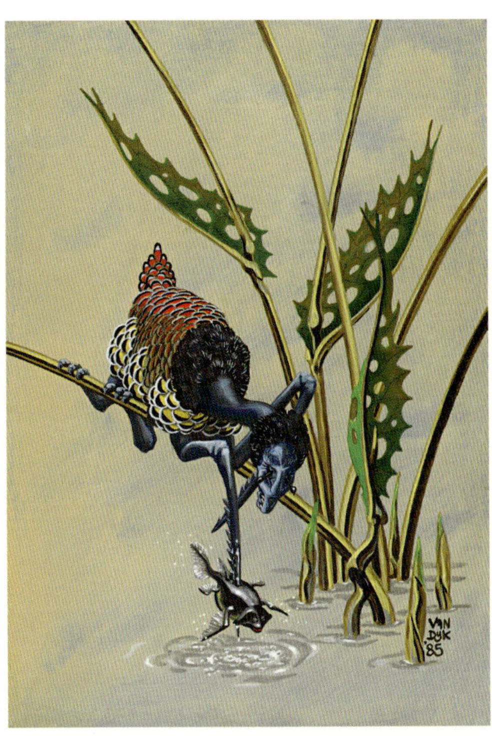

◀ The 'spietskip' ('skewering chicken') was a small hexapod marsh predator, seen here acquiring dinner. Hexapods still had very mammalian heads at this stage, and the spietskip's feathers are altogether too birdlike.

▶ The 'bleeding heart' avian was biologically the least serious of all early Furahan paintings. Many people liked it, though.

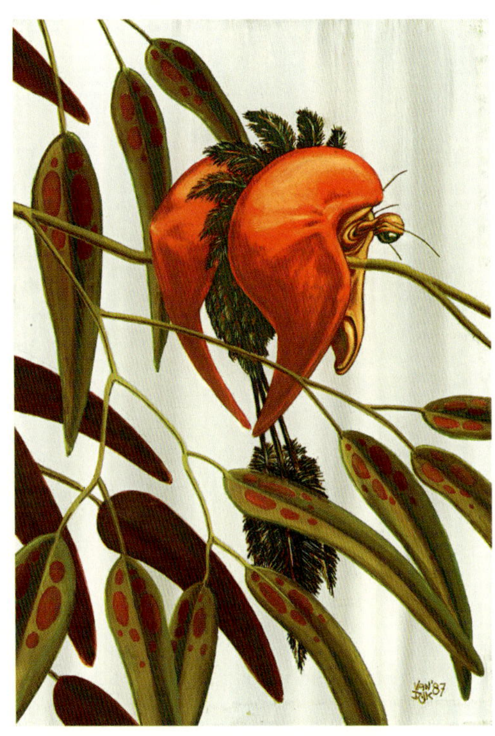

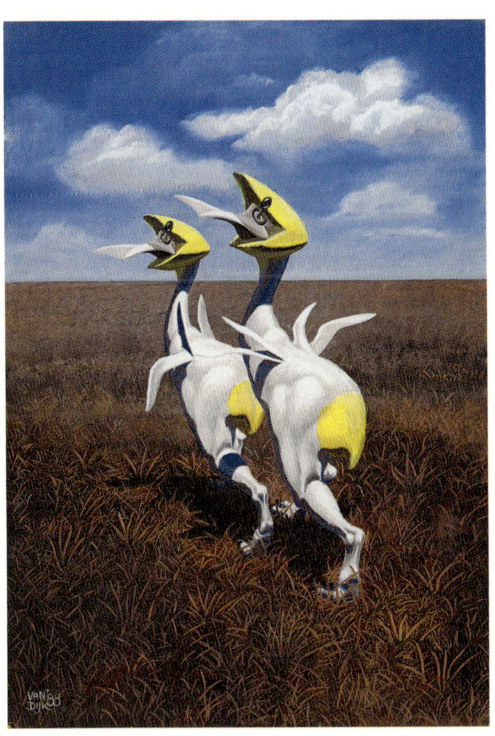

◀ These prancing animals are descendants of quadrialate flyers that lost the power of flight. Their ancestors flapped large wings to impress one another, and the animals now still flap their ever-smaller and ever-sillier wings.

▶ A herd of handlebars is seen galloping over the plains. I felt there was no biological need for horns to be brown, black or grey, so they ended up purple. The overall design of these early animals was very insect-like.

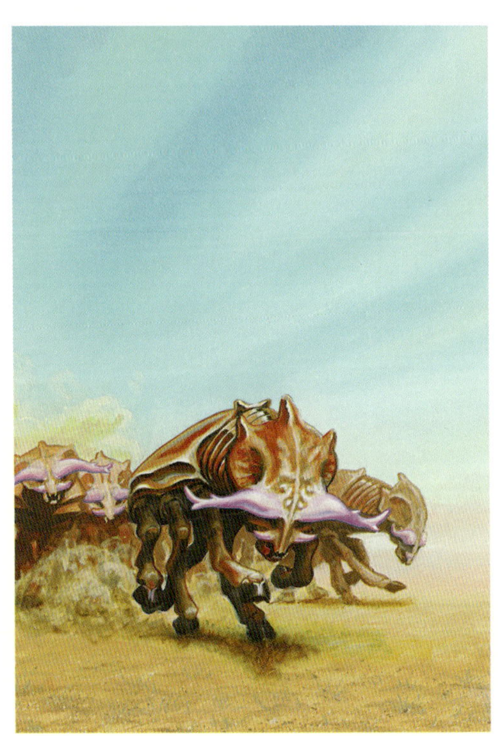

GLOSSARY OF FURAHAN BIOLOGICAL TERMS

This glossary focuses on concepts that are not already explicitly explained in the text. Most entries concern concepts that are peculiar to Furahan biology or society. To prevent confusion, the names of these entries are followed by (F) to indicate their Furahan background. Terms that are commonly used in Earth's biology are indicated with (E), but many such terms are not included.

Ab colonia condita (F) Latin phrase meaning 'from the foundation of the colony'. Used to indicate local Furahan time reckoning. Years followed by ACC indicate that the year relates to that event and is expressed in Furahan years, not Earth ones.

Aggressive Automimicry (E) Mimicry in biology indicates that one organism resembles another, with the implication that the resemblance evolved for a specific purpose, such as to avoid predation. Aggressive mimicry means that a predator is disguised to avoid recognition of its true nature by its prey (a wolf in sheep's clothing). Automimicry concerns imitation by a species of a part of itself, such as when an eye is mimicked by a spot on a tail. Taken together, aggressive automimicry describes that a body part of a species used for predation resembles a different, inoffensive, part of itself. If this is still unclear, just read the description of the nightsnare.

Autoheterotroph (E) Organism that can produce its own set of basic molecules needed for life (auto) as well as obtain them from other organisms (hetero).

Bauplan (E) German word indicating the set of anatomical features that together characterise a group of animals. For instance, the Bauplan of Furahan hexapods includes two lateral ›chordae, three pairs of locomotor limbs and a head/neck assembly consisting of, from proximal to distal, a proximal neck, the cranium, the distal neck and the jaw apparatus.

Becdacier (F) Species of brachiating hexapod, similar to the Marblebill. Name derived from the French *bec d'acier*, meaning 'beak of iron'.

Benthic (E) Adjective indicating the bottom zone of a body of water. When used in the context of oceans, benthic often indicates very deep water.

Bipterates (F) Clade of Furahan hexapods with two wings ›Dialata.

Bureaurchy (F) A system of government in which the managerial class takes on legislative roles. A person in this class is a bureaurch.

C_2H_5OH Ethanol. This simple molecule has interesting effects on Earth vertebrate brains and is by accident also produced by various Furahan life forms.

Cartouche (F) An emblem usually showing a stylised animal or plant. Cartouches are worn by higher-ranking members of Furahan society to indicate said rank and secondly to indicate their family group, that is, their ›Clan.

Centauraptor (F) The word 'raptor' originally meant a large bird of prey but came to mean a large range of predators, in particular predatory dinosaurs. The part 'centaur-' here indicates ›centaurism. The combined word centauraptor indicates terrestrial hexapod species whose front limbs are used exclusively as weapons.

Centaurism (F) In evolution, locomotor limbs may gain another function and no longer act as walking legs. In one group of predatory hexapods, the first pair of limbs evolved into hunting weapons. This feature is a defining characteristic of the ›Neocarnivores. Because the resulting animal scheme resembles centaurs from Greek mythology, this evolutionary trend was labelled centaurism. Souren Nyoroge coined this phrase.

Cephalothorax (E) Some animals do not have a head (cephalon) that is clearly separated from the next major body part, the breast (thorax). When the two parts form one whole, the word 'cephalothorax' may be used. Note that the functions situated in a human head and thorax are not necessarily the same ones found in an alien animal.

Cetiform (E,F) Whale-like

Chlorochrome (F) From the Greek words for 'green' and 'colour'. The term is reserved for the clade of Furahan plants using a pigment that does not absorb green wavelengths for photosynthesis, leaving green light to be reflected. In this aspect, these plants resemble Earth plants.

Chorda (E,F) Generally, a chorda in anatomy is a flexible but fairly stiff rod providing structural strength. In Earth vertebrates, the chorda preceded the vertebral column. In Furahan ›Scalata, there are two chordae, a left and a right one. Their later development into chains of bone (›ossochordae) on either side formed the side rails of the ladder (scala) that is the basis of the Scalate skeleton.

Clade (E,F) In the common evolutionary sense, a clade describes a group of organisms consisting of an ancestor and all its descendants. In Furahan Society, a Clade was also used to describe a cultural group to which individuals professed to belong, such as West Germanic or Perimediterranean. Clades were usually voluntarily chosen, not based on genetic ancestry.

Clan (E,F) An extended family group providing support and status in Furahan society.

Commensalism This term refers to a relationship between species in which one party benefits from the relation while the other is neither harmed nor benefitted.

Convergent Evolution (E,F) Describes that evolution tends to find similar solutions to similar problems, even if the starting conditions are different. One classic example is a streamlined form to reduce friction when moving through water, evolved in fish (Earth and Furaha), cephalopods, plesiosaurs, whales and so on.

Corpus spheroidum dentatum (F) Literally, this phrase means 'toothed spherical body'. The word 'body' should be regarded as organ in this context, meaning part of an organism. It does not therefore mean the entirety of an organism, which would be the normal sense of body. The term describes the shape of this specific mixotroph organ well. Then again, 'toothed' should be regarded as 'having indentations' rather than having actual hardened teeth. 'Indentations' here in fact only means that the margins of the slit-like opening on the top of the organ (the stoma) are not smooth but are wave-like. Perhaps the description is not that good after all.

Corpora spheroida dentata (F) Plural of *›corpus spheroidum dentatum.*

Crambologist (F) Scientist specialised in the study of rusps.

Craniognathal neck (F) Distal neck of Furahan hexapods, between the ›neurosensory cranium and the ›orognathes.

Deluge (F) Synonym with 'diluvium', 'collapse' or 'disaster'. Term indicating the time of climatic, environmental and societal collapses of the final industrial period of Old Earth. Because of its impact, the article is usually also written with a capital, as in 'The Deluge' or 'The Collapse' ›Prediluvian, ›Postdiluvian.

Dialata (F) Clade of Furahan hexapods with two wings ›Bipterates, ›Quadrialata.

Epitheton Laborans Byname related to one's work.

Erythrochrome The combination of the Greeks words for red and colour describe a specific type of photosynthesis, or for the clade of plants with that specific photosynthesis pathway. The leaves are red in colour.

Firing the Feng (F) Engaging the spatial displacement engines, operating on the Feng-Malparaiso-Kerkrade principle. The Feng drive is, rather obviously, only effective as long as spacetime is unmoored.

Frugivore. Fruit-eating.

Fungol (F) A medium to large ›mixotroph with characteristic fuzzy globes.

Furfeathery (F) Colloquial term that indicates the general appearance of one type of integument of hexapods, in between Earth fur and feathers.

Galea (F) In Furahan biology, a galea is a covering of a head or of the part of the body that functions as a head. The term is derived from the Latin word for helmets as worn by Roman soldiers.

Gnathoskeleton (F) Grinding plates and associated bone elements that function as mechanical food grinders inside the stomach equivalents of ›Scalates.

Hab (F) Colloquial abbreviation of 'habitable zone' around a star. Mainly used by spacers.

Hexameric (F) ›Trimeric.

Hexamera (F) Clade of animals with a ›Hexameric ›Bauplan.

Hexapoda/Hexapodes Clade within the larger ›Scalate ›Clade, in principle containing terrestrial scalates. 'Terrestrial' should not be taken too literally here, as the clade contains groups that secondarily returned to an aquatic existence and also contains flying animals.

Lamarckian relates to Jean-Baptiste Lamarck, who proposed that a physical capability acquired by an individual organism during life can be passed to its descendants. This is a view that contrasts with classical genetics.

Maker factory (F) In Furahan society, as elsewhere in the ›oikos, many objects are made one at a time. The production is usually overseen by ›Minds without human interference. For small maker factories (Makershops), Small Minds suffice.

Mdudu (F) Singular of ›wadudu. The word 'mdudu' is used to indicate an individual specimen of a 'bug', for example one that is trod upon or one that is witnessed while biting, piercing or lancing a human.

Mdudulogue (F) A zoologist specializing in the biology of ›wadudu.

Mesoskeleton (F) Skeletons are either found on the inside of animals, such as Earth vertebrates, or on the outside, as in Earth arthropods. A mesoskeleton exists partially on the outside of animals and partially on the inside. The mesoskeleton is a feature of Furahan wadudu, in particular large ones. The mesoskeleton seems to be a mark of late development of an exoskeleton, in which the outermost integumentary later could be detached from inner layers. The flexible outer layers provided protection against water loss and some mechanical protection, while the stiff inner layers served as mechanical stress-resistant elements. Once separate, the stress-bearing inner tissues became free to evolve towards the middle of limbs or bodies. Note that an external cylindrical stiff structure serves well if the limb is small in size, so for small animals there was little advantage in moving the skeleton towards the inside of a limb. For larger limbs and bodies, an inner skeleton provided significant advantages, such as savings of mass and protecting the skeleton from direct impact.

Minds (F) Term used in ›postdiluvian times to indicate machines with artificial intelligence. Minds are typically employed to control tasks with complex interactions, as Minds are better at tracking myriad interactions than humans. Minds are typically classified as 'Microminds', 'Small Minds', 'Broad Minds' and 'Overlords'.

Mixotrophs (F) Some life forms are neither simply heterotroph nor autotroph. Heterotrophs cannot produce essential biochemical substances themselves but acquire them by absorbing the products of other organisms. Autotrophs can produce such substances themselves. Earth plants are typical examples, producing basic chemicals through photosynthesis. Furahan mixotrophs can produce biochemical substances themselves (auto-) but can also absorb ready-made substances (hetero-). They can derive their energy from light (photo-), but also from chemical processes (chemo-). To avoid awkward combinations such as 'autoheterophotochemotrophy', they are called 'mixotrophs'.

Mixotrophologist (F) A biologist specialised in ›mixotrophs. For most of the history of the Institute of Furahan Biology, the field of mixotrophology was peopled by enthusiasts who started as zoologists or as botanists. Although differences in prior training highlighted strengths and weaknesses in both groups, both sides tended to emphasise the strength of their own background and the weaknesses of their counterparts. Initially, neither side agreed with the introduction of a specific mixotrophology training program.

Multirobur (F) This adjective, meaning 'many stems', is derived from Latin roots and describes plants large enough to be called trees that never have just one trunk or stem. The number varies but is usually at least five.

Mutualism A relation between two species in which both benefit from the association.

Naming ceremony (F) Even though finding a species new to science was a common event throughout Furahan history, doing so is still regarded as something worth celebrating, provided the species is fairly conspicuous (microscopic species are often ignored). From the earliest days of the Institute of Furahan Biology, people tended to use a Presentation to create interest in their work, engage in departmental politics, have a party, or all of these. Over time, the habits and rules of Presentations varied greatly, but the basic elements remained present. The Recitator is the scientist who formally describes a species, proposes its binomen and whose name will forever be linked to that species. The Quaestor, usually a professor, has the job of questioning various aspects of the description, such as whether the separation from other species is large enough to warrant a new species. A Iudex from the Institute has the last word. At times, acceptance was a foregone conclusion, but at other times proposals could just as easily be turned down.

Neurosensory Cranium (F) Anatomical term to indicate the proximal part of the hexapod head/neck assembly. The word indicates that this skull (cranium) contains major sense organs (sensory) as well as important brain functions (neuro) ›truncocranial, ›craniognathal and ›orognathes.

Nonameric (F) ›Trimeric.

Nuntsjuk Species of brachiating hexapod, similar to the Marblebill.

Oikos (F) A term commonly used to describe the community of humans in space, but only in the oikos.

Orognathes (F) The jaw assembly of Furahan hexapods, separated from the skull containing the major senses and the brain by the distal neck.

Ossochorda (F) Chain of bones on the left and right sides of Furahan scalates.

Poliochrome (F) From the Greek words for 'grey' and 'colour'. This describes the clade of plants using a specific photosynthetic pathway that makes use of a large part of the spectrum of visible light, so no specific colour band is reflected more than others. As a result, the photosynthesis surfaces are overall grey in colour. Various subgroups reflect purple a bit more than other colours, giving the surfaces an overall grey-purple aspect, sometimes called 'grurple' by Furahan citizens.

Polymery/Polymeric Meaning many parts ›Trimeric.

Polypremny (F) This noun, based on Greek, indicates the presence of multiple trunks in a tree-like large plant. *See* multirobur for the corresponding adjective. In fact, 'polypremny' may be defined as the defining characteristic of multirobur trees. The term applies to erythrochrome plants only.

Postdiluvian (F) Adjective indicating the period after the collapse of Old Earth ›Deluge.

Prediluvian (F) Adjective indicating the period before the collapse of Old Earth ›Deluge.

Prostak trees (F) Name of Furahan trees of unknown origin. The literal meaning is 'oaf' or 'rude person', which does not explain why any species of tree should be named for a concept that does not apply to trees. But it is.

Quadrialata (F) Clade of Furahan hexapods with four wings. Also indicated by 'tetrapterates', which means the same thing, but based on Greek instead of Latin ›Dialata.

Rail bone (F) A tubular lengthwise bone forming part of a series that together form the long side of the ladder skeleton in ›Scalates.

Recitator (F) The scientist presenting a species at a ›Naming ceremony.

Rung Bone (F) A tubular transverse bone connecting the left and right sides of the ladder skeleton in ›Scalates.

Ruspveldt (F) A habitat comprising grassland to open savannah, usually kept from becoming woodland or dense forest by megarusp populations whose rough foraging prevents tree growth.

Sardoon (F) Medium-sized species of Fishes VI.

Sarissa (F) Spike-like mixotroph outgrowth, named after Macedonian infantry pike.

Scalata/scalates (F) The Furahan clade characterised by a ladder-like skeleton. Contains the subset ›Hexapoda.

Slipdrifting (F) Spaceship movement method while spacetime is moored.

Terraforming (E,F) Large-scale action to change the overall environmental features of a planet, such as increasing global temperature and changing the composition of the atmosphere. The term was originally meant to indicate intentional actions to make a planet more Earth-like. Making a planet less Earth-like unintentionally proved equally feasible ›The Deluge.

Tetrapterates (F) Clade of Furahan hexapods with four wings ›Quadrialata and ›Bipterates.

Thagomizer (E,F) Originally a late twentieth-century term for stegosaur tail spikes, the term has come to indicate any array of spikes at the back of animals.

Trimeric (F) Composite word consisting of words meaning 'three' and 'part'. In Furahan biology, the word is used to indicate parts of an animal, in particular when it comes to radial symmetry. In this context, a trimeric animal would indicate a radial animal with three identical parts, like a hypothetical orange with just three parts. Variants concern 'di-' , 'tri-', 'tetra-', 'penta-', 'hexa-', 'hepta-', 'octo-' and 'nona-' for the numbers 1 to 9.

Truncocranial Neck (F) Proximal neck of Furahan hexapods, between the trunk and the ›neurosensory cranium.

Voronoi Tree (F) Species of tree with a characteristic partition of the bark into polygons. Named after an Earth mathematician. It appears likely that the shape of the polygons is governed by Voronoi principles.

Wadudu (F) Plural of the noun 'mdudu', a Swahili word indicating small animals. 'Bug' might be a good translation in the original Earth context, but on Furaha the meaning has shifted to something more formal when used by citizen-scientists, referring to small mesoskeletal animals. The general public completely ignores formal divisions between clades that zoologists view as unbridgeable chasms and use the word just as original Swahili speakers intended it to be used: 'bugs'. Perhaps even 'creepy-crawlies'.

Wurm (F) Colloquial term for small soft animals without limbs.

Xyloblastic (F) Cells that build wood are called xyloblastic. The deposition of new wood is important not only during tree growth but also as a response to changing circumstances. Xyloblastic cells are spurred into action by mechanical stresses that therefore result in new wood. Wood can however also be taken away actively, if circumstances changed to the effect that a particular part of wood is no longer under stress, so it has become mechanically useless. For the dissolution of wood ›xyloclastic.

Xyloclastic (F) An adjective indicating the dissolution of wood tissue in Furahan plants. Xyloclastic cells dissolve wood from the inside of a tree, helping to shape wooden structures into the optimal shape for a branch or trunk. Without xyloclastic cells, all wood formed at any age of a tree would remain present throughout the life of the tree, resulting in much superfluous wood, exactly in the manner of Earth trees. For tissues producing wood ›xyloblastic.

Zwam (F) Species of tropical fungus.

INDEX

adaptive radiation 117
Altanero, Katarzyna 125
amphibious 117
automimicry 40, 152
autotroph 27, 35, 155
avian 30, 130, 133, 153
baignac 110, 127
ballont 9, 154, 156
balore 148
barrelfish 65
Bauplan 91, 104, 146, 152, 154–155
becdacier 113, 128, 154
beetle fish 99
berbie 127–128
big yellow blob 52
bilateral 65, 69, 77, 81
bioluminescence 40
blackgammon 117
blue leviathan 95
blue moon 65
bobberpalm 151
bobbuck 122–123
botoro 39
boxfish 65
brachiate 127–128, 135, 152
brachiator 154–155
brachiologue 128
brontorusp 45, 140–144, 151
brownfish 98
Bruyningh 7, 20
bureaurch 20–22, 61, 154
camera eye 68–69, 84, 144
camouflage 49, 77, 81
capaespada 65, 67
cartouche 17, 27, 35, 81, 91, 139, 154
cataphract 46
centauraptor 125, 154
centaurism 109, 122, 125, 154
cephalisation 140
cephalothorax 77, 154
chemotroph 35, 155
chlorochrome 28–29, 81, 154
cinder 7–11
citizen-scientist 22, 27, 30, 33, 49, 51, 57, 61, 79, 97, 156
cladogram 93
clan 17, 22, 33, 81, 139, 155
clap and fling 88, 133
cloakfish 55, 65–71 , 93, 135
colour contrast 87
column tree 120
common cloak 65, 68
compound eye 69, 77, 97, 127
continental drift 14–15
convergent evolution 95, 155
counter colouring 98
cowfish 99
crambologist 142, 144, 146, 155
cross flow filter 69
cthulhu 62, 100
cuirassier 147
dandy 113
dax 144
Deluge 20–21, 153, 155–156
desert ghost 118–119
dialate 130, 133, 154–156
dipeau 136
disruptive colouration 65–67
dragon 130, 133
endoskeleton 142
erythrochrome 30, 81, 155–156
exoskeleton 46, 82, 142, 155
farfalderol 87
Felsacker, Sigismunda 22, 28–29, 72, 75, 151
flachkopf 57, 59, 62
flimb 92–93
flortle 59
flounder 65
flyfoam 43
flyfoam 43
funk 36
Gaean 8, 11, 18
giraffatitan 144
gna 117
goppy 100
grade 17, 22
grapple 102
greater snook 92–93
green snakefish 95
grouillard 75
habitable zone 10, 155
haplodiplontic 36
hermaphroditism 79, 135
heterotroph 27, 35, 154–155
Homo semisapiens 20–21
honeybun 151
Horizonists 8-9, 20–22, 25, 87
Hrvatsky, Sébastien 22, 151
Institute of Furahan Biology 7, 17, 22, 49, 51, 61-62, 79, 91, 139, 142, 156
intelligence 18, 21, 59, 61, 109, 155

iridescence 65, 81, 92, 99
jove 10
Jua 8–12
kermitoid 130
korongo 133
kwal 52, 55
kwals 52–56
lausfresser 49
lighter than air 43, 152
lone star 71
lorica 72
louse/lice 46, 49
mackleral 92–93
mad sickle 77
marblebill 110, 113, 127–128
marshwallow 117
mdudu (*see also* wadudu) 30, 35, 39, 40, 46–51, 62, 81, 155–156
mdudulogue 51, 155
megarusp 30, 140–145
mesoskeleton 46, 155
milky petal pedal 53
minds 20–22, 156
mixotroph 35–43, 75, 109, 148, 156
moon 9, 13, 18
naming ceremony 49, 57, 156
Nastrarruzzo 27, 33, 65, 97, 139
    Caradoc 33
    Caspar 27
    David 27
    Friedrich 27
    Grover 139
    Seamus 65
    Sean 97
Ngonjera 8–13, 18–20, 30, 35, 130
nightsnare 40, 152
nuntsjuk 128, 155
Nyoroge 7, 11, 17, 20–21, 27, 35, 45, 91, 139
    Souren 7, 11, 17, 20–21, 27, 35, 45, 91, 139
    Tigran 91
Odin's spear 33
oikos 18, 155–166
ommatidia 127
Palaeo Days 22, 72, 75, 151
phalanx 36
photosynthesis 27–28, 33, 39, 110, 154–156
phunque 36
pied stickler 125
play 27, 67
plumfish 71
poliochrome 29, 49, 156
polypods 72–76, 146
porcelain seasoar 79, 135
predator 14, 49, 57, 62, 67, 84, 92, 100, 102, 110, 122, 128, 133, 140–141, 153–154
prigoon 81
prober 122–123
purple flapper 52
quadrialate 130, 154, 156
radial 52, 65, 68–69, 77, 81–82, 87–88, 156
rail bones 104, 156
recitator 49, 156
red baron 84, 89
red floater 43
righteous phleph 57, 59
romanarc 30
rostrum 144, 151, 153
rung bones 104, 156
runrusp 146
sawjaw 97, 117
scala 93, 104–105, 154
sea wheel 52
sexual dimorphism 84, 133
shadowfisher 130
shimmerusp 144
snafe 29, 109
Snaiad 18
sneeth 51
snigel 61–62
spidrid 35, 49, 51, 72, 77–83, 151
spidridologist 79
spidronomer 79
spotty woodrustler 77
starfish 65, 93
strandsprab 77
streamlining 57, 99–102, 107, 154
strider 100
sundancer 79, 83
swobbler 110
tenterhook 128
tetrapter 27, 40, 84–89, 139, 151
tetrapterate 157
thagomizer 147, 157
thresher 114
treehugger 151
trench gobbler 100
tribune 52, 55
umbrella tree 109
Uytterwaerde 17, 57, 59, 62
    Estrelinha 62
    Grover 17
    Ottokar 57, 59, 62
voronoi tree 128, 157
wadudu, *see* mdudu
wadudu castle 30, 35
wardens 57–63
wetsloth 107
white-headed seasoar 135
witvis 99
woolly-haired shuffler 120
Worlds Apart 7, 11, 17, 21, 27, 35, 43, 45, 91, 139
wurm 51, 77, 157
xyloblastic 33, 157
xyloclastic 33, 157
yellow brick 43
yellow zoom 46

First published in 2025 by
The Crowood Press Ltd
Ramsbury, Marlborough
Wiltshire SN8 2HR

enquiries@crowood.com
www.crowood.com

This impression 2026

British Library Cataloguing-in-Publication Data
A catalogue record for this book is available from the British Library.

For product safety-related questions, contact:
productsafety@crowood.com

ISBN 978 0 7198 4571 0

The right of Gert van Dijk to be identified as author of this work has been asserted by him in accordance with the Copyright, Designs and Patents Act 1988.

Typeset by Simon & Sons
Cover design by Sergey Tsvetkov
Printed and bound in India by Thomson Press India Private Limited

Acknowledgements

I would like to express my gratitude to those who encouraged me to paint and write this book over the years. The late Sytske Looijen suggested adding layers of content, such as people and their oddities. Mark Fogg designed a solar system that would be stable over enough gigayears to allow complex life to develop. Thijs van Ebbenhorst Tengbergen provided advice on publishing and preparing paintings for printing. Many readers of my blog provided very insightful and useful comments, including Evan Black, J.W. Bjerk, Rodlox, Anthony Docimo, Davide Gioia, Petr, Keavan, Andrew Broeker, Zerraspace, Abbydon, Idle Speculation.

Various people provided names for Furahan life forms or lent their faces to Furahan dignitaries, including Hans Goossens, Marca van Dijk, Tais Teng, Dougal Dixon, Ben Blansjaar and Roelien Bastiaanse.

Charlotte Lemmens and Vincent Icke provided a strong impetus to take my art seriously. I thank Dougal Dixon, Jean-Sébastien Steyer and Adrian Tchaikovsky for support to get the book into print. Finally, Roelien Bastiaanse deserves special mention for overall enthusiastic support and setting up a shuffler support centre.

Special thanks go to the people at The Crowood Press for making this book real.

The author
https://planetfuraha.blogspot.com
https://www.planetfuraha.nl/